STRUCTURE AND EVOLUTIONARY HISTORY
OF THE SOLAR SYSTEM

GEOPHYSICS AND
ASTROPHYSICS MONOGRAPHS

AN INTERNATIONAL SERIES OF FUNDAMENTAL TEXTBOOKS

VOLUME 5

STRUCTURE AND EVOLUTIONARY HISTORY OF THE SOLAR SYSTEM

by

HANNES ALFVÉN AND GUSTAF ARRHENIUS

Univ. of Calif., San Diego, La Jolla., Calif., U.S.A.

D. REIDEL PUBLISHING COMPANY

DORDRECHT-HOLLAND / BOSTON-U.S.A.

Library of Congress Cataloging in Publication Data

Alfvén, Hannes, 1908–
    Structure and evolutionary history of the solar system.

    (Geophysics and astrophysics monographs; v. 5)
    'Based on four papers which have been published in Astrophysics and
    space sciences 1970–1974.'
    Includes bibliographies and index.
    1.   Solar system.   I.   Arrhenius, Gustaf, joint author.   II.   Title.
    III.   Series.
QB501.A533   523.2   75–29444

Cloth edition: ISBN 90 277 0611 5
Paperback edition: ISBN 90 277 0660 3

Published by D. Reidel Publishing Company,
P.O. Box 17, Dordrecht, Holland

Sold and distributed in the U.S.A., Canada, and Mexico
by D. Reidel Publishing Company, Inc.
306 Dartmouth Street, Boston,
Mass. 02116, U.S.A.

Printed in The Netherlands by D. Reidel, Dordrecht

TABLE OF CONTENTS

PREFACE

This monograph is based on four papers which have been published in Astrophysics and Space Sciences 1970–1974. They contain the results of our joint work started in 1968 at the University of California, San Diego, in La Jolla. The work was based on the belief that the complicated processes by which our solar system was formed can only be clarified by close collaboration between representatives of the physical and chemical sciences.

Our investigations have also been strongly supported by work at other institutions, especially by a group at the Royal Institute of Technology, Stockholm, where a number of plasma experiments have been made in order to clarify basic processes which are relevant to cosmogonic problems. These experiments were, in their turn inspired by theoretical work on primordial processes carried out during the last thirty-five years.

We especially want to acknowledge the contributions by Drs N. Herlofson, B. Lehnert, C.-G. Fälthammar, and Lars Danielsson in Stockholm and by Drs J. Arnold, W. Thompson, A. Mendis, B. De, and W. Ip in La Jolla, J. Trulsen in Tromsö, Norway and D. Lal in Ahmedabad, India. The subject index was compiled by D. Rawls. The work has been supported by the U.S.A. National Aeronautics and Space Administration through the Planetology Program Office, Office of Space Science under grant NASA NGL 05-009-110, the Lunar and Planetary Program Division under grant NASA NGL 05-009-002 and the Apollo Lunar Exploration Office under grant NASA NGL 05-009-154; the support of the Swedish National Research Council is also gratefully acknowledged.

We wish to thank Prof. Zdeněk Kopal and his associates for permission to reprint these articles with some alterations from the journal *Astrophysics and Space Science*.

Part I was originally printed in Volume **8**, pp. 338–421, with received date 20 March 1970.

Part II was originally printed in Volume **9**, pp. 3–33, with received date 20 March 1970.

Part III was originally printed in Volume **21**, pp. 117–176, with received date 28 December 1972.

Part IV was originally printed in Volume **29**, pp. 63–159, with received date 6 December 1973.

(a)

Aurora borealis over Alaska

(b)

Comet

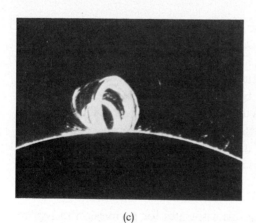

(c)

Solar prominences

(d)

Solar eclipse

Magnetohydrodynamics and plasma physics are the keys to understanding the early phases of solar system evolution. The basic cosmogonic processes can be understood as extrapolations of present day phenomena of the types we observe (a) in the magnetosphere-ionosphere, (b) in comets, (c) in solar prominences and (d) in the solar corona. The Aurora borealis is Courtesy of Lee Snyder, University of Alaska, and the solar prominences are Courtesy of Sacramento Park Observatory, Air Force Cambridge Research Laboratories.

INTRODUCTION

The present state of astrophysics is remarkable. During the last decades there has been an explosion of observational facts, largely due to space research. Matching this there has also been an explosion in the number of theoretical papers, but this has been a quantitative expansion, which in many fields has been decoupled from the new observational material. In fact much of the theoretical work is based on hypotheses which may have seemed reasonable when they were introduced, but which have never been subject to careful critical examination in the light of present knowledge. When this is done it turns out that many of the 'generally accepted' theories lack a valid foundation. In several cases it is easily shown that they are in conflict with well-established laws of physics. Hence in astrophysics today one of the most urgent tasks is to subject the theoretical foundations to a critical analysis, and to give the field a firm base.

We devote a large part of this monograph to analysis of some existing theories. In part III the well-known criticism of certain developments of cosmical plasma physics is recapitulated. The 'first approach' to cosmic electrodynamics does not describe a real plasma but a fictitious medium, a 'pseudoplasma' with properties which sometimes drastically differ from those of a real cosmic plasma.

Another theory which cannot stand a critical examination is the Laplacian concept of the formation of the Sun and the solar system by non-hydromagnetic processes. Although in most other fields of cosmic physics it was already realized 25 years ago that electromagnetic processes have a dominating influence on the dynamics of cosmic gas clouds (plasmas), the majority of cosmogonic papers published today are still based on the assumption that such forces can be neglected. It is claimed that if the primeval Sun was dark or its light absorbed by dust, the degree of ionization in its surroundings should be negligible. This argument is not correct. In the present solar system very little of the ionization is produced by solar light – in fact mainly in the D-region of the day-side ionosphere of planets. If the sunlight were switched off we would still have a high degree of ionization in the solar corona, interplanetary space, in the magnetosphere and in the upper layers of the ionosphere. In fact, the high degree of ionization in space above the solar photosphere and above the F-layers of the ionosphere is produced by a hydromagnetic transfer of kinetic energy (initially due to convection in the Sun) into electromagnetic energy. There is no reason to doubt that similar processes were active during the formation of the solar system. Indeed they are likely to have been much more intense because of the large release of gravitational energy which necessarily accompanied the formation of the solar system.

The recent discovery of strong magnetic fields in dark clouds shows that much energy must be stored as magnetic energy there. Hence if solar systems are produced in dark clouds – which is possible but not certain – hydromagnetic effects cannot be neglected for the origin of such systems. This is further shown by the presence in the dark clouds of a wide range of organic molecules manifest from their microwave emission. Their formation at the observed rates is considered to be possible only by ion-molecule reactions and also requires that ultraviolet radiation is practically absent. The radio astronomical observations thus provide a direct insight into the plasma effects in dark clouds. Another example of the development of a whole new literature based on questionable assumptions is the theory of equilibrium condensation of solid matter out of a gas cloud which has been made the basis for interpretation of meteorite data. It has been assumed that the process takes place in thermodynamic equilibrium in spite of the fact that it has been known for a long time that the departure from such a state is large under cosmic conditions of condensation. A number of other similar misunderstandings are also criticized in the present treatise.

However, our criticism of certain hypotheses should not be interpreted as a negative attitude to all speculation. Speculation – of course – is essential to all scientific progress. The reader will find several rather speculative theories in this monograph. However, we do not attribute more than preliminary value to these. Our aim is not to substitute for other hypotheses our own ones. Instead we consider the most important part of our work, aside from the critical analysis of basic concepts, to be the establishment of a *framework of boundary conditions* which all theories of different processes have to fit. Such a framework must be acceptable both from physical and chemical points of view.

The solar system of today is a result of a large number of complicated processes and it is likely that we need a large number of partial theories to account for its evolution. But because of the limited amount of observational data and the still incomplete understanding of the basic processes many of these theories necessarily must remain speculative. However, what can be established with some degree of certainty is their internal relations and the general criteria they have to fit.

There is also a third important feature of our approach. Warned by the failure of the pseudoplasma theories we do not believe that it is possible to derive realistic models of the cosmogenic processes by purely theoretical work starting from first principles. Instead we should observe that the cosmogonic conditions cannot have been drastically different from the conditions we find today either in the ionosphere, photosphere, magnetosphere, solar corona or interplanetary space. Hence by studying processes occurring in these regions we could learn a good deal about the processes which formed our solar system. Many of the cosmogonic processes may be considered as extrapolations of processes we observe today. This approach has a higher chance of leading to reliable results than a purely theoretical deduction from first principles. The frontispiece suggests which phenomena we should study to understand the processes which were important in the formative stage of our solar system.

PART I

GENERAL PRINCIPLES AND OBSERVATIONAL FACTS

Abstract. (1) *Introduction and Survey*. The method for studying the structure and evolution of the solar system is discussed. It is pointed out that theories that account for the origin of planets alone are basically insufficient. Instead one ought to aim for a general theory for the formation of secondary bodies around a central body, applicable both to planet and satellite formation.

A satisfactory theory should not start from assumed properties of the primitive Sun, which is a very speculative subject, but should be based on an analysis of present conditions and a successive reconstruction of the past states.

(2) *Orbits of Planets and Satellites*. As a foundation for the subsequent analysis, the relevant properties of planets and satellites are presented.

(3) *The Small Bodies*. The motion of small bodies is influenced by non-gravitational forces. Collisions (viscosity) are of special importance for the evolution of the orbits. It is pointed out that the focusing property of a gravitational field (which has usually been neglected) leads to the formation of jet streams. The importance of this concept for the understanding of the comet-meteoroid relations and the structure of the asteroidal belt is shown.

(4) *Resonance Structure*. A survey is given of the resonances in the solar system and their possible explanation. It is concluded that in many cases the resonances must already be produced at the times when the bodies formed. It is shown that resonance effects put narrow limits on the post-accretional changes of orbits.

(5) *Spin and Tides*. Tidal effects on planetary spins and satellite orbits are discussed. It is very doubtful if any satellite except the Moon and possibly Triton has had its orbit changed appreciably by tidal effects. The isochronism of planetary and asteroidal spins is discussed, as well as its bearing on the accretional process.

(6) *Post-accretional Changes in the Solar System*. The stability of the solar system and upper limits for changes in orbital and spin data are examined. It is concluded that much of the present dynamic structure has direct relevance to the primordial processes.

1. Introduction and Survey

1.1. STRUCTURE OF THE SOLAR SYSTEM

We plan to publish a series of papers aiming at clarification of the evolutionary history of the Solar System. For this purpose it is essential to start from an analysis of the present state of the system. There are already a number of excellent surveys of this topics, first of all Middlehurst-Kuiper's *The Solar System*. However a large amount of new physical and chemical information has accumulated in recent years. Data with particular relevance to our problem comprise direct observations in space and on the Moon and new investigations of chronology, chemistry and irradiation history of lunar and meteoritic matter. Furthermore new basic knowledge has been supplied by experimental and theoretical work in plasma chemistry and plasma physics and theoretical studies of resonances between orbiting bodies, and of the focusing of orbits in a Coulomb field.

A complete description of the evolution of the system – if such a description is at all possible – should account for every observed fact. Our present task must necessarily be much less ambitious since we find that only a relatively small fraction of the total body of knowledge concerning the present state can be used for reconstruction of the past. A survey of the relevant material is an essential part of this paper.

When analyzing the 'generally accepted' current views of the structure of the Solar System, one is struck by the fact that a surprisingly large fraction of these concepts derives from old, foggy theories, accepted even if there is little observational support. Observations are often arranged in the framework of such theories, even when the result is contradictory. We believe that much of the traditional text book material should not be perpetuated without scrutiny of its foundations.

To be fruitful, an analysis of the Solar System must combine both the chemical and the physical information in an internally consistent way; this is one of the underlying principles in our joint attempt.

1.2. THE APPROACH TO THE EVOLUTIONARY HISTORY

Physics and chemistry are essentially products of observations made on objects under human control and the theoretical treatment of such observations. Astrophysics is an application or extrapolation of these results to objects which usually cannot be observed under such favorable conditions as on Earth. The question arises whether the laws found valid in our present environment hold in a different time and in other parts of the universe. This is a controversial question in the cosmological field; it has brought about the occasional claim of new fundamental laws and changes in universal constants. For the purpose of the present treatise, it is not necessary to involve phenomena unaccountable under the well-known laws of physics; consequently, our basic assumption is that these laws are valid for our own and other solar systems, now and throughout their past history.

Astronomy has been concerned mainly with phenomena which are distant in space. The available information is conveyed to us by means of light and other forms of radiation which are very much diluted before reaching us. Much of the information is hard to interpret, but has sufficed to give us an increasingly clear picture of space around us.

Our problem is different. The phenomena which we are investigating are astronomically speaking close in space; but they are distant in time. We cannot 'see' what happened in our immediate environment some billion years ago. Thus we must obtain information about the history of the solar system by other means which are not necessarily less accurate.

Radiochemistry has given us a powerful means of investigating ancient events When applied to the Earth, meteorites, the Moon, and hopefully soon to other celestial bodies, it can supply us with detailed information about the evolution of the solar system. In certain respects the *dynamical state* of the solar system has been conserved from early times in a state which allows us to reconstruct phenomena connected with its formation. A review of the dynamical data considered to be of importance to our study will be given in Sections 2–5, and a review of the chemical data will follow in a later paper.

1.3. PLANETARY SYSTEM AND SATELLITE SYSTEMS

The solar system consists of the Sun surrounded by a number of planets (including asteroids), comets, meteoroids, and dust; in addition, there are a number of satellites orbiting some of the planets. The satellite systems of Jupiter, Saturn, and Uranus are particularly well developed. Most theories dealing with the formation of the Solar System focus their attention exclusively on the planetary system, ignoring the satellite systems. This is an unfortunate approach. Indeed, we should not aim for a theory of the formation of planets; rather we should arrive at a general theory of the formation of secondary bodies around a central body. This theory should account for both the satellite systems and the planetary systems. Since we have three well developed satellite systems and only one planetary system the emphasis is naturally placed on the satellite systems. This is also important because we know very little about the formation and early life of the Sun. Treating the planetary system alone it is difficult to exclude the possibility that planetary matter was ejected from the Sun, or that solar radiation, or other uniquely solar phenomena played a decisive role in the development of the system. Neither Jupiter, Saturn, nor Uranus – and in any case not all three – has produced phenomena analogous to those of the Sun. We can consequently surmise that such phenomena are not essential for the formation of secondary bodies around a central body. Furthermore, since a theory of the formation of secondary bodies must account for the formation of planets, the formation of the central bodies of the satellite systems is necessarily included in the theory. This correspondence is not necessarily true for the Sun, which may have originated differently.

Hence a theory of the formation of secondary bodies around central bodies must *primarily account for the satellite systems*. If we find that such a theory is also applicable to the planetary system – and this is the case – then we can draw important conclusions about the early state of the Sun, and by implication about stellar evolution and the formation of solar systems of other stars. Throughout this text we frequently rely on this important principle.

We continually refer to the concept of the formation of the solar system emphasizing the formation of satellites around planets as much as the formation of planets around the Sun. No suitable brief term which succinctly refers to the formation of the solar system exists; although earlier literature resorts to terms such as cosmogony or cosmology. These are correctly used for only the more general concepts of the creation of the cosmos as a whole respectively the science of the cosmos. Since the origin of the solar system is a question of the formation of *accompanying bodies* we propose and use as a suitable term in this text the term *hetegony* from ἑταῖρος or ἔτης, the Greek words for companion.

1.4. INFRARED STARS

Applying the theory of the development of secondary bodies around a central body to other stars, we find that many of them should form solar systems and that possibly some of these systems are at present in the state of formation. It is tempting to identify the T Tauri stars as such objects. If this is correct, which is by no means sure, we may be able to profit from direct observation of the process of formation. Analysis of the radiation from these objects has provided information on the properties of circumstellar matter in the low temperature range including chemical and structural characteristics of the solid grains. Such radiation evidence from the actual processes of transformation of primitive plasma – gas – grain systems offers a valuable comple-ment to the fossil information obtained from the chemical and dynamic properties of our own system.

1.5. PLASMA PHENOMENA

The processes involved in the formation of the celestial bodies in our solar system requires us to use not only the methods of ordinary chemistry and ordinary celestial mechanics, but also those of plasma chemistry and magnetohydrodynamics. The latter phenomena have generally been ignored by cosmogonists or applied incorrectly. As we shall find the plasma phenomena turn out to be very important, both to the chemical differentiation and the dynamical state of the system.

1.6. THE ACTUALISTIC APPROACH

Most theories concerning the origin of the solar system start with some reasonable assumptions about the primeval Sun. This is unfortunate. Since our knowledge of the formation and early development of the Sun is rudimentary, such theories are necessarily speculative.

Instead, here we approach the hetegonic problem using the same actualistic principle that Hutton successfully introduced for the reconstruction of the history of the Earth. Accordingly, we draw on information from the present state of the solar system to reconstruct successively more ancient states. In this manner the more recently observed phenomena are considered to be known with a higher degree of accuracy than the more ancient ones. Even so we find that some of the most ancient processes have left unmistakable records.

1.7. THE MAIN FEATURES OF EVOLUTIONARY PROCESSES

The major stages considered here in the evolution of a planetary or satellite system are represented by Figure 1.1 in essential agreement with what is often referred to as the planetesimal approach.

In considering the origin of asteroids we discuss two different possibilities. These bodies may be produced by the disruption of one or several protoplanets, or they may be formed directly by accretion from the intermediate embryonic stage.

Accordingly, the evolution of a system of secondary bodies is believed to have passed through four main phases:

(1) In connection with the formation of the central body, or in the final phase of its formation, plasma is accumulated around the body and brought into partial corota-

tion with it. During this phase a transfer of angular momentum from the central body determines the dynamic state of the secondary bodies formed later. We can use some of our knowledge of the present state of the terrestrial magnetosphere in the analysis of the plasma distribution around the central bodies.

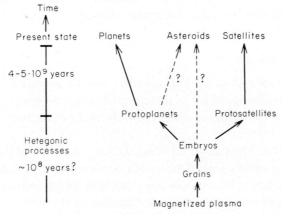

Fig. 1.1. The major stages in the evolution of a planetary or satellite system according to the planetesimal approach.

(2) The plasma condenses, at least partly, to grains whose dynamical state is determined by that of the plasma. Even the primeval plasma may contain a large quantity of grains. Chemical differentiation takes place during the first and second phases of evolution. In order to understand this differentiation it is necessary to consider chemical effects which differ in several important aspects from those which have been conventionally considered in discussions of the formation of primordial matter.

(3) The grains accrete to larger bodies. It appears likely that the development of jet streams of gases and solids is an intermediate step in the formation of these bodies. During the accretion a state is reached which in some respects is similar to that which is now observed in the asteroidal region. The study of the spins of planets and asteroids provide us with valuable information for the understanding of the accretion process. Phases (1)–(3) are not strictly consecutive. They overlap and proceed simultaneously, covering a time period of perhaps 10^8 yr.

(4) The later development of our solar system has been much less dramatic. Tidal effects have braked the spin of all satellites and transformed the Moon from a retrograde to a prograde satellite. With these and a few other exceptions, the state resulting from the first three phases appears to be well preserved. Thus it is possible to trace some of the early phenomena from the present state with surprisingly high accuracy. The fine structure of the Saturnian ring system is one of the most striking examples. The dimensions of the markings seem to have changed less than one percent since the formation of the rings some billion years ago. Similar conclusions are reached from studies of the resonance structure of the solar system (See Section 4).

To take advantage of the principle outlined in Section 1.6, we are treating the evolution through the four phases in reversed order; i.e., backward from the present time. The first six sections consequently center on the present state and the late evolution whereas Sections 7–9 deal with the accretion. Following sections will be devoted to the condensation of primordial matter and its chemical differentiation, and further the injection of plasma and the mass distribution. Finally the general theory will be applied to a number of special problems.

1.8. PLANETARY COSMOGONIES

The number of publications dealing with the various aspects of the evolutionary processes is overwhelming. Many of these contributions discuss only isolated aspects or are entirely speculative. In the present context, many of them are of only historical importance.

For reasons outlined in Section 1.3, a satisfactory treatment of the evolution of the solar system must contain a general theory of the formation of secondary bodies around primary central bodies. Although this principle remains unchallenged, new contributions continue to appear which postulate theories of the formation of a *planetary* system which are obviously inapplicable to the formation of *satellite* systems. A recent review (Herczeg, 1968), for example, hardly mentions the existence of satellite systems; still less does it try to explain their origin. This type of approach to the hetegonic problem is unfortunately common. Such theories would have been quite legitimate before Galileo's discovery of the Jovian satellites since at that time it was possible to consider the formation of planets without being burdened by the existence of complex satellite systems. In the following sections, we refer to cosmogonic theories only in cases where satellites are also considered, and where the fundamental problem of the formation of secondary bodies around a primary body have been realistically considered.

2. The Orbits of Planets and Satellites

2.1. DIFFERENT PRESENTATIONS OF CELESTIAL MECHANICS

The dynamic state of a celestial body can be represented by nine quantities. Of these, three give the position of the body (e.g., its center of gravity) at a certain moment, three give its three-dimensional velocity, and three its spin (around three orthogonal axes). These quantities vary more or less rapidly in a way which is tabulated in the Nautical Almanac. In order to study the origin and the long-time evolution of the dynamic state of the solar system, we are mostly interested in those dynamical quantities which are invariant or vary very slowly.

The typical orbits of satellites and planets are circles in certain preferred planes. In the satellite systems, the preferred planes tend to coincide with the equatorial planes of the central bodies. In the planetary system, the preferred plane is essentially the orbital plane of Jupiter (because this is the biggest planet), which is close to the plane of the ecliptic. The circular motion with period T is usually modified by super-imposed oscillations. Radial oscillations (in the preferred plane) with period $\approx T$

change the circle into an ellipse with eccentricity e. Axial oscillations (perpendicular to the preferred plane), also with a period $\approx T$, make the orbit inclined at an angle i_0 to this plane.

The traditional way of presenting celestial mechanics has the aim of making it useful for the preparation of the Nautical Almanac, and nowadays also for the calculation of spacecraft trajectories. This way is not very suitable if we want to study the interaction of grains with a plasma or with any viscous medium, or the mutual interaction between grains. It is more convenient to use an approximate method, the essence of which is that an elliptical orbit is treated as a perturbation of a circular orbit. This method is obviously applicable only for orbits with small eccentricities. From a formal point of view the method has some similarity to the guiding-center method of treating the motion of charged particles.

2.2. CIRCULAR ORBITS

If a body with negligible mass is moving around a central body, an important quantity is the angular momentum $\mathbf{C}$ (per mass unit) of the small body with reference to the central body (or, strictly speaking, to the center of gravity). This is defined as

$$\mathbf{C} = \mathbf{r} \times \mathbf{v},$$ (2.1)

where $\mathbf{r}$ is its distance to the center, and $\mathbf{v}$ is its orbital velocity. $\mathbf{C}$ is an invariant vector during the motion.

The body is acted upon by the gravitational attraction f_g of the central body and by the centrifugal force

$$f_c = v_\psi^2/r = C^2/r^3,$$ (2.2)

where v_ψ is the tangential velocity component.

The simplest type of motion is the motion with constant velocity v_0 in a circle with radius r_0. The gravitational force f_g is exactly compensated by the centrifugal force. We have

$$v_0 = \frac{C}{r_0} = (r_0 f_g)^{\frac{1}{2}} = \frac{r_0^2 f_g}{C}.$$ (2.3)

The angular velocity of the motion is

$$\omega_K = \frac{v_0}{r_0} = \left(\frac{f_g}{r_0}\right)^{\frac{1}{2}} = \frac{r_0 f_g}{C} = \frac{C}{r_0^2},$$ (2.4)

with the period $T_K = 2\pi/\omega_K$.

2.3. OSCILLATION

The body can perform oscillations around the circular orbit in both the radial and the axial directions.

If the body is displaced radially from r_0 to $r = r_0 + \Delta r$, it is acted upon by the force

$$\Delta f_r = f_c - f_g = \left[\frac{C^2}{r^3} - f_g(r)\right]_0.$$ (2.5)

Because the force is zero for $r=r_0$ we obtain

$$\mathrm{d}f_r = -\left(\frac{3C^2}{r^4} + \frac{\partial f_g}{\partial r}\right)_0 \mathrm{d}r . \tag{2.6}$$

This means that the body oscillates around the circle with the angular velocity

$$\omega_r = \left(-\frac{\mathrm{d}f_r}{\mathrm{d}r}\right)_0^{\frac{1}{2}} = \left(\frac{3C^2}{r^4} + \frac{\partial f_g}{\partial r}\right)_0^{\frac{1}{2}} = \left(\frac{3f_g}{r} + \frac{\partial f_g}{\partial r}\right)_0^{\frac{1}{2}} . \tag{2.7}$$

If it is displaced in the z-direction (axial direction), it is acted upon by the force f_z, which because div $f = 0$ is given by

$$\frac{\partial f_z}{\partial z} = -\left[\frac{1}{r}\frac{\partial}{\partial r}(rf_g)\right]_0 = \left[-\frac{1}{r}\left(f_g + r\frac{\partial f_g}{\partial r}\right)\right]_0 . \tag{2.8}$$

The angular velocity of the oscillation is

$$\omega_z = \left(-\frac{\partial f_z}{\partial z}\right)^{\frac{1}{2}} = \left(-\frac{f_g}{r} - \frac{\partial f_g}{\partial r}\right)_0^{\frac{1}{2}} . \tag{2.9}$$

From (2.7), (2.9), and (2.4) we find

$$\omega_r^2 + \omega_z^2 = 2\omega_K^2 . \tag{2.10}$$

We place a moving coordinate system with the origin at a point moving along the circle r_0 with the angular velocity ω_K (Figure 2.1). The x-axis points in the radial direction and the y-axis in the forward tangential direction. We have

$$x = r\cos(\psi - \omega_K t) - r_0 , \tag{2.11}$$
$$y = r\sin(\psi - \omega_K t) , \tag{2.12}$$

where ψ is the angle measured from a fixed direction. The radial oscillation can be written

$$r = r_0\left[1 - e\cos(\omega_r t - \psi_r)\right], \tag{2.13}$$

where $e\, r_0$ is the amplitude ($e \ll 1$) and ψ_r is a constant. Because C is constant, we have

$$\frac{\mathrm{d}\psi}{\mathrm{d}t} = \frac{C}{r^2} \approx \frac{C}{r_0^2}\left[1 + 2e\cos(\omega_r t - \psi_r)\right]. \tag{2.14}$$

As $x \ll r_0$ and $y \ll r_0$ we find from (11), (12), (13), (14), and (4):

$$x \approx r - r_0 = -r_0 e\cos(\omega_r t - \psi_r)$$
$$= -r_0 e\cos(\omega_K t - \omega_\pi t - \psi_r), \tag{2.15}$$

where we introduce

$$\omega_\pi = \omega_K - \omega_r . \tag{2.16}$$

$$\frac{\mathrm{d}y}{\mathrm{d}t} \approx r_0\left(\frac{\mathrm{d}\psi}{\mathrm{d}t} - \omega_K\right) \approx \frac{2eC}{r_0}\cos(\omega_r t - \psi_r); \tag{2.17}$$

or, after integration,

$$y \approx 2r_0 e \left(1 + \frac{\omega_\pi}{\omega_r} \right) \sin \left(\omega_K t - \omega_\pi t - \psi_r \right). \tag{2.18}$$

The pericenter (point closest to the center) is reached when x is a minimum; that is, when

$$\omega_K t - \omega_\pi t - \psi_r = 2\pi n \quad (n = 0, 1, 2 \ldots). \tag{2.19}$$

If we put

$$\psi_\pi = \omega_\pi t + \psi_r, \tag{2.20}$$

the pericenter is reached when $t = (\psi_\pi + 2\pi n)/\omega_K$. This shows that the pericenter moves (has a 'precession') with the velocity ω_π, given by (2.16).

In a similar way we find the axial oscillations:

$$z = r_0 i \sin (\omega_z t - \psi_z) = r_0 i \sin (\omega_K t - \omega_\theta t - \psi_z) \tag{2.21}$$

where $i \, (\ll 1)$ is the inclination and

$$\omega_\theta = \omega_K - \omega_z. \tag{2.22}$$

The angle ψ_θ of the 'ascending node' (point where z becomes positive) is given by

$$\psi_\theta = \omega_\theta t + \psi_z. \tag{2.23}$$

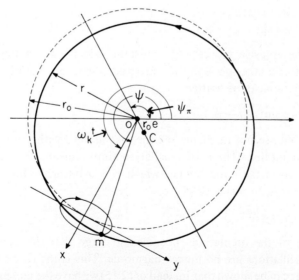

Fig. 2.1. The central body is at the origin 0, which is the center of the dashed (---) circle with radius r_0. Radial oscillations change the orbit of a point mass m into an ellipse, which almost coincides with the solid circle (———) which has its center at C. The distance OC is $r_0 e$. The position of the pericenter is given by ψ_π. The difference between the full circle and the exact Kepler ellipse is less than the thickness of the line. Let the origin of a coordinate system (x, y) move with constant velocity along the dashed circle. In this coordinate system the point mass m moves in an 'epicycle' which is an ellipse with the axis ratio 2:1. The epicycle motion is retrograde.

2.4. COULOMB FORCE

If the mass of the orbiting body is taken as unity, then

$$f_g = \mu/r^2,\tag{2.24}$$

with $\mu = \kappa M_c$ is the mass of the central body, and κ the gravitational constant. As $f_c = f_g$ for the undisturbed motion we have from (2.2) and (2.24)

$$C = (\mu r_0)^{1/2}.\tag{2.25}$$

From (2.24) we get

$$\partial f_g/\partial r = -2f_g/r.\tag{2.26}$$

Substitution of (2.26) into (2.7) and (2.9) shows

$$\omega_r = \omega_z = \omega_K,\tag{2.27}$$

with

$$\omega_K = \mu/Cr_0 = (\mu/r_0^3)^{1/2}.\tag{2.28}$$

The significance of (2.27) is that the frequencies of radial and axial oscillation coincide with the fundamental angular velocity of circular motion. Consequently we have $\omega_\pi = \omega_\theta = 0$ and there is no precession of the pericenter or of the nodes. According to (2.15) and (2.18) the body moves in the 'epicycle'

$$x = -r_0 e \cos(\omega_K t - \psi_r),\tag{2.29}$$
$$y = 2r_0 e \sin(\omega_K t - \psi_r).$$

The center of the epicycle moves with constant velocity along the circle r_0. The motion in the epicycle takes place in the retrograde direction. See Figure 2.1.

The axial oscillation can be written

$$z = r_0 i \sin(\omega_K t - \psi_z).\tag{2.30}$$

We still have an ellipse but its plane has the *inclination i* with the plane of the undisturbed circular motion. The axial oscillation simply means that the plane of the orbit is changed from the initial plane, which was arbitrarily chosen because in a Coulomb field there is no preferred plane.

2.5. LARGE ECCENTRICITY

If the amplitude of the oscillations becomes so large that the eccentricity is not negligible, the oscillations are no longer harmonic. This is the case for most comets and meteoroids. It can be shown that instead of (2.15) we have the more general formula

$$x \approx r - r_0 \approx -\frac{r_0 e \cos\varphi}{1 + e \cos\varphi},\tag{2.31}$$

where r_0 is the radius of the undisturbed motion, defined by (2.25) and φ the angle between the vector radius of the satellite and of the pericenter of its orbit (the point where it is closest to the central body). The relation (2.27) is still valid, but the period changes

$$T = T' = 2\pi (\kappa M)^{-1/2} a^{3/2}, \tag{2.32}$$

with

$$a = r_0/(1 - e^2). \tag{2.33}$$

It can be shown that geometrically the orbit is an ellipse with $a =$ semi-major axis and $e =$ eccentricity.

2.6. MOTION IN THE FIELD OF A ROTATING CENTRAL BODY

According to (2.27), the motion in a Coulomb field is degenerate in the sense that $\omega_r = \omega_z = \omega_K$. This is due to the fact that there is no preferred direction.

In the planetary system and in the satellite systems the motions are *perturbed* because the gravitational fields deviate from Coulomb fields. This is essentially due to the effects discussed in this Section and Section 2.7.

The axial rotations (spins) of the planets change the shape of these bodies from spherical to ellipsoidal. We can consider their gravitation to consist of a Coulomb field from a sphere, on which is superimposed the field from the 'equatorial bulge'. The latter contains higher order terms, but has the equatorial plane as the plane of symmetry. We can write the gravitational force *in the equatorial plane*

$$f_g = \frac{\mu}{r^2}\left(1 + \frac{\alpha}{r^2}\right), \tag{2.34}$$

taking account only of the first term from the equatorial bulge. The constant α is always positive. As in this case

$$|\partial f_g/\partial r| > 2f_g/r, \tag{2.35}$$

we have from (2.7), (2.9), and (2.4)

$$\omega_z > \omega_K > \omega_r. \tag{2.36}$$

According to (2.16) and (2.22) this means that the pericenter moves with the velocity

$$\omega_\pi = \omega_K - \omega_r > 0, \tag{2.37}$$

i.e., in the prograde direction, and the nodes move with the velocity

$$\omega_\theta = \omega_K - \omega_z < 0 \tag{2.38}$$

i.e., in the retrograde direction.

Further we obtain from (2.10), (2.37), and (2.38)

$$\omega_\pi + \omega_\theta = \frac{\omega_\pi^2 + \omega_\theta^2}{2\omega_K}. \tag{2.39}$$

As the right-hand term is very small we find in a first approximation

$$\omega_\pi = -\omega_\theta, \tag{2.40}$$

which is a well-known result in celestial mechanics.

Introducing (2.40) into (2.39) we get in a second approximation

$$\Delta\omega = \omega_\pi + \omega_\theta = \omega_\pi^2/\omega_K.$$ (2.41)

Brouwer (1945) has treated the case when a satellite moves so close to a rotating central body that the quadrupole term of the equatorial bulge becomes important, and applied the result to the motion of Amalthea (Jupiter V), for which we have from observations

$$\omega_K = 722 \text{ deg day}^{-1} = 722 \cdot 365 \text{ deg yr}^{-1};$$ (2.42)
$$\omega_\pi = 917.4 \text{ deg yr}^{-1}; \quad \omega_\theta = -914.7 \text{ deg yr}^{-1}.$$

From an analysis according to the usual methods of celestial mechanics, Brouwer finds

$$\Delta\omega = 3.2 \text{ deg yr}^{-1}$$ (2.43)

(in reasonable agreement with the observational value 2.7 deg).

Our formula (2.41) gives

$$\Delta\omega = \frac{917^2}{722\,365} = 3.16 \text{ deg yr}^{-1}.$$ (2.44)

Hence even for this case where the equatorial bulge effect is large, our simple formula seems to give a result in good agreement with the elaborate treatment by the usual methods. It should be observed, however, that it is valid only for small inclinations (see Öpik, 1958).

2.7. PLANETARY MOTION PERTURBED BY OTHER PLANETS

The motion of the body we are considering is perturbed by other bodies orbiting in the same system. Except when the motions are commensurable so that resonance effects become important, the main perturbation can be computed from the average potential produced by the other bodies.

As most satellites are very small compared to their central bodies, the effects described in Section 2.6 dominate in the satellite systems. On the other hand the perturbation of the planetary orbits is almost exclusively due to the gravitational force of the planets, among which the gravitational effect of Jupiter dominates. In order to calculate this one smears out Jupiter's mass along its orbit and computes the gravitational potential from this massive ring. It produces a perturbation which both outside and inside Jupiter's orbit obeys (2.35). Hence (2.36), (2.37), and (2.38) are also valid.

The dominating term for the calculation of the perturbation of the Jovian orbit derives from a similar effect produced by Saturn.

2.8. ORBITAL PROPERTIES OF PLANETS AND SATELLITES

The most important invariants of the motion are the absolute values of the *orbital angular momenta* and of the *spins* of the celestial bodies. Although the space orientation of these vectors is not constant with time, but changes with a period of a few years up to 10^6 yr (the first figure referring to close satellites and the latter to the outer planets), there are reasons to believe that – with some noteworthy exceptions – the absolute values have remained essentially constant since the formation of the bodies.

There are exceptions to this general rule. Tidal effects have changed the orbital momentum and spin of the Moon and the spin of the Earth in a drastic way, and have produced a somewhat similar effect in the Neptune-Triton system (Section 5). The spins of Mercury and Venus were possibly slowed down until they were captured in spin-orbit resonances. The spins of all satellites have been braked to synchronism with the orbital motion. It is a controversial question to what extent the orbits of satellites other than the Moon and Triton have been changed by tides. As we shall see in Sections 4.5 and 4.11, it is likely that the changes have been very small.

Tables 2.1, 2.2, and 2.3 provide those physical elements of the planets and satellites which are relevant to our discussion. Particularly important is the first column, which gives the specific angular momentum (that is, the angular momentum per unit mass) of the orbiting body, defined by

$$\mathbf{C} = \mathbf{r} \times \mathbf{v}, \tag{2.45}$$

where $\mathbf{r}$ is the radius vector from the central body (ideally from the common center of gravity) and $\mathbf{v}$ is the orbital velocity. The absolute values of $\mathbf{C}$ are listed, as are those of the total angular momentum $C_M = M \cdot C$, where M is the mass of the orbiting body.

If M_C is the mass of the central body and κ is the gravitational constant (6.67×10^{-8} dyne cm^{-2} g^{-2}), then the semi-major axis a and the eccentricity e of the orbital ellipse are connected with C through

$$C^2 = \mu a (1 - e^2). \tag{2.46}$$

Of the planets and the prograde satellites, all except Nereid have $e < 0.25$. Most of them, in fact, have $e < 0.1$ (exceptions are the planets Mercury and Pluto, the satellites Jupiter 6, 7, and 10, and Saturn's satellite Hyperion). Hence the approximation

$$C \approx (\mu a)^{1/2} \tag{2.47}$$

is correct within 3.1% in the first case and within 0.5% in the second.

From the value for the semi-major axis, a, is calculated the sidereal period of revolution

$$T = 2\pi\mu^{-\frac{1}{2}}a^{\frac{3}{2}} = 2\pi C^3 \mu^{-2}(1 - e^2)^{-\frac{3}{2}},$$

$$T \approx \frac{2\pi}{\mu^2} C^3; \tag{2.48}$$

and the average orbital velocity

$$v \approx 2\pi a/T = \mu^{1/2}a^{-1/2} = \mu C^{-1}. \tag{2.49}$$

In Table 2.1 the orbital inclination of the planets, i_0, refers to the orbital plane of the earth (the ecliptic plane). It would be more appropriate to reference it to the invariant plane of the solar system, the so-called Laplacean plane. However, the difference is small and will not seriously affect our treatment.

For the satellites, the orbital plane is referred to the most relevant reference plane as shown in the tables. For close satellites this is the equatorial plane of the planet because the precession of the orbital plane is determined with reference to this plane. For distant satellites the orbital plane of the planet is more relevant because the influence from the Sun's gravitational field is more important.

2.9. PHYSICAL PROPERTIES OF PLANETS AND SATELLITES

Having dealt with the orbital characteristics, we devote the remainder of each table to the orbiting body itself. Given the mass and the radius of the body, its *mean density* is calculated from

$$\theta = 3M(4\pi R^3)^{-1}. \tag{2.50}$$

Following a list of observed *periods of axial rotation* (spin periods), the planetary *moments of inertia* are tabulated.

If R_i is the radius of inertia and R the radius of the body, the ratio $\alpha = R_i/R$ is a measure of the mass distribution inside the body. For a homogeneous sphere we have $\alpha^2 = 0.4$. A smaller value indicates that the density is higher in the central parts than in the outer layers.

Next, the *inclination of equator to orbit plane*, i_e, is tabulated in the planetary table.

The velocity which is necessary for shooting a particle from the surface of a celestial body to infinity is called the *escape velocity* v_e. It is also the velocity at which a particle hits the body if falling from rest at infinity. We have

$$v_e = (2\mu/R)^{1/2}. \tag{2.51}$$

If a satellite is orbiting very close to the surface of the planet, such a *grazing satellite* has $a = R$. Its orbital velocity is $v_e/\sqrt{2}$.

We define the 'time of escape' through

$$\tau_e = R/v_e, \tag{2.52}$$

which means

$$\tau_e = (8\pi\kappa\theta/3)^{-1/2} = 1340\,\theta^{-1/2}\ \sec\ (\text{g/cm}^3)^{1/2}. \tag{2.53}$$

It is easily shown that if a particle is shot up vertically from a body with velocity v_e, it reaches a height

$$h = [(\tfrac{5}{2})^{2/3} - 1]\,R_0 = 0.84\,R_0 \tag{2.54}$$

after the time τ_e. This time is related to the period τ_g of a 'grazing satellite' in Table 2.1 through

$$\tau_g = 2\pi\sqrt{2}\tau_e = 8.9\,\tau_e. \tag{2.55}$$

For the Earth ($\theta = 5.5$) we have $\tau_e = 10$ min.

2.10. OTHER QUANTITIES OF HETEGONIC IMPORTANCE

We have two columns listing the values of q and Q, the *ratio of the orbital distances* and the *ratio of the masses of consecutive bodies* respectively. The quantity q takes the place of the number magic of Titius-Bode's 'law' (Section 2.13).

All the planets and most of the satellites orbit in the same sense ('prograde') as their central body spins. This is probably the result of a transfer of angular momentum from the spin of the central body to orbital momentum of the secondary bodies at the time when the system was formed.

However, there are a few satellites which orbit in a *retrograde* direction. With the exception of Triton, their orbits differ from those of the prograde satellite also in the respect that their eccentricities and inclinations are much larger. As their origin is likely to be different (they are probably captured bodies) they are listed separately (Table 2.3). Being initially a captured satellite, the Moon is also included in Table 2.3 (See Section 5.4).

The heading '*grazing planet (satellite)*' refers to the dynamical properties of a body moving in a Kepler orbit grazing the surface of the central body. Similarly, *synchronous planet (satellite)* refers to a body orbiting with a period equal to the spin period of the central body. The data of such fictitious bodies are merely useful references for the orbital parameters of the system.

Some of the relations given in Tables 2.1 and 2.2 are plotted in the diagrams of Figures 2.2, 2.3, 2.4, and 2.5.

2.11. MASS DISTRIBUTION IN THE SOLAR SYSTEM

Practically all cosmogonic theories assume that the celestial bodies have condensed from a primeval dilute gas or plasma. If we find a group of secondary bodies orbiting at a certain distance from a central body, it is reasonable to assume that they derive from a plasma or gas cloud which once was located in the same region. As we shall show in a later part of this work, the relation between a primeval plasma and the bodies formed out of it is more complicated, but relevant conclusions can be drawn from this simple picture.

If we accept this picture, we can reconstruct the primeval plasma density in the solar system by smearing out the masses of the planets and satellites. It is reasonable to assume that the mass M_n of a planet or satellite was initially distributed over a toroidal volume around the present orbit of the body. Let us further assume that the small diameter of the toroid is defined by the intermediate distances to adjacent orbiting bodies; that is, the diameter will be the sum of half the distance to the orbit of an adjacent body closer to the central body and half the distance to the orbit of one farther from it.

We find

$$M_n = 2\pi r_n \times \pi \left[\frac{r_{n+1} - r_{n-1}}{4} \right]^2 \times \varrho = \frac{\pi^2}{8} \left[q_n^{n+1} - \frac{1}{q_{n-1}^n} \right]^2 r_n^3 \varrho \qquad (2.56)$$

where r_n is the orbital distance of the n-th body from the central body and ϱ is the smoothed-out density. Further $q_n^{n+1} = r_{n+1}/r_n$. Numerical values of q are given in

TABLE 2.1

Planets	Specific Angular Momentum $C = [\mu a(1-e^2)]^{1/2}$ 10^{19} cm² sec⁻¹	Semi-Major Axis a 10^{13} cm	Sidereal Period of Revolution $T = 2\pi\mu^{-1/2}a^{3/2}$ 10^8 sec	Average Orbital Velocity $v = 2\pi a/T$ 10^5 cm sec⁻¹	Eccentricity e	Inclination of Orbit to Ecliptic i_0 °	Total Angular Momentum $C_M = M \cdot C$ 10^{46} g cm² sec⁻¹	Mass M 10^{27} g	Equatorial Radius R 10^9 cm
Grazing Planet	0.304	0.00696	0.0001	437.					
Synchronous Planet	1.83	0.253	0.0219	72.5					
Mercury	2.77	0.579	0.0760	47.9	0.206	7 00	0.922	0.333	0.244
Venus	3.79	1.08	0.194	35.0	0.00682	3 23	18.5	4.87	0.605
Earth	4.46	1.50	0.316	29.8	0.0168	0 00	26.6	5.97	0.638
Mars	5.50	2.28	0.593	24.1	0.0933	1 51	3.53	0.642	0.339
Jupiter	10.2	7.78	3.74	13.1	0.0483	1 18	19300.	1900.	7.14
Saturn	13.8	14.3	9.30	9.65	0.0559	2 29	7830.	569.	6.04
Uranus	19.5	28.7	26.5	6.80	0.0471	0 46	1690.	86.8	2.35
Neptune	24.4	45.0	52.0	5.43	0.00853	1 46	2520.	103.	2.23
Pluto	28.0	58.9	78.0	4.75	0.249	17 06	31.	1.1	
							$\Sigma \approx 31420.$		

TABLE 2.1 (*Continued*)

Planets	Average Density $\theta = 3M/4\pi R^3$ g cm^{-3}	Sidereal Spin Period t 10^5 sec	Relative Moment of Inertia $\alpha^2 = R_1^2/R^2$	Inclination of Equator to Orbital Plane i_e °	Escape Velocity $v_e = [2\mu/R]^{1/2}$ 10^6 cm sec^{-1}	Distributed Density M/a^3 10^{-13} cm^{-3}	Ratio of Semi-Major Axes $q_n = a_{n+1}/a_n$	Ratio of Masses $Q_n = M_{n+1}/M_n$
Grazing Planet								
Synchronous Planet								
Mercury	5.50	50.7			0.427	17.	} 1.87	} 14.6
Venus	5.25	210.0		~180	1.04	38.	} 1.38	} 1.23
Earth	5.50	0.864	0.335	23 27	1.12	18.	} 1.52	} 0.108
Mars	3.92	0.886	0.389	25 12	0.502	0.54	} (3.42)	} (2960)
Jupiter	1.25	0.354	0.250	3 07	5.06	40.	} 1.83	} 0.299
Saturn	0.617	0.368	0.220	26 45	3.54	2.0	} 2.01	} 0.153
Uranus	1.59	0.389	0.230	97 59	2.22	0.037	} 1.57	} 1.19
Neptune	2.22	0.504	0.290	29 00	2.49	0.011	} 1.31	} 0.011
Pluto		5.52				~0.00005		

TABLE 2.2A

Prograde Satellites	Specific Angular Momentum $C=[\mu a(1-e^2)]^{1/2}$ 10^{16} cm^2 sec^{-1}	Semi-Major Axis a 10^{10} cm	Sidereal Period of Revolution T 10^6 sec	Average Orbital Velocity $v=2\pi a/T$ 10^6 cm sec^{-1}	Eccentricity e	Inclination of Orbit Degrees
Jupiter						
Grazing Satellite	3.01	0.714	0.106	4.22		
Synchronous Satellite	4.76	1.64	0.354	2.91		
5 Amalthea	4.79	1.81	0.430	2.64	0.0030	0.4
1 Io	7.31	4.22	1.53	1.73	0.0000	0.0
2 Europa	9.24	6.72	3.07	1.38	0.0003	0.0
3 Ganymede	11.7	10.7	6.18	1.09	0.0015	0.0
4 Callisto	15.5	18.8	14.4	0.821	0.0075	0.0
6	38.2	115.	217.	0.333	0.1580	28.4
10	38.2	115.	219.	0.331	0.1405	28.4
7	38.6	117.	224.	0.329	0.2072	27.8
Saturn						
Grazing Satellite	1.51	0.604	0.152	2.51		
Synchronous Satellite	2.16	1.12	0.368	1.92		
10 Janus	2.41	1.58	0.647	1.53	0.000	0.0
1 Mimas	2.66	1.86	0.814	1.43	0.0201	1.52
2 Enceladus	3.00	2.38	1.18	1.26	0.0044	0.03
3 Tethys	3.35	2.95	1.63	1.14	0.0000	1.10
4 Dione	3.78	3.77	2.36	1.00	0.0022	0.03
5 Rhea	4.46	5.27	3.90	0.848	0.0010	0.34
6 Titan	6.81	12.2	13.8	0.557	0.0291	0.33
7 Hyperion	7.48	14.8	18.4	0.506	0.1042	0.43
8 Iapetus	11.6	35.6	68.6	0.326	0.0283	18.4
Uranus						
Grazing Satellite	0.369	0.235	0.0943	1.57		
Synchronous Satellite	0.628	0.624	0.389	1.01		
5 Miranda	0.832	1.27	1.22	0.654	0.0000	0.0
1 Ariel	1.06	1.91	2.18	0.553	0.0000	0.0
2 Umbriel	1.26	2.68	3.58	0.470	0.0000	0.0
3 Titania	1.60	4.38	7.52	0.366	0.0000	0.0
4 Oberon	1.86	5.86	11.6	0.317	0.0000	0.0
Neptune						
Grazing Satellite	0.392	0.223	0.0798	1.76		
Synchronous Satellite	0.770	0.786	0.504	0.980		
2 Nereid	4.14	55.6	311.	0.112	0.7493	27.8
Mars						
Grazing Satellite	0.0121	0.0339	0.0600	0.355		
Synchronous Satellite	0.0315	0.211	0.886	0.150		
1 Phobos	0.0201	0.0938	0.276	0.214	0.0170	0.95
2 Deimos	0.0318	0.235	1.09	0.135	0.0031	1.73

TABLE 2.2B

	Total Angular Momentum	$C_M = M \cdot C$ 10^{11} g cm² sec⁻¹	Mass M 10^{24} g	Equatorial Radius R 10^7 cm	Distributed Density M/a^3 10^{-9} g cm⁻³	Ratio of Semi-Major Axes $q_n = a_{n+1}/a_n$	Ratio of Masses $Q_n = M_{n+1}/M_n$
Jupiter							
Grazing Satellite							
Synchronous Satellite							
5 Amalthea				0.70?		(2.33)	
1 Io		53.5	73.	16.7	074.	1.59	0.65
2 Europa		43.9	48.	14.6	157.	1.60	3.3
3 Ganymede		180.	15.5	25.5	126.	1.76	0.62
4 Callisto		147.	95.	23.6	14.2	(6.09)	
6				0.50?		1.00	
10				0.07?		1.02	
7		$\Sigma \approx 424$		0.10?			
Saturn							
Grazing Satellite							
Synchronous Satellite							
10 Janus							
1 Mimas		0.0101	0.038	3.0?	5.94	1.18	
2 Enceladus		0.0214	0.071	3.0?	5.28	1.28	1.9
3 Tethys		0.219	0.65	5.0?	25.6	1.24	9.2
4 Dione		0.390	1.10	5.0?	19.3	1.28	1.6
5 Rhea		1.02	2.3	7.0	15.6	1.40	2.2
6 Titan		93.4	137.	24.4	75.1	(2.32)	(60.)
7 Hyperion		0.0851	0.11	2.0?	0.351	1.21	0.00083
8 Iapetus		1.32	1.1	5.0?	0.0252	2.41	10.
		$\Sigma \approx 96$					
Uranus							
Grazing Satellite							
Synchronous Satellite							
5 Miranda		0.00832?	0.1?	1.0?	47.7?	1.51	12.?
1 Ariel		0.127?	1.2?	3.0?	170.?	1.40	0.42?
2 Umbriel		0.0629?	0.5?	2.0?	26.3?	1.64	8.0?
3 Titania		0.0640?	4.0?	5.0?	47.6?	1.34	0.65?
4 Oberon		0.484?	2.6?	4.0?	12.9?		
		$\Sigma \approx 0.7$					
Neptune							
Grazing Satellite							
Synchronous Satellite							
2 Nereid		0.00188	0.03?	1.0?	0.000174		
Mars							
Grazing Satellite							
Synchronous Satellite							
1 Phobos				0.06?		2.50	
2 Deimos				0.03?			

TABLE 2.3

Selected Physical Elements of the Retrograde (Captured) Satellites of the Solar System

Planet	Satellite	Specific Angular Momentum $C = [\mu a(1-e^2)]^{1/2}$ 10^{16} cm² sec⁻¹	Semi-Major Axis a 10^{10} cm	Sidereal Period of Revolution T 10^5 sec	Average Orbital Velocity $v = 2\pi a/T$ 10^6 cm sec⁻¹	Eccentricity e	Inclination of Orbit degrees	Mass M 10^{24} g	Equatorial Radius R 10^7 cm
Jupiter	12	51.9	212.	545.	0.244	0.1687	146.7		0.06?
	11	53.5	226.	598.	0.237	0.2068	163.4		0.08?
	8	50.0	235.	637.	0.232	0.4000	147.		0.10?
	9	51.9	237.	655.	0.228	0.2750	157.		0.08?
Saturn	9 Phoebe	22.2	130.	476.	0.171	0.1632	173.9		1.0?
Neptune	1 Triton	1.54	3.53	5.08	0.437	0.0000	159.9	138.	20.0
Earth	Moon	0.394	3.84	23.6	0.102	0.0549	5.15	73.5	17.4

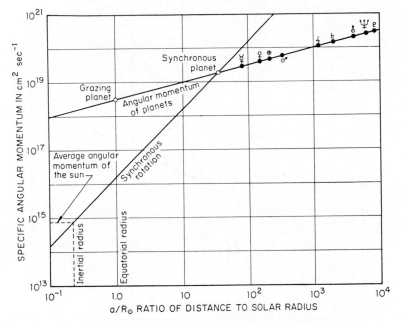

Fig. 2.2. Specific angular momenta of the Sun and Planets.

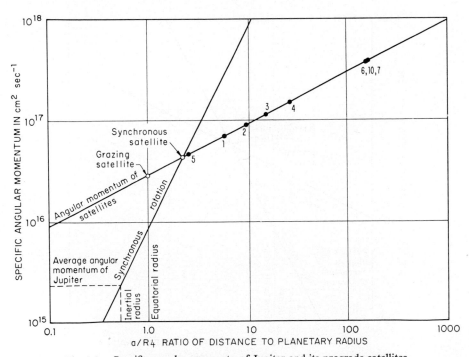

Fig. 2.3. Specific angular momenta of Jupiter and its prograde satellites.

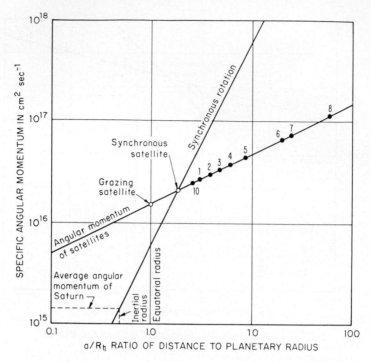

Fig. 2.4.   Specific angular momenta of Saturn and its prograde satellites.

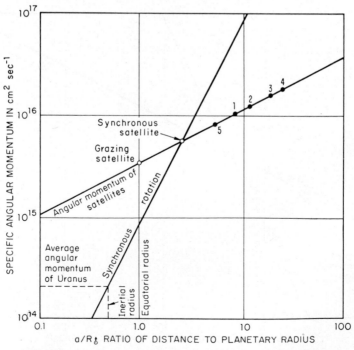

Fig. 2.5.   Specific angular momenta of Uranus and its satellites.

Tables 2.1 and 2.2. The equation has no physical meaning at the inner or outer edge of a group of bodies. Inside the groups q is about 1.2–1.6, which means that the square of the term in parenthesis varies between 0.1 and 1.0. Hence in order to calculate an order of magnitude value we can put

$$\varrho = M_n r_n^{-3}, \tag{2.57}$$

which is the formula employed for the distributed density column in Tables 2.1 and 2.2. These values are plotted in Figures 2.6–2.9, and curves are drawn to suggest a possible primeval mass density distribution.

It should be kept in mind, however, that for example the terrestrial planets contain mostly non-volatile substances, presumably because volatile substances were not able to condense in this region of space. As the primeval plasma probably contained mainly volatile substances, its density may have been a few orders of magnitude larger than indicated in the diagram.

2.12. LAPLACIAN MODEL VS. REAL CONDITIONS

According to the well-known Laplacian model (to which posterity has ascribed much more importance than did Laplace himself) the primeval Sun was surrounded by a uniform disk which later broke up into rings which condensed to planets. The model idealizes the planetary system as consisting of a uniform sequence of bodies, the orbital radii of which obeyed a simple exponential law (or Titius-Bode's law). Also the model requires that the planetary masses be a simple function of the solar distance. It is natural that there should be an outer limit to the sequence of planets, presumably determined by the outer limit of the original disc. Furthermore, it is natural that no planets could condense very close to the Sun because of its high temperature. But unless a number of ad hoc hypotheses are introduced the theory does not predict that the density should vary in a non-monotonic way inside these limits.

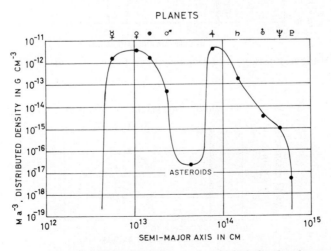

Fig. 2.6. Distributed density versus semi-major axis for the planets.

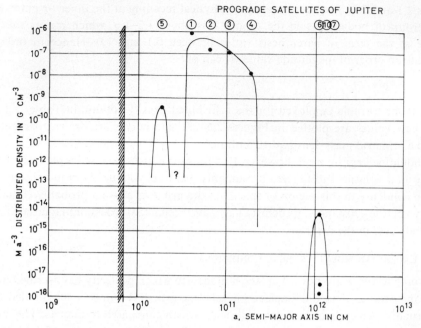

Fig. 2.7.   Distributed density versus semi-major axis for the prograde satellites of Jupiter.

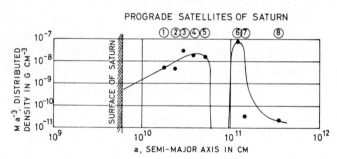

Fig. 2.8.   Distributed density versus semi-major axis for the prograde satellites of Saturn.

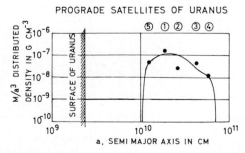

Fig. 2.9.   Distributed density versus semi-major axis for the prograde satellites of Uranus.

As we see from Figure 2.6, the Laplacian model is very far from reality. The density in the region between Mars and Jupiter is lowered by five orders of magnitude. Laplacian type theories have postulated one or more broken-up planets, the fragments of which should now be the asteroids. Even if it were correct, it could not explain the very low density of matter in this region. This requires a systematic transport outwards or inwards, and no very plausible mechanism for this has been proposed. (The difficulties inherent in this view are discussed further in Section 7).

If we try to look at Figure 2.6 without centuries of Laplacian brainwashing, we are more inclined to describe the mass distribution in the planetary system in the following way.

There were two clouds, one associated with the terrestrial, or inner, planets, and a second with the giant, or outer, planets. These clouds were separated by a vast, almost empty region. The inner cloud covered a radial distance ratio of $q(\female - \male) = 3.9$ (where q is the ratio between two orbital radii.) For the outer cloud the corresponding distance ratio is $q(\Psi - 2\!\!\!\!\iota) = 5.8$, or if Pluto is taken into account, $q(\mathrm{P} - 2\!\!\!\!\iota) = 7.6$. cf. Table 2.4. The clouds were separated by a gap with a distance ratio of $q(2\!\!\!\!\iota - \male) = 3.4$. The bodies deriving from each of the two clouds differ very much in initial chemical composition.

As always, the analysis of a single specimen like the planetary system is necessarily somewhat arbitrary; thus it is important to study the satellite systems before we can make objections to the Laplacian model with confidence. We find in the Jovian system that the four Galilean satellites form a group with $q = 4.5$. Similarly, for the five Uranian satellites $q = 4.6$. These values fall within the range of those in the planetary system.

In the case of the planetary system, one could argue that the reason why there are no planets inside the orbit of Mercury is that solar heat prevented a condensation very close to the Sun. This argument is invalid for the inner limit of the Galilean satellites, as well as for the Uranian satellites. Neither Jupiter nor Uranus can be expected to have been so hot as to prevent a formation of satellites close to its surface. We see that Saturn, which both in solar distance and in size is intermediate between Jupiter and Uranus, has satellites (including the ring system) virtually all the way to its surface. Hence the Saturnian system inside Rhea would be reconcilable with a Laplacian uniform disc model, but both the Jovian and the Uranian systems are not in agreement with this picture.

Furthei, in the Saturnian system (Figure 2.8) the fairly homogeneous sequence of satellites out to Rhea is broken by a large void (between Rhea and Titan $q = 2.3$). Titan, Hyperion, and possibly also Iapetus may be considered as one group ($q = 2.9$). The inner satellites including the ring should be counted as a group; q(Rhea-Janus) $= 3.3$.

It is not justifiable to ignore the evident groupings in both planetary and satellite systems and the large gaps which demarcate the groups. The groups are listed with their orbital ratios in Table 2.4.

TABLE 2.4

Groups of planets and satellites

Central Body	Name	Secondary bodies	Orbital ratio	Remarks
Sun	Terrestrial planets	Mercury Venus Earth Moon? Mars	$q = 3.9$	Irregularity: Moon-Mars problem
Sun	Giant planets	Jupiter Saturn Uranus Neptune Triton? Pluto	$q = 5.8$ $q = 7.6$	Doubtful whether Pluto and Triton belong to this group
Jupiter	Galilean satellites	Io Europa Ganymede Callisto	$q = 4.5$	A very regular group: $e \approx 0$, $i \approx 0$. Amalthea is too small and far away to be a member
Uranus	Uranian satellites	Miranda Ariel Umbriel Titania Oberon	$q = 4.6$	Also very regular: $e \approx 0$, $i \approx 0$. The satellites move in the equatorial plane of Uranus, not in its orbital plane ($i_e = 98°$).
Saturn	Inner Saturnian satellites	Janus Mimas Enceladus Thetys Dione Rhea	$q = 3.3$	The satellites form a very regular sequence down to the associated ring system.
Saturn	Outer Saturnian satellites	Titan Hyperion Iapetus	$q = 2.9$	Irregular because of the smallness of Hyperion.
Jupiter	Outer Jovian satellites	VI X VII	$q = 1.0$	Very irregular group consisting of three small bodies in eccentric and inclined orbits

Other prograde satellites: Amalthea, Nereid, Phobos and Deimos.

Thus we find that the celestial bodies in the solar system occur in groups, each having 3–6 members.

It is reasonable that the outer Jovian (prograde) satellites should be considered a similar group consisting of closely spaced small members.

Amalthea may be regarded as the only observed member of another less massive group.

Nereid is perhaps the only remaining member of a regular group of Neptunian satellites which was destroyed by the retrograde giant satellite during the evolution of its orbit.

Phobos and Deimos form a group of extremely small Martian satellites.

2.13. TITIUS-BODE'S 'LAW'

This 'law' has been almost as misleading as the Laplacian model. In spite of the criticism of it by Schmidt (1946) it still seems to be sacrosanct in all textbooks. In its original formulation it is acceptable as a device for memorizing the planetary distances. It is not applicable to Neptune and Pluto, and had they been discovered at the time, the 'law' would probably never have been formulated.

The arguments for the Titius-Bode law are essentially of two different – and incompatible – kinds. According to the first kind there is an excellent numerical agreement between the observed planetary distances and the formula

$$r = 0.4 + 0.3 \cdot 2^n \, \text{AU} \tag{2.58}$$

(with n = integer numbers). This is true only under the following conditions: (a) For Mercury the ad hoc value $n = -\infty$ is inserted. (b) The asteroids are supposed to have been produced from an explosion of a planet between Mars and Jupiter with $n = 3$. (c) Neptune and Pluto are neglected (because they do not fit). (d) No attempt is made to derive the formula theoretically.

According to the second kind of arguments the 'law' is an expression for the supposed fact that the planetary distances follow an exponential law, such that the ratio q between consecutive orbital distances should be a constant.

As is seen from Table 2.2 this is approximately correct within some of the groups of bodies but the q-values vary from one group to another. In fact the values are scattered between the limits $q = 1.2$ and $q = 2$, with a few still higher values.

If Titius-Bode's law is interpreted as q = constant, the constant term in (2.58) should be removed. The result is, however, that the excellent numerical agreement is lost, and hence the charm of the formula.

Attempts have been made to find similar 'laws' for the satellite systems. This is possible only by postulating a distressingly large number of 'missing satellites'.

3. The Small Bodies

3.1. FORCES ACTING ON SMALL BODIES

Besides planets and satellites there are a large number of smaller bodies in interplanetary space; i.e., *asteroids*, *comets*, and *meteoroids*. Further, there are the constituents of the interplanetary plasma; *atoms* or *molecules*, *ions* and *electrons*. The total mass spectrum covers about 50 orders of magnitude. See Figure 3.1. The behavior of the bodies depends in a decisive way on their mass.

(a) For masses $m > 10^{-10}$ g (including asteroids, comets and most meteoroids) the solar gravitation

$$f = \kappa M_\odot m / r^2 \tag{3.1}$$

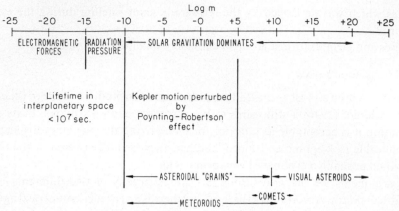

Fig. 3.1.   Forces acting on interplanetary bodies with different masses. (The diagram refers to the present solar radiation field.)

dominates. The basic motion is in a Kepler ellipse. This motion is perturbed in several ways:

(1) *Mass-independent perturbations* are produced by the gravitation of the planets. This effect consists of a secular precession of the nodes and the perihelion of the orbits. See Section 2.3.

(2) *Mass-dependent perturbations* are produced by *light pressure and the Poynting-Robertson* effect. These effects become larger (compared to gravitation) the smaller the body. They are insignificant for $m > 10^5$ g.

(3) Another mass-dependent effect is due to *viscosity*, i.e., to collisions with other bodies. See Section 3.4.

(b) For $m < 10^{-10}$ g, radiation pressure becomes comparable to the solar gravitation. The dynamics of such small particles are very complicated and depend on their size and chemical composition in a critical way. Their lifetime in interplanetary space is often shorter than 10^7 sec. They contribute to the zodiacal light and gegenschein. Recent reviews are given by Biermann (1967) and Belton (1967).

(c) For masses $< 10^{-15}$ g, electromagnetic forces dominate the motion. In the presence of an electric field E and a magnetic field B a particle with a mass m, charge e, and velocity v is acted upon by the force

$$f = e \left(\mathbf{E} + \frac{1}{c} \mathbf{v} \times \mathbf{B} \right). \tag{3.2}$$

For electrons, and also for ions, electromagnetic forces are usually orders of magnitude larger than any other force. In a plasma the number of electrons equals the number of ions and **f** for electrons is almost exactly antiparallel to and equally as large as **f** for ions, with the result that the electromagnetic force on plasmas may be orders of magnitude smaller than given by (3.2). Hence mechanical forces on plasmas often are as important as electromagnetic forces (Alfvén and Fälthammar, 1963).

Furthermore a solid body in space is usually electrically charged. There are two competing effects, the photo-electric effect which tends to give it a positive charge and the ambipolar diffusion in the surrounding plasma, which tends to give it a negative charge. Under the conditions we are considering, the result is usually a charge of the order of a few volts, positive if the grain is situated in a low-density plasma and strongly exposed to solar radiation, but negative if the grain is situated in a dense plasma and receiving little light from the Sun. Suppose that a spherical grain with radius R and density θ has a voltage V esu ($=300$ V). Its charge is $q=RV$, and its mass $m=\frac{4}{3}\pi\theta R^3$. It is subject to a gravitational force f_g from a body with mass M_c

$$f_g = \kappa(mM_c/r_c^2), \tag{3.3}$$

where r_c is the distance to the body. If it moves with the velocity v in a magnetic field B the electromagnetic force f_m is

$$f_m = q\frac{v}{c}B. \tag{3.4}$$

The ratio $\alpha = f_m/f_g$ is

$$\alpha = \frac{3}{4\pi}\frac{VB}{\theta}\frac{v}{c}\frac{r_c^2}{\kappa M_c}\frac{1}{R^2}. \tag{3.5}$$

Suppose that $V=\frac{1}{300}$ esu ($=1$ V), $\theta=3$ g cm^{-3}, $v/c=10^{-4}$ and $\kappa M_c/r_c^2=1$ cm sec^{-2} (for the order of magnitude the solar attraction near the Earth's orbit). Then we have

$$\alpha = 2.5 \times 10^{-8} B/R^2. \tag{3.6}$$

From a theoretical point of view, particles with $m<10^{-10}$ g are of interest as a transient stage between the condensation from a plasma and the accretion to larger bodies. The presence of such small particles, as revealed e.g. by the zodiacal light, give us in principle an opportunity to study the behavior of such particles under present conditions. The results may be applicable to the hetegonic process, especially to the problem of condensation nuclei.

However, the particles in this mass range, which are observed today, are possibly not directly relevant to our problem. As their lifetime is very short, they are not likely to convey any important message to us about the hetegonic process.

Our knowledge of the motion of very small particles is based on theories, some of which are rather speculative. There is little direct evidence that the small particles really behave as they are supposed to.

3.2. PRODUCTION OF SMALL BODIES: FRAGMENTATION, CONDENSATION AND ACCRETION

There are two different ways of accounting for the existence of the small bodies:

(1) They may be produced by *fragmentation* of larger bodies. The asteroids have traditionally been regarded as splinters from one or more planets which have exploded or become fragmented by mutual collisions. In a similar way the meteoroids may have been emitted from comets, or possibly from other bodies like the asteroids.

There is no doubt that collisions at hypervelocities occur in interplanetary space, and that a number of small bodies are fragments from larger bodies.

(2) Small bodies may also be produced by *condensation* of a plasma or gas – existing now or in hetegonic times. *Accretion* of such grains to larger grains which finally become planets or satellites is a basic process of all 'planetesimal' theories. For clarifying these processes it would be very important if we could find and identify these surviving primeval grains in interplanetary space.

Hence, an important problem for our whole analysis is to decide to what extent the small bodies are produced by fragmentation and by condensation-accretion.

There are two different ways to approach the small-body problem:

(a) The study of the *distribution of their orbits*. The theory of their motions will be given in Section 3.7. The observational evidence will be discussed in a later section of this work.

(b) The study of their *size spectra*. This will be discussed in Section 3.8 and 3.9.

3.3. RADIATION PRESSURE. POYNTING-ROBERTSON EFFECT

Besides the gravitation there are essentially two effects which influence the motion of grains $> 10^{-10}$ g. The first one which is due to the interaction between the grains and solar radiations (essentially solar light) will be the subject of this paragraph. The second one is produced by the mutual interaction between the grains due to collisions (viscosity) and will be studied in the following paragraph.

If a body with the cross-section σ is hit by radiation with an energy density w, it will be acted upon by the force

$$f_l = \sigma w / c. \tag{3.7}$$

If the body is a perfect mirror reflecting all light in an antiparallel direction, the force f_l is doubled. If the energy is re-emitted isotropically (seen from the coordinate system of the body), this emission produces no resultant force on the body.

The corpuscular radiation, solar wind, etc. gives a pressure of the same kind.

As shown by Robertson, a black body moving with the velocity v_r, v_θ in the environment of the Sun is acted upon by the radiation pressure with the components

$$f_r = f_l(1 - 2v_r/c), \tag{3.8}$$
$$f_\vartheta = - f_l(v_\theta/c). \tag{3.9}$$

The tangential component (3.9), called the Poynting-Robertson effect, is due to the motion in relation to the radiation field of the Sun.

As f_l decreases in the same way as the gravitational force, it has the same effect as a reduction in $\mu = \kappa M$. For the Sun we have $w = 1.35 \times 10^6$ erg sec^{-1} cm^{-2} and $w/c = 0.45 \times 10^{-4}$ g cm^{-1} sec^{-2} at the distance of the Earth.

Hence we find for the reduced gravitation

$$\mu' = \mu(1 - \gamma) \tag{3.10}$$

with

$$\gamma = \sigma w r^2 / c\kappa M_\odot m.$$ (3.11)

As $\kappa M_\odot / r^2 = 0.59$ cm sec^{-2} we find

$$\gamma = 0.76 \times 10^{-4} (\sigma/m).$$ (3.12)

For a black sphere with density θ and radius R we have

$$\gamma = 0.57 \times 10^{-4} (\theta \times R)^{-1}.$$ (3.13)

For θ of the order of one, this means that the Sun will repel particles with $R < 10^{-4}$ cm (see Lovell, 1954, p. 406). The corresponding mass is of the order of 10^{-10} or 10^{-11} g. This is one of the effects putting a limit to the dominance of the gravitation. It so happens that the size of the particles at this limit is of the same order as the wavelength of maximum solar radiation. The behavior of particles with $m < 10^{-11}$ g is extremely complicated and not very well known.

3.4. Viscosity (collisions)

The evolution of an assembly of particles of different size is in a decisive way governed by their mutual interaction. For the small bodies under consideration, the gravitational interaction is usually negligible and the main interaction is due to viscosity or – to use another term – mutual collisions.

The small grains of different sizes together with the plasma constitute what is referred to as the *interplanetary medium*. The presence of this means that the motions of all bodies in interplanetary space are affected by viscous effects. For planets and satellites these effects are so small that they have not yet been discovered. For the smaller bodies we are studying in this chapter viscosity plays a role which increases as the mass of the body decreases.

The study of the motion of comets has revealed that forces other than gravitation are sometimes important (Marsden, 1968; Hamid *et al.*, 1968). More critical views on these deviations are expressed by Roemer (1963, p. 545). Especially for bodies in the asteroid belt, *viscous effects* (which is a synonym for *collisions*) are sometimes important. All views on the asteroids agree in the respect that collisions have been decisive for the evolution of the asteroidal belt. Kiang (1966) finds that a correlation between proper eccentricities and proper inclinations of asteroids suggest the existence (or former existence) of a resisting mechanism. However, even down to the smallest observed asteroids ($r = 10^5$ cm) gravitation is by far the main force, and viscosity only enters as a small correction. This applies to the present state of the solar system, but not to earlier states. When the matter was dispersed in space, collisions were more frequent than now, and viscosity a more important phenomenon.

For an order-of-magnitude estimate, let us assume that the viscous effect consists of collisions of the grain with other grains of comparable mass, and that such collisions take place with the frequency $1/\tau$. If a celestial body (a grain) moves in a Kepler orbit with period T_K this type of motion is not changed very much if

$$\tau \gg T_K.$$ (3.14)

However, because of perturbations a Kepler orbit is usually subject to secular variation with periods T_s which in a typical case may be as large as $10^4 T_K$ or even larger (Brouwer and Clemence, 1961). See Section 2.7. In order not to interfere with the secular variations of the orbit we must require

$$\tau \gg T_s. \tag{3.15}$$

No viscous effect has been observed to disturb the motion of real planets and satellites. However, in the asteroid belt viscous effects interfering with the secular perturbations may exist. As we shall see, it seems difficult to explain the formation of the observed asteroidal jet streams on any other basis. If this interpretation is accepted, viscous effects can not be neglected in the present state of the Solar System for bodies smaller than 10^6 cm. It is probable that viscous effects were of decisive influence during the early state of the Solar System.

3.5. A BASIC MISTAKE

Most of the discussion of the mutual interaction between grains is based on a misleading conception which we shall discuss in this paragraph.

Suppose that a parallel beam of particles is shot through a medium at rest. Then collisions between the beam particles and the particles at rest will scatter the moving particles. The beam is diffused so that its particles will spread in space.

Almost all treatment of the motion of grains (including meteoroids and asteroids) in interplanetary space is based on this model without realizing that it is applicable only under certain conditions, which are usually *not* satisfied for the following reasons:

(a) It is often implicitly assumed that in interplanetary space there is a 'resistive' medium which is essentially at rest. We know that in interplanetary space there is a radial outward motion of a very thin plasma (solar wind), but its density is too low to affect the motion of grains appreciably (it is smaller than the Poynting-Robertson effect). Hence the resistive medium must necessarily consist of grains. However an assembly of grains cannot possibly be at rest, because the grains are attracted by the Sun. The only possibility is that they are supported by the centrifugal force, i.e., they must move in Kepler orbits.

Hence a 'resistive medium' affecting the motion of the asteroids can be 'at rest' only in the sense that on the average there are an equal number of grains moving in all directions. As we shall see in the following paragraphs, observations do not support the existence of a resistive medium at rest in interplanetary space. There is rather good evidence showing that on the average, meteoroids are *moving in the prograde sense*.

(b) The second mistake is the belief that collisions – either with a 'medium' or with other moving grains – will lead to an increased spread of the velocities and the orbits. As stated above, this is true for a beam of particles which are not moving in a gravitational field. It is also true for particles in a gravitational field under the condition that the collisional frequency is larger than the Kepler frequency. The most important case, however, is when the grains make many Kepler revolutions between collisions. In this case collisions lead to an equalization of the orbits of the colliding particles, with the result that *the spread* in space and in velocity *is reduced*.

Suppose that two particles with masses m_1 and m_2 move in orbits with angular momenta C_1 and C_2 $(C_2 > C_1)$. Around circles with radii r_1 and r_2 they perform oscillations with amplitudes $r_1 e_1$, $r_1 i_1$; and $r_2 e_2$, $r_2 i_2$, respectively (see Section 2.3).

If $r_1 (1 + e_1) > r_2 (1 - e_2)$ they have a chance of colliding. If the precession rates of their nodes and perihelia are different and not commensurable, they will sooner or later collide at a point at the central distance r_3. At the collision their tangential velocities will be $v_1 = C_1/r_3$ and $v_2 = C_2/r_3$. If the collision is perfectly inelastic, their common tangential velocity v_3 after the collision will be $v_3 = (m_1 v_1 + m_2 v_2)/(m_1 + m_2)$. Hence each of them will have the angular momentum

$$C_3 = \frac{m_1 C_1 + m_2 C_2}{m_1 + m_2}, \tag{3.16}$$

which means $C_1 < C_3 < C_2$. Collisions which are at least partilly inelastic will tend to equalize the angular momenta of colliding particles. It is easily seen that collisions also will tend to make the particles oscillate with the same amplitude and phase. This means that the *general result of viscosity is to make the orbits of particles more similar*. In other words, the effect of viscosity is to produce what we may call an apparent attraction between the orbits.

3.6. JET STREAMS

Like many types of electric and magnetic fields a gravitational Coulomb field has a *focussing effect*. We shall in this paragraph discuss some of the consequences of this.

Suppose that a body moves with velocity v_0 and semi-major axis r_0 in a Kepler orbit which is sufficiently close to a circle to allow us to treat it according to Section 2.1. Suppose further that the field is an unperturbed Coulomb field. Hence the orbit of the body will remain an ellipse which does not precess.

Let the body emit particles in all directions with the velocity v. The particles emitted in the radial direction will oscillate around the orbit of the body with the amplitude

$$x = r_0 v/v_0. \tag{3.17}$$

Particles emitted in the axial direction will oscillate with the same amplitude. Further, particles ejected in the tangential (forward) direction will have the angular momentum $C = r_0 (v_0 + v)$, which because $C = \sqrt{(\mu r)}$, is the same as a particle oscillating around the circle

$$r' = \frac{(r_0 v_0)^2}{\mu} \left(1 + \frac{v}{v_0}\right)^2 \approx r_0 \left(1 + \frac{2v}{v_0}\right). \tag{3.18}$$

Hence it will oscillate with the amplitude $2x$ and its maximum distance from the orbit of the body is $4x$.

The particles emitted with a velocity v will remain inside a torus with the small radius $= x' = \beta x$ where x is given by (3.17) and β is between 1 and 4 depending on the angle of emission. This result also applies to the case when the body does not move exactly in a circle.

If a body, or a number of bodies in the same orbit, emit gas molecules with an r.m.s. thermal velocity $v = (3kT/m)^{1/2}$, the gas will be confined within a torus with the typical thickness of $4x$, with

$$x = vr_0/v_0 = (3kTr_0^3/m\mu)^{\frac{1}{4}}. \tag{3.19}$$

As a typical example, the thermal velocity of hydrogen molecules at $T = 300$ K is of the order 10^5 cm/sec. If we put $v_0 = 3 \times 10^6$ cm/sec ($=$ the Earth's orbital velocity), we have $x/r_0 = \frac{1}{30}$.

Hence the body, or bodies, may be surrounded by an atmosphere. The gravitation which prohibits the evaporation of the atmosphere of, say, the Earth, may in effect be substituted for by the forces f_r and f_z from Section 2.3, Equations (2.6) and (2.8) (although there is not a perfect analogy). The outer parts of the torus move slower, the inner parts swifter than the body in the central orbit.

The jet streams differ from the rings in Laplacian theories in the respect that the mean free path of particles in a jet stream is long compared to the dimensions. Further, they need not necessarily be circular. In fact, the phenomena we are discussing will take place even in jet streams with high eccentricity.

Suppose that there is a jet stream, consisting of a large number N of particles all confined to move inside a torus with small radius $= x$. In relation to particles moving in the central orbit their velocity is v. If all of the particles are spheres with the same radius R, the collisional cross-section is $4\sigma = 4\pi R^2$, and each particle in the torus will collide with a frequency which is of the order

$$1/T_v = v\sigma n, \tag{3.20}$$

where T_v is the average time between two collisions and

$$n = N/2\pi^2 r_0 x^2 \tag{3.21}$$

is the particle density. If the average mass density of a particle is θ, its mass is $m = 4\pi\theta R^3/3 = 4\theta\sigma R/3$. The density of the grains may be about 3 g cm^{-3}. Hence, for the order of magnitude we may put $m = 4\sigma R$. If we put the space density $\varrho = mn$ we have

$$T_v = 4R/v\varrho. \tag{3.22}$$

If we consider $v = 10^5$ cm sec^{-1} as a typical relative velocity we find the values for $T_v\varrho$ given in Table 3.1.

In order to keep the jet stream together, the collisional time T_v must be smaller

TABLE 3.1

$R =$	10^{-3}	1	10^3	10^6	cm
$T_v\varrho =$	4×10^{-8}	4×10^{-5}	4×10^{-2}	$4 \times 10^{+1}$	sec g cm^{-3}
$\varrho \geqslant$	1.3×10^{-21}	1.3×10^{-18}	1.3×10^{-15}	1.3×10^{-12}	g cm^{-3}
$n \geqslant$	1.1×10^{-12}	1.1×10^{-18}	1.1×10^{-24}	1.1×10^{-30}	cm^{-3}

than the time constant for the dispersive processes. Most important of these are the differential precession of the different orbits in the jet stream, and the Poynting-Robertson effect. For the order of magnitude we may put $T_v = 10^5$ yr $= 3 \times 10^{12}$ sec (see Whipple, 1968). This value gives the values of ϱ in Table 3.1.

The contraction of a jet stream is produced by inelastic collisions between the particles. The time constant for contraction should be a few times T_v.

This also means that a jet stream can be formed only when there is no disruptional effect with a shorter time constant. For example, the differential precession of the pericentre and the nodes of an elliptic orbit will disrupt a jet stream, unless T_v is smaller than the period of the differential precession.

A more refined model should take account of the size distribution of the particles. Since the smallest particles are usually the most numerous, these will be the most efficient in keeping a jet stream together.

As the relative velocities in the interior of a jet stream decrease, the accretion of grains to larger bodies will be more and more efficient. Hence, for the order of magnitude, T_v is a sort of *coagulation* time of the jet stream. However, if larger bodies are formed the result is that T_v will increase and the contractional force will be smaller. Eventually the jet stream may no longer keep together.

3.7. COLLISIONS BETWEEN A GRAIN AND A JET STREAM

We shall study what happens if a grain with mass m_g collides with a jet stream.

Suppose that the grain moves in an orbit which at one point P crosses the jet stream. See Figure 3.2. (In principle, its orbit could cross the jet stream at two points, but we confine ourselves to the simplest case.) We are considering motions in an unperturbed

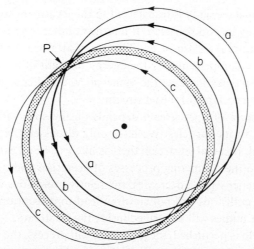

Fig. 3.2.   The capture of a grain by a jet stream. The shaded area represents the jet stream. The orbit of a grain (thick curve) intersects the jet stream at P. Collisions lead to fragmentation and fragments are ejected, for example, along the thin curves. All these orbits carry them back to the point P. Subsequent collisions at P may lead to further fragmentation, but if the collisions are at least partly inelastic, the final result is that all the fragments will be captured by the jet stream.

Coulomb field, which means that the orbits remain unchanged unless the particles collide.

In the region where the grain crosses the jet stream, it will sooner or later collide with one of its constituent particles. The collision likely is partially inelastic; in other words, part of the kinetic energy due to the relative motion is dissipated. The collision may result in a disruption of one of the colliding grains, or of both, into a number of fragments.

After the collision each of these fragments will move in a new orbit which in general differs from the initial orbits of the grains. The orbit may be situated inside the jet stream, but it may very well be mostly outside it. However, all possible orbits of the fragments will necessarily bring them back again to the point where the collision took place. Since by definition this point was situated inside the jet stream, all the fragments will repeatedly cross the jet stream. (An exception to this rule occurs when the collision has taken place near the surface of the jet stream and the latter has had time to contract so much before the next collision that the point of the first collision is outside the stream.) Sooner or later this orbital intersection will lead to new collisions with the particles in the jet stream.

As on the average the collisions smooth out the relative velocities, the fragments will finally be captured by the jet stream. At the same time this capture will change the shape of the jet stream so that the new orbit is a compromise between its original orbit and the orbit of the colliding grain.

Hence a grain which collides with a jet stream will be 'eaten up' by it, with or without fragmentation. In the former case the jet stream 'chews' before it 'swallows'. This again can be considered as a consequence of the focussing effect of a Coulomb field.

In this way new kinetic energy is transferred to the jet stream so that the decrease in its internal energy may be compensated. If a large number of grains are colliding with the jet stream, a temporary state of equilibrium is attained when the losses due to internal collisions are compensated by the energy brought in by the 'eaten' particles. However, the new particles increase the value of N, and hence the losses. The final destiny is in any case a collapse of the jet stream.

The internal structure of a jet stream depends on the size distribution and on the velocity distribution of its particles. We have only discussed the ideal case where all particles are identical. In a real jet stream there is likely to be an assortment of particles of all sizes, subject to the competing processes of accretion and disruption. When the internal energy of the jet stream decreases, the relative velocities will become smaller. This means that the collisions will not so often lead to disruption. The accretion will dominate, and larger bodies will be formed inside the jet stream.

If the Coulomb field is perturbed, the jet stream will precess, the nodes moving in the retrograde and the pericentre in the prograde sense. However, the rate of precession depends on the orbital elements; and these are slightly different for the particles inside the jet stream. Hence the perturbations tend to disrupt the jet stream. The permanence of the jet stream depends upon whether the viscosity, which keeps the jet stream together, is strong enough to dominate. In general, large bodies will leave the jet stream more readily than small bodies.

We conclude:

(1) If a large number of grains are moving in a Coulomb field, its focussing effect may lead to the formation of jet streams. These are kept together by viscosity (mutual collisions).

(2) The jet streams have a tendency to capture all grains which collide with them.

(3) The jet streams have a tendency to contract.

(4) Inside a jet stream the grains will aggregate to larger bodies.

(5) Large bodies formed in a jet stream may break loose from it.

3.8. SIZE SPECTRA

The size spectra of meteoroids, asteroids, and other bodies are of basic importance for the understanding of the origin and evolution of the bodies. A spectrum can be expressed as a function of the radius r (assuming spherical bodies), the cross section $\sigma = \pi r^2$ or the mass $m = \frac{4}{3}\pi\theta r^3$ (where θ is the average density). Furthermore it can be given as a function of the astronomical magnitude which is $g = \text{const.} - 5 \log r$.

The number of particles in the interval between r and $r + dr$ is denoted by $N(r)$, and the functions $N(\sigma)$ and $N(m)$ are defined in similar ways.

We have

$$N(r)\, dr = N(\sigma)\, d\sigma = N(m)\, dm, \tag{3.23}$$

and, consequently,

$$N(r) = 2\pi r N(\sigma) = 4\pi\theta r^2 N(m). \tag{3.24}$$

It is often possible to approximate the distribution functions as power laws valid between certain limits. As the variable can be either r, σ, g, or m and as sometimes differential spectra and sometimes integrated spectra are considered, the literature is somewhat confusing. We put

$$N(r) = v_r r^{-\alpha}, \tag{3.25}$$
$$N(\sigma) = v_\sigma \sigma^{-\beta}, \tag{3.26}$$
$$N(m) = v_m m^{-\gamma}, \tag{3.27}$$

where v_r, v_σ, v_m, and α, β, and γ are constants. We find

$$v_r r^{-\alpha} = 2\pi r v_\sigma \sigma^{-\beta} = 4\pi\theta r^2 v_m m^{-\gamma},$$
$$v_r r^{-\alpha} = 2\pi v_\sigma r \pi^{-\beta} r^{-2\beta} = 4\pi\theta r^2 v_m \left(\tfrac{4}{3}\pi\theta\right)^{-\gamma} r^{-3\gamma},$$

which gives the relations

$$\alpha - 1 = 2(\beta - 1) = 3(\gamma - 1) \tag{3.28}$$

and

$$v_r = 2\pi^{1-\beta} v_\sigma = (4\pi\theta)^{1-\gamma}\, 4^\gamma\, v_m. \tag{3.29}$$

Integrating (3.25) between r_1 and $r_2 (> r_1)$ we obtain

$$I = \int_{r_1}^{r_2} N(r)\, \mathrm{d}r = \frac{v_r}{a} \left[r_1^{-a} - r_2^{-a} \right], \tag{3.30}$$

with $a = \alpha - 1$. In case $a = 0$ we obtain instead a logarithmic dependance.

If $a > 0$ the smallest particles are most numerous and we can often neglect the second term.

The total cross-section of particles between $\sigma_1 = \pi r_1^2$ and $\sigma_2 = \pi r_2^2 > \sigma_1$ is

$$I(\sigma) = \int_{\sigma_1}^{\sigma_2} \sigma v_\sigma \sigma^{-\beta}\, \mathrm{d}\sigma = \frac{v_\sigma}{2 - \beta} \left[\sigma_1^{-(\beta - 2)} - \sigma_2^{-(\beta - 2)} \right], \tag{3.31}$$

$$I(\sigma) = \frac{v_\sigma}{b} (\sigma_1^{-b} - \sigma_2^{-b}),$$

with $b = \beta - 2$.

If $b > 0$ (which often is the case) the smallest particles determine the total cross-section.

The total mass between m_1 and m_2 $(> m_1)$ is

$$I(m) = \int_{m_1}^{m_2} m v_m m^{-\gamma}\, \mathrm{d}m, \tag{3.32}$$

$$I(m) = \frac{v_m}{-c} (m_2^{-c} - m_1^{-c}),$$

with $c = \gamma - 2$.

If $c < 0$ (which often is the case) the largest particles have most of the mass.

If the magnitude g is chosen as variable, we have for the differential spectrum $\log N(g) = \mathrm{const.} + 0.2(\alpha - 1)g$.

3.9. THREE SIMPLE MODELS

In order to get a feeling for the relation between different physical processes and the resulting size spectra, we shall derive such spectra for three very simple models. See Table 3.2.

In the first two models we assume that embryos are accreting mass by capturing all small particles (ions, atoms, or dust particles much smaller than the embryo) which hit it. Hence the mass of the embryo increases at the rate

$$\mathrm{d}M/\mathrm{d}t = \varrho v \sigma, \tag{3.33}$$

where ϱ is the mass density of the particles to be accreted, and v the velocity which they have before they are accelerated by the gravitation of the embryo. The capture cross-section is

$$\sigma = \pi r^2 \left[1 + v_e^2/v^2 \right], \tag{3.34}$$

TABLE 3.2
Survey of spectra and models

		Non-grav. Accretion		Grav. Accretion	Fragmen- tation	
Differential Spectra		↓	↘	↓	↙	↘
Number of bodies in radius interval dr	α	0	1	2	3	4
Number of bodies in surface interval dσ	β	0.5	1	1.5	2	2.5
Number of bodies in mass interval dm	γ	0.67	1	1.33	1.67	2.
Number of bodies in magnitude interval dg	$0.2(\alpha-1)$	−0.2	0	+0.2	+0.4	+0.6
Integrated Spectra						
Number of bodies r and 0 or ∞	a	−1	0	+1	+2	+3
Total cross-section of bodies between σ and 0 or ∞	b	−1.5	−1	−0.5	0	+0.5
Total mass of bodies between m and 0 or ∞	c	−1.33	−1	−0.67	−0.33	0

Large grains most numerous	Small grains most numerous	
Most cross-section due to large grains	Most cross- section due to small grains	
Most mass in large bodies		Most mass small bodie

Note: Öpik's 'population index' S is identical with a.

where

$$v_e = r/t_e \tag{3.35}$$

is the escape velocity, and

$$t_e = \left(\frac{3}{8\pi\kappa\theta}\right)^{\frac{1}{2}} = 1340\ \theta^{-\frac{1}{2}}\ \text{sec}\ (\text{g/cm}^3)^{1/2} \tag{3.36}$$

θ is the mean density of the embryo.

3.9.1. *Non-Gravitational Accretion*

In our first model we assume that the embryo is so small that the gravitation can be neglected ($v \gg v_e$). Hence

$$4\pi\theta\ r^2\ \frac{\mathrm{d}r}{\mathrm{d}t} = \pi r^2 \varrho v,$$

which means

$$\frac{\mathrm{d}r}{\mathrm{d}t} = \text{const}.$$

If we further assume a time constant injection of infinitely small particles, we obtain the spectrum

$$N(r) = \text{const.} \tag{3.37}$$

Hence, we have $\alpha = 0$.

3.9.2. *Gravitational Accretion*

In our second model the embryo is supposed to be so large that its gravitation focuses grains towards it. This means $v_e \gg v$. Hence

$$dr/dt = 2\pi\kappa\varrho r^2/3v, \tag{3.38}$$

which gives

$$r = A/(t_1 - t) \tag{3.39}$$

with

$$A = 3v/2\pi\kappa\varrho. \tag{3.40}$$

If we have a time constant injection of small particles, we have for $t < t_1$

$$N(r)\, dr = \text{const}\, dt,$$

which from (16) gives

$$N(r) = \text{const}\, r^{-2}. \tag{3.41}$$

Hence, $\alpha = 2$.

If the injection of small particles has started at $t = 0$, the first particle reaches $r = \infty$ at $t = t_1$.

3.9.3. *Fragmentation*

In our third model we assume that whenever a body is hit by a particle, it is split up into n smaller bodies which all are equal. Hence the cross-section for fragmentation is proportional to σ or to $m^{2/3}$. This means that bodies in the interval m to $m + \Delta m$ are leaving this interval at a rate proportional to $m^{2/3}$. At the same time bodies are injected into the interval by the splitting of bodies in the interval $n \cdot m$ to $n(m + \Delta m)$, and this occurs at a rate which is proportional to $(n\,m)^{2/3}$. If particles are continuously injected at the upper end of the spectrum, we get a time-independent distribution if

$$N(m) \cdot m^{2/3} \Delta m = N(nm)(nm)^{2/3} n\Delta m. \tag{3.42}$$

Introducing (27) we find

$$m^{-\gamma + 2/3} = (nm)^{-\gamma + 2/3} n, \tag{3.43}$$

which is satisfied if

$$\gamma = \tfrac{5}{3} \tag{3.44}$$

and

$$\alpha = 3, \quad \beta = 2.$$

Hence, the integrated mass spectrum has the exponential $c = -\frac{1}{3}$.

The mass of the largest particles dominates, but the cross-section is equal in all logarithmic intervals.

Piotrowski (1953) has worked out a model which is essentially the same as given here. The power law with $\alpha = 3$, $a = 2$ is often referred to as Piotrowski's law.

There are a number of alternative models taking account of the fragmentational process in a more exact way. The a-values are usually found to be $2 > a > \frac{5}{3}$.

Dohnanyi (1969) takes account of both the fragmentation and the erosion at hypervelocity impacts and finds $\gamma = \frac{11}{6}$ which means $\alpha = 3.5$, $a = 2.5$.

All the theoretical models seem to agree that the result of fragmentation is that most of the mass remains in the largest bodies, whereas most of the cross section is due to the smallest particles.

3.10. ASTEROIDS

3.10.1. *General Survey*

Semi-major axes and periods. The asteroids, of which more than 1700 are tabulated in *Ephemeridy malych planet*, move mostly in the region between Mars and Jupiter. A few specimens are also found beyond Jupiter or inside the orbit of Mars. The orbits of asteroids have in general higher eccentricity and larger inclination than those of the real planets. The average eccentricity is 0.14. For less than 1% of the orbits the eccentricity exceeds $e = \frac{1}{3}$. The average inclination is 9.7°. Graphs showing statistical aspects of asteroids have been published by Brown *et al.* (1967).

The semi-major axis, a, of the orbits of most of the asteroids is between 2 and 3.5 AU. However, there is a group called the 'Trojans' which moves at the same orbital distance ($a = 5.20$ AU) as Jupiter. There is also another group called the Hilda group, with $a = 3.98$ AU.

If we plot the number of known asteroids as a function of a, we obtain Figure 3.3. The diagram shows a series of sharp gaps where no, or very few, asteroids are found. These are due to resonance effects from Jupiter. As the period T is proportional to $a^{3/2}$, all bodies with a certain a-value have the same period. The gaps correspond to $T/T_{2\!\!\!\perp} = \frac{1}{2}, \frac{1}{3}, \frac{2}{5}$, and $\frac{3}{7}$, the gap for $\frac{1}{2}$ being very pronounced. Gaps corresponding to $\frac{2}{7}, \frac{3}{8}, \frac{3}{10}, \frac{4}{11}, \frac{5}{12}, \frac{6}{13}$ have also been traced (Hirayama, 1918). Mars also produces a resonance at $T/T_{\delta} = 2$, but no resonance with the period of Saturn or any other planet has been found. Figure 3.4 shows the number of asteroids as a function of angular momentum, Figure 3.5 as a function of eccentricity and Figure 3.6 of inclination.

The motion of the Trojans is coupled to Jupiter's motion because the periods are equal. These asteroids oscillate around certain positions (libration points) situated in Jupiter's orbit 60° on both sides of Jupiter. A resonance effect of a similar kind is active in the Hilda group. The period of the Hilda asteroids is $\frac{2}{3}$ of the period of Jupiter (Schubart, 1968; Chebotarev, 1970). The motion is coupled to the motion of Jupiter in such a way that a Hilda asteroid never comes very close to Jupiter. The

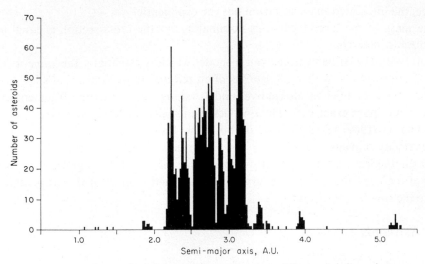

Fig. 3.3.   The number of asteroids as a function of the semi-major axis a.

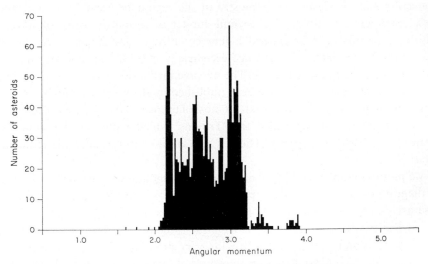

Fig. 3.4.   The number of asteroids as a function of their specific angular momentum,
$C = [\mu a(1 - e^2)]^{1/2}$ for $2.14 < a < 4.00$ AU.

resonance phenomena active in the Trojan and the Hilda groups will be discussed in Section 4.

There are a few asteroids with very large eccentricities. Some of these (Eros, Apollo, Icarus) form a group the orbits of which cross the Earth's orbit. Another asteroid, Hidalgo, goes out far beyound the orbit of Saturn. These asteroids form a transition between the real asteroids and the periodic comets. There is no well-defined limit between the orbits of comets and of asteroids. The main difference

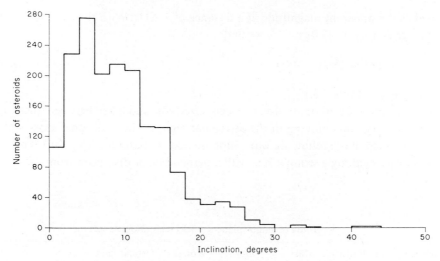

Fig. 3.5.   Distribution of the inclination of asteroid orbits for 1697
asteroids where $2.14 < a < 4.00$ AU.

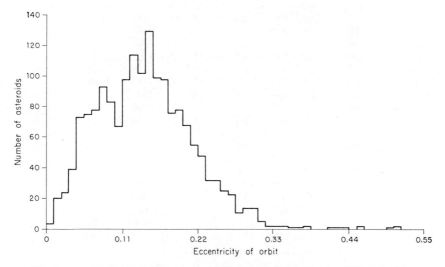

Fig. 3.6.   Distribution of the eccentricities of asteroid orbits for 1697 asteroids
where $2.14 < a < 4.00$ AU.

between these two types of bodies is that comets evaporate gases which form their tails.

The diameters of asteroids can usually not be measured, nor can their masses be determined directly. The sizes and masses are therefore estimated from their apparent magnitudes with reasonable assumptions about the albedo and average density.

According to Blanco and McCuskey (1961), we have

$$\log R_0 = 8.27 - 0.2g ,\tag{3.45}$$

where R_0 is the radius of the asteroid in cm and g its absolute visual magnitude (defined as the apparent magnitude at a distance of 1 AU). With the assumption that the average density θ is $3.0\,\mathrm{g\,cm^{-3}}$, we find

$$\log M = 25.91 - 0.6g \tag{3.46}$$

for M in gm (see Table 3.3.)

Collisions must occur mutually between asteroids and also between these and smaller bodies (grains) moving in the asteroidal region. The collisional cross section can be computed in the following way. Suppose that a small body (a grain) is moving with a velocity v_0 along a straight line with a perpendicular distance r from the origin.

TABLE 3.3
Asteroids or minor planets – Selected minor planets

Number and name	Radius	Mass	m_{pg} at $r\varDelta = 1$	Rot. period		Orbital data Period	a	e	i
	km	g		h	m	d	AU		°
1 Ceres	350	60×10^{22}	4.0	9	05	1681	2.767	0.079	10.6
2 Pallas	230	18×10^{22}	5.1			1684	2.767	0.235	34.8
3 Juno	110	2×10^{22}	6.3	7	13	1594	2.670	0.256	13.0
4 Vesta	190	10×10^{22}	4.2	5	20	1325	2.361	0.088	7.1
6 Hebe	110	20×10^{21}	6.6	7	17	1380	2.426	0.203	14.8
7 Iris	100	15×10^{21}	6.7	7	07	1344	2.385	0.230	5.5
10 Hygiea	160	60×10^{21}	6.4	18?		2042	3.151	0.099	3.8
15 Eunomia	140	40×10^{21}	6.2	6	05	1569	2.645	0.185	11.8
16 Psyche	140	40×10^{21}	6.8	4	18	1826	2.923	0.135	3.1
51 Nemausa	40	9×10^{20}	8.6			1330	2.366	0.065	9.9
433 Eros	7	5×10^{18}	12.3	5	16	642	1.458	0.223	10.8
511 Davida	130	3×10^{22}	7.0			2072	3.182	0.177	15.7
1566 Icarus	0.7	5×10^{15}	17.7			408	1.077	0.827	23.0
1620 Geographos	1.5	5×10^{16}	15.9			507	1.244	0.335	13.3
Apollo	0.5	2×10^{15}	18			662	1.486	0.566	6.4
Adonis	0.15	5×10^{13}	21			1008	1.969	0.779	1.5
Hermes	0.3	4×10^{14}	19			535	1.290	0.475	4.7

Radius of planet R: $\log R = 2.95 - \frac{1}{2}\log p - 0.2 m_{pg}$ (at $r\triangle = 1$); R in km, $p =$ albedo factor. The table is based on $p = 0.16$. From Allen (1963).

If we place a bigger body with the radius R_0 at the origin, its gravitation will deflect the orbit of the grain so that it will hit this 'embryo' under the condition (compare Section 3.9).

$$r < R_0 \left(1 + v_e^2/v_0^2\right)^{\frac{1}{2}}, \tag{3.47}$$

where

$$v_e = \left[2\kappa M/R_0\right]^{1/2} = \left(\frac{8\pi}{3}\kappa\theta\right)^{1/2} \cdot R_0 = 1.3 \times 10^{-3} R_0 \tag{3.48}$$

TABLE 3.4

Relation between magnitude, mass and cross section

Absolute visual magnitude g	Radius in cm R_0	Mass in g	Escape velocity in cm sec^{-1} v_e	Geometric cross section in cm^2 σ_g	Collisional cross-section in cm^2 for various approach velocities σ		
					$v_0 = 10^4$ cm/sec	$v_0 = 10^5$ cm/sec	$v_0 = 10^6$ cm/sec
3	4.68×10^7	1.29×10^{24}	6.06×10^4	6.87×10^{15}	2.59×10^{17}	9.39×10^{15}	6.89×10^{15}
4	2.95×10^7	3.24×10^{23}	3.82×10^4	2.74×10^{15}	4.28×10^{16}	3.14×10^{15}	2.75×10^{15}
5	1.86×10^7	8.13×10^{22}	2.41×10^4	1.09×10^{15}	7.42×10^{15}	1.15×10^{15}	a
6	1.18×10^7	2.04×10^{22}	1.52×10^4	4.34×10^{14}	1.44×10^{15}	4.44×10^{14}	
7	7.41×10^6	5.13×10^{21}	9.61×10^3	1.73×10^{14}	3.33×10^{14}	1.75×10^{14}	
8	4.68×10^6	1.29×10^{21}	6.06×10^3	6.87×10^{13}	9.39×10^{13}	6.89×10^{13}	
9	2.95×10^6	3.24×10^{20}	3.82×10^3	2.74×10^{13}	3.14×10^{13}	2.75×10^{13}	
10	1.86×10^6	8.13×10^{19}	2.41×10^3	1.09×10^{13}	1.15×10^{13}	a	
11	1.18×10^6	2.04×10^{19}	1.52×10^3	4.34×10^{12}	4.44×10^{12}		
12	7.41×10^5	5.13×10^{18}	9.61×10^2	1.73×10^{12}	1.75×10^{12}		
13	4.68×10^5	1.29×10^{18}	6.06×10^2	6.87×10^{11}	6.89×10^{11}		
14	2.95×10^5	3.24×10^{17}	3.82×10^2	2.74×10^{11}	2.75×10^{11}		
15	1.86×10^5	8.13×10^{16}	2.41×10^2	1.09×10^{11}	a		
16	1.18×10^5	2.04×10^{16}	1.52×10^2	4.34×10^{10}			
17	7.41×10^4	5.13×10^{15}	9.61×10	1.73×10^{10}			
18	4.68×10^4	1.29×10^{15}	6.06×10	6.87×10^9			

a Remaining figures in each column do not differ from those for σ_g to 3 significant figures.

is the escape velocity of the embryo in c.g.s. units. Hence, its collisional cross section is

$$\sigma = \sigma_g [1 + v_e^2/v_0^2], \tag{3.49}$$

where $\sigma_g = \pi R_0^2$ is the geometrical cross section.

If $v_0 \ll v_e$, we have

$$\log \sigma = 27.8 - 0.8g - 2 \log v_0. \tag{3.50}$$

All these parameters are tabulated in Table 3.4 in the range of absolute magnitudes observed for asteroids and with appropriate approach velocities.

It should be observed that (3.49) is derived under idealized conditions, among others that the solar gravitational field can be neglected. See further Sections 8 and 9.

3.11. HIRAYAMA FAMILIES AND JET STREAMS

As discovered by Hirayama there are '*families*' of asteroids with almost the same values of a, i, and e.

As Brouwer (1951) has pointed out, both i and e are subject to secular variations with periods of the order 10^4 or 10^5 yr. From a hetegonic point of view we want to eliminate these. This can be done by introducing the 'proper elements'.

The eccentricity e and the longitude π of the perihelion of a Kepler orbit are subject to secular variations. The same is the case for the inclination i and the longitude of the ascending node θ. Following Brouwer (Brouwer, 1951; Brouwer and Clemence, 1961) we write:

$$e \cos \pi = E \cos \pi_1 + p_0, \tag{3.51}$$
$$e \sin \pi = E \cos \pi_1 + q_0, \tag{3.52}$$
$$\sin i \cos \theta = I \cos \theta_1 + P_0, \tag{3.53}$$
$$\sin i \sin \theta = I \cos \theta_1 + Q_0; \tag{3.54}$$

E means the 'proper eccentricity' and I the 'proper inclination'; π_1 and θ_1, the corresponding longitudes. Furthermore, p_0, q_0, P_0 and Q_0 are the forced oscillations produced by planetary disturbances which have been calculated by Brouwer. The quantities p_0 and q_0 are a function of the planets' eccentricities and perihelia; P_0 and Q_0, of their inclinations and nodes. See also Kiang (1966).

Figure 3.7 and Table 3.5 show the relation between the 'osculating' elements

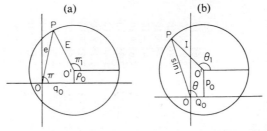

Fig. 3.7. (a) Illustrating the relation between (e, π) and (E, π_1). (b) Illustrating the relation between $(\sin i, \theta)$ and (I, θ_1) (after Kiang, 1966).

(referring to the present orbits) and the 'proper' elements according to Kiang (1966). The vectors E and I rotate around center O', with periods given in Table 3.5. The distance of the vector from origin gives the numerical value of e and i, and the angles of these lines the longitudes of the perihelion and the node. The center O' is essentially

TABLE 3.5

a (AU)	P_1 (yr)	q_0	ϱ_0	Q_0	P_0
2.15	41 400	$+0.0567$	-0.0363	$+0.0108$	$+0.0031$
2.60	26 300	$+0.0302$	-0.0056	-0.0006	$+0.0181$
3.15	14 400	$+0.0374$	$+0.0058$	-0.0029	$+0.0210$
4.00	4 400	$+0.0421$	$+0.0093$	-0.0038	$+0.0222$

given by the eccentricity and inclination of Jupiter, and varies with a period of 300 000 yr.

Brouwer has given the values of E, I, π_1, and θ_1, for 1537 asteroids (*loc. cit.*). Based on this material he treats the Hirayama families. He demonstrates that in the (E, I) diagram the points belonging to a Hirayama family show a somewhat stronger concentration than in the (e, i) diagram. This made it possible for him to detect a number of new families. For example, it is evident that the largest of the families, the Flora family, consists of at least two, and possibly four families, called Flora I, II, III, and IV.

The proper longitude of the perihelion π_1, and the proper longitude of the ascending node θ_1 change systematically because of planetary disturbances, mainly from Jupiter, according to the formulae

$$\pi_1 = \pi_0 + b(t - t_0), \tag{3.55}$$
$$\theta_1 = \theta_0 - b(t - t_0). \tag{3.56}$$

It is a consequence of celestial mechanics that the sum $\pi_1 + \theta_1$ is an invariant to a first approximation (see Section 2.6). Brouwer shows that for some families there is a maximum of $\pi_1 + \theta_1$ for certain directions. He also uses the quantity $\pi_1 + \theta_1$ for the detection of a number of new families or groups.

Members of the same family generally have different values of θ and π. This means that the space orientation of their orbits differs. In some cases, however, there are a number of orbits with the same θ and π, so that all the five orbit parameters (a, i, e, θ, π) are similar. Hence their orbits almost coincide and the asteroids are said to be members of a '*jet stream*' (Alfvén, 1969; Danielsson, 1969).

Subjecting the whole asteroid material to computer analysis, Arnold (1969) has revised Brouwer's analysis of the families. He has confirmed the existence of all the Hirayama families and some but not of all the Brouwer families. Further, he has discovered a number of new families. Table 3.6 gives a summary of the families according to Arnold. The families are arranged after increasing a-values.

TABLE 3.6
Asteroid families

Family Name Arnold's Nomenclature	Total Number in Family	Average Semi-Major axis in AU $\bar{a}$	Range of $\bar{a}$ in AU	Average Eccentricity $\bar{e}_1$	Range of $\bar{e}_1$	Average Sine of Inclination with Respect to Ecliptic Plane $\sin i_1$	Range of $\sin i_1$	Number Expected in Three-Dimensional Volume Defined by $\bar{a}_1$, $\bar{e}_1$, $\sin i_1$ R_m
6	37	2.2166	2.1746–2.2553	0.1265	0.0922–0.1525	0.0476	0.0164–0.0643	3.8
7	73	2.2208	2.1800–2.2641	0.1380	0.0475–0.1731	0.0794	0.0635–0.0948	3.2
8	10	2.2228	2.1974–2.2618	0.1458	0.1340–0.1617	0.0971	0.0948–0.0993	0.1
9	36	2.2452	2.1732–2.3150	0.1528	0.1232–0.1801	0.1126	0.0980–0.1305	4.2
C-31	6	2.2528	2.2272–2.2996	0.2484	0.2211–0.2878	0.0699	0.0284–0.1053	3.0
A-77	16	2.3566	2.3294–2.3784	0.1974	0.1791–0.2187	0.1724	0.1423–0.2029	1.9
5	34	2.3667	2.2879–2.4529	0.2418	0.1369–0.3631	0.3985	0.3234–0.4599	1.8[a]
A-75	6	2.3956	2.3637–2.4177	0.1630	0.1522–0.1733	0.0903	0.0844–0.0973	0.24
A-81	9	2.4373	2.4143–2.4541	0.1745	0.1507–0.1946	0.1306	0.1127–0.1483	1.0
A-82	7	2.4396	2.4291–2.4491	0.1311	0.1191–0.1418	0.0451	0.0287–0.0597	0.23
A-76	11	2.4409	2.4177–2.4710	0.0769	0.0597–0.0962	0.0686	0.0461–0.0892	1.3
A-74	10	2.4457	2.4199–2.4794	0.1649	0.1439–0.1779	0.0535	0.0460–0.0563	0.3
B-25	7	2.4670	2.4344–2.4899	0.1587	0.1307–0.1739	0.0837	0.0769–0.0895	0.5
4	20	2.5480	2.5284–2.5798	0.0989	0.0737–0.1246	0.2591	0.2463–0.2652	0.8
A-73	10	2.5504	2.5350–2.5735	0.1995	0.1895–0.2113	0.1231	0.0986–0.1457	0.7
A-72	18	2.5955	2.5644–2.6222	0.1720	0.1392–0.2059	0.1708	0.1472–0.1937	2.9
B-29	7	2.5979	2.5521–2.6216	0.3397	0.2821–0.4537	0.2943	0.2774–0.3129	0.6[a]
A-70	31	2.6281	2.6106–2.6595	0.1636	0.1115–0.2239	0.2231	0.1874–0.2578	6.2

(Table 3.6 continued)

A-83	8	2.6610	2.6424–2.6884	0.1197	0.1079–0.1347	0.0629	0.0470–0.0860	0.8
A-69	11	2.6677	2.6523–2.6861	0.1690	0.1637–0.1778	0.2111	0.1920–0.2342	0.3
B-18	12	2.6752	2.6529–2.7127	0.1880	0.1259–0.2325	0.2655	0.2573–0.2742	1.8
C-30	4	2.6932	2.6123–2.7691	0.4501	0.4105–0.5002	0.4593	0.4430–0.4770	0.5[a]
A-84	9	2.6963	2.6795–2.7166	0.2233	0.2078–0.2340	0.1313	0.1066–0.1589	0.8
B-28	6	2.7169	2.6303–2.8021	0.2513	0.1900–0.3318	0.5537	0.5322–0.5719	0.2[a]
A-85	17	2.7281	2.6941–2.7622	0.1020	0.0763–0.1351	0.1361	0.1166–0.1546	2.5
A-87	14	2.7312	2.6787–2.7854	0.0984	0.0814–0.1173	0.0840	0.0732–0.0931	1.2
A-86	17	2.7479	2.7024–2.7876	0.0517	0.0327–0.0734	0.0736	0.0561–0.0899	1.9
B-17	11	2.7577	2.7228–2.7847	0.2584	0.2179–0.3168	0.2794	0.2622–0.2966	3.4
A-66	23	2.7724	2.7339–2.8032	0.1719	0.1451–0.1964	0.1340	0.1005–0.1619	3.6
A-67	6	2.7860	2.7675–2.8070	0.1007	0.0884–0.1097	0.1684	0.1600–0.1760	0.2
A-91	7	2.8020	2.7916–2.8103	0.1390	0.1255–0.1593	0.1557	0.1277–0.1794	0.6
A-65	6	2.8665	2.8604–2.8735	0.1538	0.1385–0.1667	0.2082	0.1903–0.2224	0.2
3	37	2.8743	2.8346–2.9042	0.0489	0.0379–0.0672	0.0370	0.0338–0.0424	0.3
B-13	7	2.9670	2.9228–3.0161	0.0566	0.0141–0.0985	0.0522	0.0355–0.0698	4.2
B-14	7	2.9810	2.9239–3.0185	0.1859	0.1515–0.2335	0.1971	0.1803–0.2195	4.9
2	66	3.0151	3.0003–3.0297	0.0743	0.0560–0.0969	0.1758	0.1661–0.1872	0.4
B-12	6	3.0635	3.0386–3.1076	0.1178	0.0990–0.1516	0.2112	0.1883–0.2336	2.6
A-88	26	3.1360	3.1038–3.1772	0.1680	0.1288–0.2045	0.2400	0.2070–0.2725	4.4
1	67	3.1408	3.0874–3.1953	0.1527	0.1210–0.11873	0.0249	0.0120–0.0430	3.5
A-89	10	3.1571	3.1442–3.1706	0.1215	0.1040–0.1406	0.1738	0.1445–0.1956	0.8
A-90	12	3.1914	3.1659–3.2115	0.0575	0.0421–0.0715	0.1596	0.1383–0.1773	0.9
B-11	5	3.3821	3.3673–3.3930	0.1066	0.0432–0.1617	0.3555	0.3123–0.3862	0.2[a]

[a] Determined from local density. These values are uncertain (poor statistics).

TABLE 3.7

Jet streams

Stream	Members	$\langle a \rangle$	Range	$\langle e_1 \rangle$	Range	$\langle \sin i_1 \rangle$	Range
J-1	32	2.2044	2.1746–2.2485	0.1234	0.0725–0.1704	0.0632	0.0101–0.1044
J-2	18	2.3596	2.3043–2.4025	0.1991	0.1729–0.2381	0.1618	0.1231–0.2112
J-3	18	2.6146	2.5514–2.6706	0.1952	0.1406–0.2526	0.2408	0.1798–0.2864
J-4	9	2.6526	2.5801–2.7163	0.0900	0.0662–0.1121	0.1991	0.1476–0.2487
J-5	6	2.7466	2.7288–2.7725	0.2617	0.2368–0.2887	0.2892	0.2735–0.3204
J-6	10	3.1271	3.0656–3.1779	0.0552	0.0286–0.0925	0.2002	0.1515–0.2765
J-7	7	2.9993	2.9800–3.0181	0.1079	0.0927–0.1319	0.2411	0.2302–0.2671

Stream	Members	$\langle \pi_1 \rangle$	Range	$\langle \theta_1 \rangle$	Range	$\langle \pi_1 + \theta_1 \rangle$	Range
J-1	32	0.030	0.771–0.345	0.595	0.332–0.786	0.625	0.219–0.928
J-2	18	0.971	0.781–0.152	0.820	0.578–0.025	0.791	0.438–0.057
J-3	18	0.965	0.780–0.165	0.631	0.475–0.769	0.596	0.398–0.832
J-4	9	0.163	0.093–0.223	0.880	0.811–0.963	0.044	0.904–0.137
J-5	6	0.851	0.546–0.138	0.163	0.987–0.364	0.014	0.745–0.125
J-6	10	0.174	0.044–0.289	0.544	0.453–0.616	0.718	0.615–0.801
J-7	7	0.451	0.286–0.691	0.519	0.363–0.661	0.970	0.800–0.204

(Table 3.7 continued)

Jet Streams – Listing of Members

Family	J-1	J-2	J-3	J-4	J-5	J-6	J-7
Total number of members	32	18	18	9	6	10	7
Members	149	12	85	15	36	57	256
	244	84	258	407	387	251	478
	270	115	397	524	564	490	482
	352	192	687	829	599	509	611
	496	220	726	1098	735	589	816
	685	249	737	1187	1484	904	1614
	703	284	881	1287		1023	
	707	437	923	1406		1114	
	736	584	989	1471		1408	
	810	646	995			1469	
	831	916	999				
	836	1224	1168				
	901	1507	1346				
	951	1538	1441				
	960	1694	1473				
	1037	1709	1499				
	1058	1710	1554				
	1120	1718	1688				
	1130						
	1150						
	1153						
	1335						
	1422						
	1486						
	1494						
	1496						
	1536						
	1636						
	1666						
	1699						
	1703						
	1733						

From Arnold (1969).

A very important result of Arnold's analysis is his discovery of seven jet streams in the asteroid belt. Table 3.7 lists these jet streams.

Arnold's results have far-reaching consequences. If one calculates the differences in precession of nodes and perihelia inside a jet stream assuming exclusively gravitational forces one finds that the members of a jet stream should be spread out more or less uniformly in θ and π after periods of the order of only 10^4 or 10^5 yr. The existence of as many as seven jet streams (at least!) indicate that there must be some focussing mechanism keeping the jet streams together. It is very important to decide whether this may be the viscosity focussing discussed in Section 3.4 or some other mechanism.

There is some similarity between the asteroidal jet streams and meteor streams. In several cases one member of an asteroidal jet stream is much larger than the other members. This is an analogy to the association of a comet with a meteor stream. Hence the discovery of asteroidal jet streams makes the difference between asteroids and comets-meteoroids less sharp. We shall discuss these problems further in a later section.

3.12. Subvisual asteroids

There are good reasons to suppose that the asteroid size spectrum is continued down to very small bodies which we may call 'asteroidal grains'. From observations we know nothing about the size spectra of the asteroidal grains. We may extrapolate the size spectra of visual asteroids but we do not know whether this is allowed.

The subvisual asteroids may be of a decisive importance in keeping jet streams together. They may also be important for other viscosity effects in interplanetary space. The only way of getting information about them is through space probes sent to the asteroidal belt.

3.13. Comets

The orbits of most comets have much larger eccentricities than most asteroids. There is however a small number of asteroids, e.g., Hidalgo and Apollo with large eccentricities, as comets, and there are also comets with small eccentricities.

Of the 525 comets with accurately determined orbits, 199 are elliptical, 274 almost parabolic, and 52 slightly hyperbolic (Vsekhsvyatskii, 1964, p. 2). However, if the orbits of the hyperbolic comets are corrected for planetary disturbances, all of them become nearly parabolic. Hence there is no evidence that comets come from interstellar space. As far as we know, all comets seem to belong to the solar system. Planetary disturbances, however, change the orbits of some comets so that they are ejected from the Solar System into interstellar space.

As most cometary orbits are very eccentric, the approximate methods which were developed in Section 2 are not applicable. The following relations between semimajor axis a, specific orbital momentum C, perihelion r_p, aphelion r_a, and velocities v_a at r_a and v_p at r_p are useful. We have put $\delta = 1 - e$ and $\mu = \kappa M_c$.

$$C^2 = \mu a (1 - e^2), \tag{3.57}$$
$$r_a = a(1 + e) = a(2 - \delta), \tag{3.58}$$
$$r_p = a(1 - e) = a\,\delta, \tag{3.59}$$

$$v_a = \frac{C}{a(1+e)} = \left(\frac{\mu}{r_a}\delta\right)^{1/2} = v_\oplus \left(\frac{r_\oplus}{r_a}\right)^{1/2}, \tag{3.60}$$

where $v_\oplus = 3.0 \times 10^6$ cm sec^{-1} is the orbital velocity of the Earth and $r_\oplus$ its orbital radius. Similarly,

$$v_p = \frac{C}{a(1-e)} = \left[\frac{\mu}{r_p}(2 - \delta)\right]^{1/2} = v_\oplus \left(\frac{r_\oplus}{r_p}(2 - \delta)\right)^{1/2}. \tag{3.61}$$

For $\delta \ll 1$ we have approximately

$$v_a = v_\oplus (2r_\oplus r_p / r_a^2)^{1/2}.$$ (3.62)

The almost parabolic comets are in reality elliptical with an eccentricity slightly less than 1. Their aphelia are situated in what Oort calls the 'cometary reservoir', a region extending out to at least 10^{17} cm (0.1 light-year), their orbital periods range from 10^3–10^6 yr (cf. Oort, 1963). This theory has further been discussed by Lyttleton (1968). The long period comets spend most of their time near their aphelia, but at regular intervals they make a quick visit to the regions close to the Sun. It is only in the special case where the comet's perihelion is less than a few times 10^{13} cm that it can be observed. The total number of comets in the Solar System has been estimated at 10^{11}, of which 10^5 are 'bright comets' (Porter, 1963, p. 550).

The space orientation of the orbits of nearly parabolic comets is random. Almost the same number are moving in prograde and in retrograde sense. From this we conclude that on the average the comets in the reservoir are at rest in relation to the Sun, or, expressed in other words, share the solar motion in the galaxy. From (3.62) a comet whose perihelion is at 10^{13} cm will at its aphelion have a tangential velocity of 5×10^4 cm/sec if $r_a = 10^{15}$ cm, and 5×10^2 cm/sec if $r_a = 10^{17}$ cm. As the solar velocity in relation to neighboring stars is of the order of about 20 km/sec the velocities in the cometary reservoir are remarkably small. If comets originated from the environment of other stars or from a random region in interstellar space, their orbits should be hyperbolas very easily distinguishable from the nearly parabolic orbits observed. Hence we have confirming evidence that the comets are true members of our Solar System. The cometary reservoir is an important part of the system. Oort (1963) has suggested that the comets originally were formed near Jupiter and then ejected into the cometary reservoir by encounters with Jupiter. This seems to be a very unlikely process. As we shall see in the following it is more likely that the long period comets are accreted out in the cometary reservoir.

3.14. SHORT-PERIOD COMETS

Among the comets there is a special group called short period comets, which differs from the long period comets in the respect that their orbits are predominantly prograde. In fact, there is not a single retrograde comet with a period less than 15 yr (Porter, 1963, p. 556, 557).

In all textbooks and reviews it is taken for granted that the short-period comets can be accounted for as long-period comets, originating in the cometary reservoir. No proof of this is given but sometimes a reference is made to Russell (1920). Reading the summary of this paper one gets the impression that the capture theory is in order, but reading the paper itself gives a contrary impression. Russell bases his work on the investigations of Newton (1893) who shows that bodies in parabolic orbits with perihelion inside Jupiter's orbit have a chance of being deflected by Jupiter into elliptic orbits. This chance however is very small so that out of 10^9 bodies (comets) only 839 will be captured in orbits with periods $< T_{2\!\downarrow}$. The captured bodies move with preference in the prograde direction, so that out of the 839 objects 257 or 31% should have inclinations $<30°$ and 51 objects (6%) $>150°$.

Russell shows that theoretically Jupiter should be responsible for almost all of the captures. Saturn should account for only 2.5%, Uranus and Neptune for 0.025% and the effect of the Earth (and other terrestial planets) is completely negligible.

Russell finds that the relative number of comets with periods longer than 10 yr is in agreement with the capture theory (or at least not in conflict with it). The short-period comets, on the contrary, are far more numerous than predicted. Further, the periods and inclinations of the long-period orbits agrees with theory but the short-period comets are 'glaringly different'. Of 39 short-period comets none has higher inclination than 50° and 36 (or 92%) have $<30°$ compared to the theoretical value 31%. This conclusion is fully confirmed by modern data. See Figure 3.8 and Porter (1963) p. 557, 558.

Russell believes that the difference between the capture theory and observations could be explained by the cumulative effects of Jovian perturbations, but does not demonstrate that this is true.

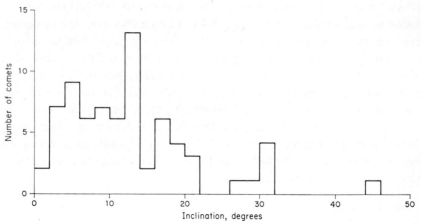

Fig. 3.8.   Distribution of the inclinations of short-period comets with aphelion less than 10 AU.

According to Öpik (1968) the present orbital distribution of short-period comets cannot be explained by gravitational perturbations. This is likely to be correct. However, Öpik claims that a 'retro rocket' effect due to anisotropic ejection of gases from comets can explain the orbital distribution. If seems very unlikely that this effect can produce any considerable change in the orbit of a comet which is supposed to be a kilometer-sized body.

Hence our conclusion is that the capture hypothesis of short period comets should be critically reviewed. We shall discuss an alternative in Section 3.16.

3.15. Meteoroids

The most striking property of meteors – easily observed even by naked eye – is the very large fluctuations in their rate of occurrence. When the Earth passes a meteor stream, the number of meteors per hour may increase by one or two orders of magnitude; during some displays even much more. Hence a large fraction of the observed meteors derive from stream meteoroids.

Radar observations have increased the number of known meteor streams. Using different statistical criteria, computer analysis of the observations has made it possible to identify a few hundred different meteor streams.

About half of all observed meteors have not been associated with any streams. They are called *sporadic* meteors. It is possible that a large fraction of them belong to streams which have not yet been identified due to the limited observational material. Many authors, however, assume that they do not have stream structure; but it is impossible to make such a decision at present.

The state of motion of the meteorites in the neighborhood of the Earth has been discussed by Ring *et al.* (1964). They conclude from a study of the Doppler shift of the zodiacal light that practically all the meteoric particles near the Earth's orbit must move in a prograde sense. Less than 10% are moving in retrograde orbits. A detailed analysis of the distribution of solid particles near the ecliptic points in the same direction. (Wall, 1967). Based on a material of 2401 photographic meteor orbits Lindblad (1970) has identified a number of meteor streams. Of 49 streams with $a < 4$ AU, all have a prograde motion.

There is an especially noteworthy meteor stream called the 'Cyclids'. The members of this traveled (before hitting the Earth) in almost circular orbits very close to the Earth's orbit. Table 3.8 gives the orbits of 30 cyclids. It seems possible that the cyclids are the remainder of the jet stream out of which the Earth once was formed (see Section 9).

3.16. Relation between comets and meteoroids

It has often been suggested that many meteorites derive from fragmented main belt asteroids. Öpik (1966), Arnold (1969), and Anders (1964) have shown that this is very unlikely or that they have evaporated from comets. There is a convincing correlation between cometary orbits and some meteor showers. This is usually interpreted as proof that meteors are debris of broken up comets. It is more doubtful whether one has actually observed a break-up of a comet into a meteor shower. It should be noted that not all meteor showers are associated with known comets. One has to postulate a number of parental bodies in order to satisfy the theory.

The view that meteors derive from broken-up comets is attractive, but only as long as one does not ask the next question: Where do the comets come from? This question, indeed, reveals some difficulties in the current views. In fact the following chain of events is assumed.

(a) *Sporadic meteors are shower meteors scattered by planetary perturbations or by*

TABLE 3.8

Parameters of Cyclids' orbits (Equinox 1950.0)

Date		Corr. rad. α	δ	V_∞ (km/sec)	V_h (km/sec)	a (AU)	e	q (AU)	q' (AU)	(')	Ω	i	π
1		2		3	4	5	6	7	8	9	10	11	12
53 I	15.19	119°	−17°	12.1	29.9	0.98	0.11	0.86	1.1	101°	115°	5°	216°
53 I	23.52	280	+55	13.5	31.0	1.06	0.09	0.96	1.1	135	303	14	78
53 II	4.15	103	+25	10.8	31.2	1.07	0.09	0.98	1.2	208	315	0	163
54 II	5.37	171	+73	11.7	30.9	1.05	0.08	0.96	1.1	226	316	6	182
53 II	7.21	129	−15	10.2	30.3	1.01	0.05	0.96	1.1	68	138	2	206
53 II	12.46	245	+45	13.6	27.7	0.86	0.14	0.74	1.0	359	323	15	322
53 II	18.44	181	+3	11.3	28.8	0.92	0.11	0.82	1.0	320	329	0	290
53 II	26.26	251	+37	12.5	27.7	0.87	0.14	0.74	1.0	1	337	9	338
54 III	1.39	184	+36	12.3	29.9	0.99	0.14	0.86	1.1	277	340	6	257
53 III	6.12	129	−64	11.9	29.8	0.99	0.02	0.97	1.0	108	165	8	274
54 IV	7.41	236	+38	11.2	29.4	0.98	0.03	0.95	1.0	324	17	3	341
53 IV	15.25	133	+11	10.9	30.4	1.05	0.05	1.00	1.1	350	205	0	195
53 IV	21.40	211	+6	11.7	30.2	1.04	0.13	0.91	1.2	261	31	3	292
54 V	3.29	252	+71	11.9	30.5	1.07	0.06	1.01	1.1	168	42	8	210
53 V	7.34	253	+32	8.2	29.1	0.98	0.04	0.93	1.0	321	46	4	7
53 V	8.34	241	+19	11.5	29.3	0.99	0.07	0.92	1.1	289	47	4	337
53 V	12.20	263	+33	9.4	29.1	0.98	0.03	0.95	1.0	343	51	2	34
53 V	12.31	260	+37	13.4	28.9	0.96	0.12	0.85	1.1	301	51	13	352
52 V	19.21	64	+37	11.4	30.3	1.06	0.13	0.92	1.2	104	58	2	162
54 V	28.27	327	+73	13.6	29.5	1.01	0.09	0.91	1.1	81	66	14	147
53 VI	5.31	265	+43	14.1	30.2	1.06	0.12	0.94	1.2	254	74	15	328
54 VI	11.42	265	−6	11.6	29.2	1.00	0.13	0.86	1.1	287	80	3	6
53 VI	20.42	289	+60	12.3	28.9	0.97	0.05	0.93	1.0	343	89	11	72
52 VI	24.21	289	+36	13.4	28.4	0.95	0.14	0.82	1.1	301	93	12	40
52 IX	17.40	6	−25	11.5	29.9	1.02	0.13	0.89	1.1	92	354	3	86
52 X	24.32	62	+58	11.2	29.1	0.95	0.07	0.88	1.0	317	211	2	168
52 XI	12.23	23	+33	11.4	30.9	1.06	0.11	0.94	1.2	239	230	2	109
52 XI	20.37	186	−19	11.5	29.3	0.94	0.05	0.90	1.0	201	58	1	259
52 XII	13.34	13	+84	13.3	31.4	1.09	0.14	0.93	1.2	235	261	13	136
53 XII	31.28	99	+64	11.8	30.8	1.04	0.12	0.92	1.2	249	279	5	168

From A. K. Terenteva in Kresak and Millman (1968).

collisions. This is not necessarily true. It has been shown in Section 3.6 that collisions will tend to group grains into jet streams. This is a process which may perhaps change the picture.

(b) *Shower meteors are fragments of evaporated or broken-up comets*. This view – isolated from other links in the chain – is quite reasonable and even attractive. There is no doubt that such a process is possible as has been worked out in detail especially by Whipple (1968).

(c) Focussing the attention to the short-period comets and the short-period meteor streams we need a mechanism producing the short-period comets. As we have seen the usual capture theory is not so well in order as is usually claimed.

If we tentatively look for other explanations of both comets and meteorites it is possible that the earlier neglected focussing properties of a gravitational Coulomb field should be taken into account. In other words, the concept of jet streams gives us perhaps a possibility of reading the usual comet-meteoroid history backwards. It would be possible that condensation from a plasma produces sporadic meteors, and that these, by the mechanism discussed in Section 3.7, are captured into meteor jet streams. In these a longitudinal instability gives rise to a collection of many of the meteoroids in one point where a comet is accreted. This picture is essentially the same as discussed in Section 9 for the formation of planets and satellites. Hence in this way we could perhaps reach a unified theory for the formation of all the different celestial bodies in the solar system.

However it is far from obvious that the mechanism is working at present. This depends upon whether the collision frequency in the meteor streams is large enough to keep them together against dispersive effects. An intermediate alternative is that comets long ago were formed by this mechanism, but today are disintegrating and feeding the meteor streams according to the usual picture.

3.17. RELATION BETWEEN ASTEROIDS AND COMETS-METEOROIDS

There are numerous suggestions that collisions between main belt asteroids may send fragments to the Earth. The Earth-crossing asteroids or the meteors hitting the Earth have been interpreted as such fragments. As Öpik (1966, 1968), Arnold (1964), and Anders (1964) have shown, such processes are very unlikely.

In fact, a body originating at a solar distance $r > r_\oplus$ can reach the Earth's orbit $r_\oplus$ only if the semi-major axis a and eccentricity e satisfy the relations

$$a(1 + e) > r, \qquad a(1 - e) < r_\oplus, \tag{3.63}$$

or

$$e > (r - r_\oplus)/(r + r_\oplus). \tag{3.64}$$

If as a numerical example we select a point in the main asteroid belt $r = 2.5r_\oplus$, we find that for an orbit reaching the Earth, e must exceed 0.42. The corresponding a is 1.75. This is a region where there are very few asteroids. From (3.60) we find that an asteroid at $2.5r_\oplus$ in a circular orbit has the orbital velocity $v = 19 \times 10^5$ cm/sec, whereas if $e = 0.43$ the tangential velocity (in aphelion $= 2.5$) is 14×10^5 cm/sec. Hence a body which shall cross the Earth's orbit must be ejected backwards with a velocity of 5×10^5 cm sec^{-1}. The distribution of tangential velocities of the main belt asteroids (which have eccentricities averaging 0.14) is shown schematically in Figure 3.9.

Mutual collisions between the asteroids will in general result in intermediate values of v. Hence even if the relative velocities at the collisions sometimes are as large as 5×10^5 cm sec^{-1} (and including the radial and azimuthal components even higher) it is

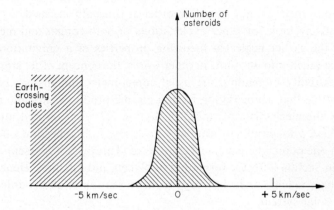

Fig. 3.9.   Tangential velocity minus circular velocity versus number of asteroids. The velocity space of asteroids and Earth-crossing bodies at $r = 2.5\, r_\oplus$.

unlikely that a transfer takes place from the velocity region occupied by main belt asteroids to the region of Earth-crossing objects. At the hypervelocity collisions most of the fragments are ejected with velocities much smaller than the impact velocity. The Earth-crossing objects include several short-period comets, meteoroids, and asteroids with high eccentricity like Icarus. (The latter objects should according to Öpik rather be classified as comets.) These objects on one hand and the main belt asteroids on the other certainly move in the same region of space, but they occupy *different regions of velocity space*. They should be counted as two different populations between which transitions are rare (even if not completely ruled out).

Hence the study of meteorites impinging on the Earth can give us little information on the main belt asteroids.

4. Resonance Structure

4.1. Resonance and the Oscillation of a Pendulum

As pointed out by Brown and Shook (1964), there is a certain similarity between the resonances in the Solar System and the motion of a simple pendulum (see Figure 4.1). Consider the motion of a mass point m, which is confined to a circle with radius l, under the action of gravitation g. If the angle with the vertical is called φ the motion

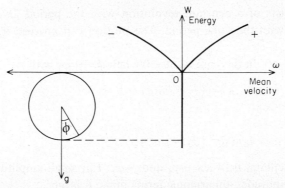

Fig. 4.1. A simple pendulum. If the energy $W > 0$, the pendulum oscillates with an amplitude $\varphi_0 < \pi$. The mean velocity $d\varphi/dt$ is zero. If the energy $W > 0$, the oscillation is superimposed by a systematic motion ω, which may be positive or negative.

is described by the equation

$$d^2\varphi/dt^2 + \kappa^2 \sin \varphi = 0, \tag{4.1}$$

where

$$\kappa^2 = g/l. \tag{4.2}$$

Equation (1) has the integral

$$(d\varphi/dt)^2 = C + 2\kappa^2 \cos \varphi, \tag{4.3}$$

where C is a constant.

Referring the energy W of the system to the potential energy at the uppermost point ($W = 0$ for $\varphi = \pi$) we have

$$W = \tfrac{1}{2}ml^2 \left(\frac{d\varphi}{dt}\right)^2 - mgl(1 + \cos \varphi) \tag{4.4}$$

and

$$C = (2W/ml^2) + 2\kappa^2. \tag{4.5}$$

Depending on the value of C we have three cases:

(a) $C > 2\kappa^2$; $W > 0$. In this case $d\varphi/dt$ never vanishes, it could be either > 0 or < 0. We have

$$t - t_0 = \int_{\phi_0}^{\phi} (C + 2\kappa^2 \cos \varphi)^{-1/2} \, d\varphi. \tag{4.6}$$

If we put

$$\frac{1}{\omega} = \frac{1}{2\pi} \int_0^{2\pi} (C + 2\kappa^2 \cos \varphi)^{-1/2} \, d\varphi, \tag{4.7}$$

we can write the solution (see Brown and Shook, 1964, p. 219)

$$\varphi - \varphi_0 = \omega t + \phi_0 - \frac{\kappa^2}{\omega^2} \sin(\omega t + \varphi_0)$$

$$+ \frac{\kappa^4}{8\omega^4} \sin 2 \,(\omega t + \varphi_0) + \cdots \quad (\varphi_0 = \text{constant}). \tag{4.8}$$

The motion consists of a constant revolution with the period $2\pi/\omega$, superimposed by an oscillation with the same period. The motion can proceed in either direction ($\omega < 0$ or $\omega > 0$).

(b) $C < 2\kappa^2$; $W < 0$. In this case $d\varphi/dt = 0$ when $\varphi = \pm\varphi_1$ with

$$\cos\varphi_1 = -C/2\kappa^2 = (-W/mlg) - 1 \qquad (4.9)$$

and the integral is

$$(d\varphi/dt)^2 = 4\kappa^2(\sin^2\tfrac{1}{2}\varphi_1 - \sin^2\tfrac{1}{2}\varphi). \qquad (4.10)$$

The value of φ oscillates between $-\varphi_1$ and $+\varphi_1$. For small amplitudes the period is $2\pi/\kappa$, for large amplitudes anharmonic terms make it larger.

(c) The case $C = 2\kappa^2$; $W = 0$ means that the pendulum reaches the unstable equilibrium at the uppermost point of the circle, with zero velocity.

The lowest state of energy is when the pendulum is at rest at $\varphi = 0$. If energy is supplied oscillations start and their amplitude grows until φ_1 approaches π. Then there is a discontinuous transition from case (b) to case (a).

In order to study the resonance phenomena in the Solar System one can start from the equations of motion of a pendulum disturbed by a periodic force (see Brown and Shook, 1964). The problems usually lead to analytically complicated formulae which are beyond the scope of this treatise. Moreover, very often only numerical solutions of a number of typical cases can clarify the situation. Instead we shall discuss qualitatively a simple case.

4.2. A SIMPLE RESONANCE MODEL

In order to demonstrate a basic resonance phenomenon let us discuss a very simple case (Figure 4.2). Suppose that a planet O is encircled by two satellites, one of not very small mass (m_2) moving in a circular orbit, and one with negligible mass (m_1)

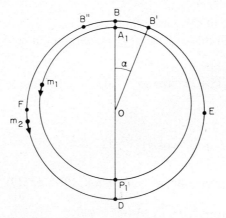

Fig. 4.2. A simple resonance model. A planet 0 is encircled by two satellites, one of not too small a mass (m_2) moving in a circular orbit, and one with negligible mass (m_1) moving in an elliptic orbit. The orbital inclinations are zero, and the ratio of average angular velocities ω_1/ω_2 is near 2 where $\omega_1 = 2\pi/T_1$ and $\omega_2 = 2\pi/T_2$. A_1 is the apocenter of the inner satellite, P_1 is its pericenter.

moving in an elliptic orbit. We denote by $\omega_1 = 2\pi/T_1$ and $\omega_2 = 2\pi/T_2$ the average angular velocities of m_1 and m_2, and we treat the case where the ratio $\omega_1 : \omega_2$ is close to 2. Orbital inclinations are put equal to zero.

If at a certain moment the longitude angles of the satellites are λ_1 and λ_2, a 'conjugation' occurs when $\lambda_1 = \lambda_2$. Consider the case when there is a conjugation between the satellites at the moment when the inner one is at its epicentre A_1 and the outer one is at $\overset{\cdot}{B}$. This implies that the outer satellite is at E when the inner satellite is at its pericentre P_1. When it moves from P_1 to A_1 it is subject to the attraction from m_2 which works in the direction of motion, hence increasing the angular momentum. When the motion continues from A_1 to P_1, m_1 is subject to a similar force from the outer satellite, which moves from B to F, but this force will diminish the angular momentum of m_1. Because of the symmetry the net result is zero (neglecting high order terms).

Suppose now that m_1 arrives at A_1 a certain time Δt before m_2 arrives at B. Because the orbits are closest together around $A_1 B$, the effects in this region predominate. If m_2 is at B' when m_1 is at A_1 the force between them will decrease the angular momentum C_1 of m_1. (The reciprocal effect on m_2 is negligible because of the smallness of m_1.) As the orbital period of a satellite is proportional to C^3, the period of m_1 will be shortened with the result that at the next conjunction it will arrive at A_1 when m_2 is still further away from B. The result is that the angle α between the bodies when m_1 is at its pericentre will increase.

If on the other hand α is negative so that m_1 arrives too late at A_1, say when m_2 already has reached B'', the angular momentum of m_1 will increase with the result that α will become still more negative.

We can compare this result with the pendulum treated in §3.1 when it is close to the upper point $\varphi = \pi$. Putting $\alpha = \varphi - \pi$ we see that the conjuction at $\alpha = 0$ represents an unstable equilibrium. We can conclude that a stable equilibrium is reached when $\alpha = \pi$, corresponding to $\varphi = 0$. This means that the inner satellite is at P_1 when the outer one is at B. (This implies that m_1 also is at P_1 when m_2 is at D. The interaction at this configuration is smaller than near A_1 because of the larger distance between the orbits.)

4.3. DEVIATIONS FROM EXACT RESONANCE

If we put the mean longitudes of the two bodies equal to λ_1 and λ_2 the resonance means that

$$\omega_1 \lambda_1 - \omega_2 \lambda_2 = 0. \tag{4.11}$$

The bodies can make oscillations around the equilibrium. (In celestial mechanics the word 'libration' is used for oscillations.) If we put

$$\varphi = \omega_1 \lambda_1 - \omega_2 \lambda_2 ; \tag{4.12}$$

φ will 'librate' with a period T_L which may be many orders of magnitude larger than the orbital period. This corresponds to the oscillations of the simple pendulum.

As the velocity of the libration is added or subtracted to the average orbital velocity, the apparent period will deviate from T_1 or T_2 by a fraction T_L/T_1 or T_L/T_2. This corresponds to the broadening of a resonance line.

The amplitude of the libration is a measure of the stability of the resonance coupling. If we increase the amplitude to π the system passes in a discontinuous way from state (b) to state (a). We no longer have a resonance, but what is called a 'near-commensurability'. The average value of φ will increase or decrease indefinitely with time.

In our simple model resonance requires that the ratio $\omega_2 : \omega_1$ is exactly 2 in case the libration is zero. In a real case this is not necessarily true because the pericentre at longitude π_1 may have a precession with the velocity $\omega_{\pi,1}$. As our discussion refers to the position of the pericentre, we have the more general resonance condition

$$2\omega_2 = \omega_1 + \omega_{\pi,1}. \tag{4.13}$$

4.4. Resonances in the Solar System

If we make a table of all the periods in the Solar System – orbital periods as well as spin periods – we find that many of them are commensurable. This indicates that there exist a number of resonance effects between mutually coupled resonators. Thus there are resonances between the orbital periods of members of the same system and there are also resonances between orbital periods and the spin periods of rotating bodies.

Such resonances seem to be very important features in the Solar System. As bodies once trapped in a resonance remain trapped indefinitely (under certain conditions), the resonance structure may stabilize the Solar System for very long periods of time.

Furthermore, a study of the resonance structure may give us relevant information about the evolution of the system. In order to draw any conclusions in this respect we much decide how the present resonance structure has been established. Two different mechanisms have been suggested:

(a) The first one envisages that the bodies were produced at the time of the hetegonic process with spin and orbital periods without resonances – except those necessarily resulting from a random distribution. A later evolution of the system, mainly by tidal effects, has changed the periods in a non-uniform way. This has resulted in the establishment of resonances.

(b) According to the second alternative, resonance effects have been important in the hetegonic process itself, so that bodies have preferentially been produced in states of resonance with other bodies. Hence the resonance structure may give us direct information about the hetegonic process.

The main purpose of this section is to decide between the two mechanisms.

4.5. Different types of resonances

In the Solar System the following types of resonances have been shown to be important (see Table 4.1):

TABLE 4.1

Different types of resonances

	Satellite Orbit	Planetary Orbit	Planetary Spin
Satellite Orbit	*Jovian satellites* Io-Europa-Ganymede *Saturnian satellites* Mimas-Tethys Enceladus-Dione Titan-Hyperion	(Jupiter 8, 9, 11 and Sun) (Phoebe and Sun) (Moon and Sun)	Tidal effects Possible coupling between Earth and Moon in the past
Planetary Orbit		(Jupiter-Saturn) Neptune-Pluto Jupiter-asteroids: Trojans Hilda asteroids Kirkwood gaps Thule	Spin-orbit of Mercury Spin of Venus- orbit of Earth

Parenthesis means a near-commensurability, (not a captured resonance).

(a) *Orbit-orbit resonances*. If two planets or two satellites have orbital periods T_1 and T_2 and the ratio between them can be written

$$T_1/T_2 = n_1/n_2, \tag{4.14}$$

where n_1 and n_2 are small integers; such periods are called commensurable. Resonance effects may be important under the condition that the gravitational attraction between the bodies is above a certain limit. There are several pronounced examples of this in the satellite systems of Jupiter and Saturn, and the effect is also important in the planetary system, especially for the asteroids.

Resonance between the orbital motion of a planet and the orbital motion of one of its own satellites has also been discussed. Seen from the point of view of the planet this is a resonance between the apparent motion of the satellite and the apparent motion of the Sun. Such resonances are sometimes referred to as 'satellite-sun resonances'.

(b) *Spin-orbit resonances*. If the density distribution in a rotating body is asymmetric, an imposed periodically varying gravitation field may influence its rotation. This effect is important in case a resonance occurs between the period of the field variations and the spin. The spin of Mercury seems to be locked in a resonance with its own orbital period, and similarly the spin of Venus is coupled with the orbital motion of the Earth (in relation to Venus).

A similar asymmetry of a planet may also affect the motion of a satellite encircling the planet. This effect is not known to be important today but it may have affected the evolution of the Earth-Moon system (see Section 4.9).

If a satellite produces tides on its primary, the tidal bulges corotate with the satellite. The coupling between the tidal bulges and the satellite may be considered as a spin-orbit resonance with $n_1 = n_2 = 1$.

4.6. Orbit-orbit Resonances in Satellite Systems

Our first example is the resonance between Enceladus and Dione. The mass of Enceladus is less than 10% of that of Dione. The period of Enceladus is half of the period of Dione. However, due to the perturbation of the Coulomb field which is caused by the spin of Saturn, the pericentre of Enceladus has a precession $\omega_{\pi, E}$. The equilibrium is constantly situated at the pericentre P_1. Hence we have

$$2\omega_D = \omega_E + \omega_{\pi, E}. \qquad (4.15)$$

Around this equilibrium Enceladus oscillates with a very small amplitude (20′) and with a period $T_p = 12$ yr.

A similar treatment as in Section 4.2 of a small satellite outside a big one give as a result that the small satellite has its equilibrium located at the epicentre.

This case is applicable to Titan and Hyperion with the resonance $n_1 = 3$, $n_2 = 4$. We have

$$4\omega_H = 3\omega_T + \omega_{\pi, H}, \qquad (4.16)$$

where $\omega_{\pi, H}$ is the precession of the pericentre of Hyperion.

A somewhat different type of resonance takes place between Mimas and Tethys. It is not connected with the peri- or epicentre but with the nodes. The orbits are inclined 1.5° and 1.1° and the conjuction between the satellites oscillates about the midpoint of these two ascending nodes. It obeys the relation

$$4\omega_T - 2\omega_M - \omega_{\theta, T} - \omega_{\theta, M} = 0, \qquad (4.17)$$

where $\omega_{\theta, T}$ and $\omega_{\theta, M}$ are the precessions of the nodes.

A fourth case of exact commensurability is found in the Jovian system, where Io, Europa, and Ganymede obey the equation

$$\omega_I - 3\omega_E + 2\omega_G = 0 \qquad (4.18)$$

within the observational accuracy (which is 10^{-9}). The mechanism is rather complicated. It has been treated in detail by the exact methods of celestial mechanics.

4.7. On the Absence of Resonance Effects in the Saturnian Ring System

The dark marking in the Saturnian ring system, especially Cassini's division, have long been thought to be due to resonances produced by Mimas and perhaps also by other satellites. It has been claimed that the Saturnian rings exhibted a number of gaps similar to the Kirkwood gaps in the asteroid distribution. This is erroneous as has been shown both observationally and theoretically (see Alfvén, 1968).

The accurate measurements of Dollfus (1961) are shown in Figure 4.3. It is obvious that there is no correlation between the observed markings and the gaps which are expected as an analogy to the Kirkwood gaps in the asteroid belt. Furthermore, from a theoretical point of view one should not expect any analogy to the Kirkwood gaps. The mass-ratio Mimas-Saturn is $1 : 8 \times 10^6$, whereas the ratio Jupiter-Sun is $1 : 10^3$. Hence the relative disturbing function is almost 10^4 times smaller in the case of the

TABLE 4.2

Interlocked resonances

Ratio	Bodies	Period	e	i	Type of resonance	Libration Period	Amplitude	References
1	Tethys	1.887802^d	0.00	1.1°	Resonances related to the nodes	70.8 yr	47°	(1)
2	Mimas	0.942422^d	0.0201	1.5°				(4)
1	Dione	2.73681^d	0.0021	0.0°	Conjunction when Enceladus at peri-saturnian	3.89 yr	11′24″	(1)
2	Enceladus	1.37028^d	0.0045	0.0°				
3	Hyperion	21.276666^d	0.104	0.5°	Conjunction when Hyperion at apo-saturnian	18.75 yr	9°	(1)
4	Titan	15.945452^d	0.0290	0.3°				(4)
2	Pluto	248.43^y	0.247	17.1°	See Figure 4.4	20000 yr	39°	(2)
3	Neptune	164.78^y	0.0087	1.46°				
2	Jupiter	11.86^y	0.048	1.38°	Largest body of a group of at least 20 bodies librating with different amplitudes and phases	270 yr	40°	(3)
3	Hilda	7.90^y	0.15	7.85°				
3	Jupiter	11.86^y	0.048	1.38°				
4	Thule	8.90^y	0.03	23.°				
1	Jupiter	11.86^y	0.048	1.38°	Two groups, one at each of the libration points of Jupiter	~900 yr	10–20°	
1	Trojans	11.86^y	~0.15	10–20°				

(1) Roy and Ovenden (1954), Goldreich (1965); (2) Cohen, Hubbard and Oesterminter (1967); (3) Schubart (1968) and (4) Brouwer and Clemence (1961).

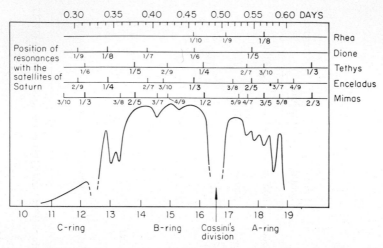

Fig. 4.3. Diagram of Dollfus' measurements of the markings of the Saturnian Rings. The diagram shows that Cassini's division (and other markings) have no relation to the points where the resonance might be produced by the satellites. Hence the structure of the Saturnian Rings is *not* a resonance phenomenon.

Saturnian rings. Such a small disturbance is not likely to produce any appreciable resonance phenomenon.

4.8. ORBIT-ORBIT RESONANCES IN THE PLANETARY SYSTEM

In the planetary system there is a resonance between Neptune and Pluto which is similar to the Titan-Hyperion resonance in the sense that the outer body is much smaller and moves in a rather eccentric orbit. The ratio of the periods is 3:2. Cohen and Hubbard (1964) have calculated numerically the relations between Neptune and Pluto. They find that the quantity

$$\delta = 3\lambda_P - 2\lambda_{\Psi} - \bar{\omega}_P - 180°, \tag{4.19}$$

where λ is the mean longitude and $\bar{\omega}$ the longitude of perihelion, oscillates with a period of 19 670 yr. The amplitude of this libration is 76° (see Figure 4.4).

The orbits are interlocked in such a way that the planets never come closer together than 18 AU ($=2.7 \times 10^{14}$ cm). In spite of the fact that the perihelion of Pluto is closer to the Sun than the aphelion of Neptune, the bodies have no chance of ever colliding.

There are a number of exact resonances between Jupiter and the asteroids. First of all a group of asteroids called the Trojans have periods which oscillate around the period of Jupiter. They move near the Lagrangean points, i.e. at the solar distance of Jupiter, 60° before and 60° after Jupiter. The Trojans perform oscillations around these points which are known to be possible equilibria.

Another group is the Hilda asteroids consisting of the minor planet 153 Hilda and 19 other asteroids. Their mean period is $\frac{2}{3}$ of Jupiter's. Contrary to the Trojans their

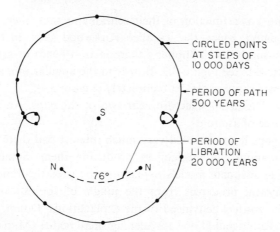

Fig. 4.4.   Path of Pluto about SN at extreme of libration (after Cohen and Hubbard, 1964). The Sun is at *S* and Neptune (*N*) librates along the dotted line.

inclinations are small but their eccentricities fairly large. Schubart (1968) has calculated their orbits with the simplifying assumption $i=0$. He finds that the eccentricity and perihelion of Hilda has a secular variation with a period of 2600 yr, superimposed by a small libration with a period of 270 yr.

An interesting and puzzling resonance phenomenon is found in the main asteroidal belt. If the number of asteroids is plotted as a function of their orbital period or –

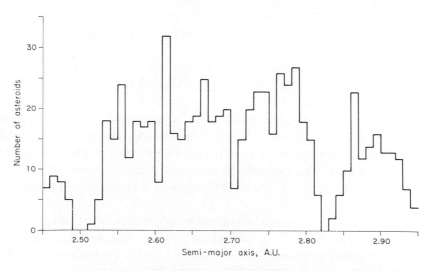

Fig. 4.5.   Detail of the structure near two of the Kirkwood gaps. The theoretical values of the 1:3 and 2:5 resonances with Jupiter are $a = 2.502$ and 2.825, respectively.

which is equivalent – as a function of their semi-major axes, there are a number of very pronounced empty zones, the so-called Kirkwood gaps. In the neighborhood of periods commensurable with Jupiter's there is an absence of asteroids. Gaps for $\frac{1}{2}, \frac{1}{3}, \frac{2}{5}$ etc. are observed. See Figure 3.3. If instead the angular momentum C is taken as abscissa the gaps disappear almost completely (Figure 3.4).

Figure 4.5 shows the detail structure near two of the gaps. An interval of about 1% in T is almost free of asteroids.

The Kirkwood gaps have attracted very much interest and there is a multitude of theoretical papers about the mechanism producing them. Some of the authors claim to have made adequate models explaining the gaps theoretically. If one tries to extract the physical principles from the jungle of sophisticated mathematical formulae, one does not feel convinced of the explanations. Doubt is due to the fact that whereas both Tethys and Dione are keeping small bodies (Mimas and Enceladus) trapped at a resonance 1:2, Jupiter produces an *absence* of small bodies at the corresponding period. Further, Jupiter keeps a number of Hilda asteroids trapped at 2:3 resonance but produces gaps at a number of other resonance points.

In absence of a clear answer to these questions, one must ask whether the Kirkwood gaps really are produced by the resonance effects which are the starting point of the theories. As we have seen in Section 3 there are reasons to believe that non-gravitational effects are of importance to the motion of comets and asteroids. It may be that viscosity effects (collisions) are essential to the understanding of the Kirkwood gaps. This would make them more interesting from a hetegonic point of view. One may hope that an explanation of them will come out from a general theory of jet streams.

4.9. Spin-orbit resonances

For all satellites with known spins these coincide with the orbital periods. This is likely to be due to tides, produced by their primaries, which have braked their synodic rotations to zero. Formally this means spin-orbit resonances with $n_1 = 1$, and $n_2 = 1$.

Mercury's spin period is 59 days which is exactly $\frac{2}{3}$ of its orbital period. This means that it is trapped in a resonance. According to Goldreich this represents the final state produced by the solar tides.

The case of Venus is very interesting. It has a retrograde spin with a period of 243 days. This value is supposed to be due to a resonance with the synodic rotation of the Earth (seen from Venus). It is very surprising that the Earth can lock Venus into such a resonance. This can only be explained if the gravitational potential of Venus has a very strong variation with the longitude. Why this should be the case is not known.

Further, the fact that the spin is retrograde is very remarkable. If we interpret the slowness of the rotation as due to the braking by solar tides, we cannot in this way account for its retrograde sense. We can hardly avoid the conclusion that Venus acquired its retrograde spin when it first accreted. This is unique among the major planets (except for Uranus the axis of which is inclined 98°).

There is another type of spin-orbit resonances, viz. the interaction between a spinning body e.g. a planet, and the satellites around it. Allan (1967) has drawn the attention to the fact that if the gravitational potential of the planet depends on the longitude, a satellite will be subject to a force in the tangential direction, which may transfer energy between the spin and the satellite orbit. In case the orbiting periods of the satellite equals the spin period of the planet we have a 1:1 resonance. The satellite will be locked at a certain phase angle around which it can perform oscillations ('librations'). Unfortunately there are no examples of synchronous natural satellites (but the theory is applicable to geostationary artificial satellites).

There are also higher resonances, but these are efficient only for satellites with high inclinations or eccentricities. A body in a circular orbit in the equational plane is not affected. It has been suggested that such resonances were of importance at the evolution of the Earth-Moon system (see Alfvén and Arrhenius, 1969).

4.10. NEAR-COMMENSURABILITIES

Besides these exact resonances there are a number of near-commensurabilities. These have attracted much attention because the perturbations become especially large. Most noteworthy is the case of Jupiter-Saturn, whose periods have a ratio close to 2:5. The near-commensurabilities have been listed by Roy and Ovenden (1954) and further discussed by Goldreich (1965).

In the case of exact resonances the relative positions of the bodies are locked at certain equilibrium positions around which they perform oscillations. At near-commensurability no such locking exists. There is a smooth transition to the near-commensurable case from a non-commensurable state.

It is doubtful whether near-commensurabilities are of hetegonic significance. If the periods of the different bodies are distributed at random, there is a certain probability that two periods should be near-commensurable. Studies by the authors cited agree that the number of observed commensurabilities is larger than expected statistically. If, however, we account for the exact resonances by a separate mechanism (see Section 11) and subtract them from the material, the remaining excess – if any – is not very large.

From a hetegonic point of view the most interesting cases are the near-commensurabilities of retrograde satellites and the 'Sun'. Also JVIII, JIX and JXI have periods which are close to 1:6 of the orbital period of Jupiter. In JXII the ratio is close to 1:7, and the same is the case for the period of Phoebe compared to the period of Saturn. There is a possibility that these commensurabilities were important at the capture of these satellites ('resonance capture').

4.11. THEORY OF RESONANCES

Goldreich (1965) has proposed that in the satellite systems *tidal effects* push out some of the satellites until they reach such orbits that they are captured into resonances. In the next section we shall discuss the tidal effects in the satellite system. The result casts some doubt on the importance of this mechanism.

In addition, there are two specific objections to the tidal theory of resonances:

(a) If a body is captured in a resonance by tidal effect, its libration amplitude immediately after capture should be close to $\pm\pi$. The amplitude can decrease due to damping, either by the viscosity of the surrounding medium or by tidal action. In some cases, however, the observed libration amplitude is very small. The libration amplitude of Hyperion is only $14°$, and in the case of Enceladus it has the extremely small value of $11'$. It is difficult to understand how the amplitude has been reduced by such a large factor.

(b) Resonances exist also in the planetary system, where tidal effects are negligible. The resonances Neptune-Pluto, Jupiter-Thule, and Jupiter-Hildas (20 asteroids!) cannot have been produced by tides. Since in the planetary system we anyhow must look for another mechanism for resonances, we should try to apply this mechanism to the satellite systems before accepting a less general mechanism.

The conclusion is that resonances are likely to be due to some effect at the accretion of the bodies. The necessary damping of the librations could easily take place when the matter was in a disperse state before or during the accretion. It seems possible that an accreting embryo can affect the accretion in an adjacent jet stream by gravitational resonance. This should follow from a general theory of the coagulation of a jet stream.

The above objections cannot be raised against the theory of the spin-orbit resonances of Mercury and Venus. Whether these were produced at the accretion of the planets or later, by tidal effects, should be considered an open question.

5. Spin and Tides

5.1. TIDES

The spins of the celestial bodies contain important information about their formation and evolutionary history.

The main effect producing changes in the spins is supposed to be tidal action. The theory of the terrestrial tides, as produced by the Moon and the Sun, has been developed especially by Jeffreys (1962), and Munk and Mac Donald (1960). The latter authors state (page 15) that 'there are few problems in geophysics in which less progress has been made'. Even if this statement is due to too high an appreciation of the progress in other fields, it shows how difficult a problem the tides offer.

For our purpose we are not only interested in the terrestrial tides but also in the tides on other celestial bodies. The internal structure of these is almost unknown. Hence, very little can be concluded theoretically. We have to look for possible effects on the orbits of satellites in order to make any conclusions.

5.2. AMPLITUDE OF TIDES

Suppose that two bodies with radii R_1 and R_2, densities θ_1 and θ_2, and masses $m_1 = 4\pi\theta_1 R_1^3/3$ and $m_2 = 4\pi\theta_2 R_2^3/3$, are situated at a center of gravity distance r_0. The gravitation from m_2 deforms the spherical shape of m_1 so that its ellipticity becomes

$$\alpha_1 = \frac{15}{4} \frac{m_2}{m_1} \left(\frac{R_1}{r_0}\right)^3 = \frac{15\theta_2}{4\theta_1} \left(\frac{R_2}{r_0}\right)^3, \tag{5.1}$$

a formula which is a good approximation for $r_0 \gg R_1$ (far outside the Roche limit). The height Z of the tides is

$$Z_1 = \frac{\alpha_1}{2} R_1. \tag{5.2}$$

Similar expressions hold for m_2. Let one of the bodies be a companion (index h) to a central body (index c) and count the distance r_0 with the radius of the central body as unit. If we abbreviate

$$\beta = r_0/R_c \tag{5.3}$$

and

$$\gamma = m_h/m_c,$$

it follows that

$$\alpha_h = \frac{15}{4} \frac{\theta_c}{\theta_h} \frac{1}{\beta^3}, \tag{5.4}$$

$$\alpha_c = \frac{15}{4} \frac{\gamma}{\beta^3}. \tag{5.5}$$

Table 5.1 shows some typical examples, the satellites of Jupiter and Saturn as well as the Moon and Triton.

As shown by these examples the tides which a satellite produces on a planet are very small. In fact it never exceeds about $\alpha = 10^{-6}$. In contrast to this the satellites are strongly deformed with α of the order 10^{-3}. If they are close to the Roche limit, equation (1) does not hold. At the Roche limit the tides become infinite.

Equation (1) can be generalized to rigid bodies by the introduction of a correction factor containing the rigidity (see Jeffreys, 1962; Munk and MacDonald, 1960).

TABLE 5.1

Central body	Companion	Central body Ellipticity α_c	Height Z_c cm	Companion ellipticity α_h
Earth	Moon	21×10^{-8}	67	3.3×10^{-5}
Jupiter	Io	67×10^{-8}	2400	4.8×10^{-3}
	Europa	11×10^{-8}	400	1.2×10^{-3}
	Callisto	9.1×10^{-8}	325	0.30×10^{-3}
	Ganymede	1.0×10^{-8}	37	0.054×10^{-3}
Saturn	Mimas	0.83×10^{-8}	25	33×10^{-3}
	Enceladus	0.74×10^{-8}	22	16×10^{-3}
	Tethys	3.6×10^{-8}	110	8.3×10^{-3}
	Dione	27×10^{-8}	81	4.0×10^{-3}
	Titan	10.6×10^{-8}	320	0.12×10^{-3}
Neptune	Triton	1.24×10^{-6}	1380	0.25×10^{-3}

5.3. TIDAL BRAKING

When a body spins with a period T in the presence of a tide-producing field, it is deformed – seen from a rotating coordinate system – periodically with the period $\frac{1}{2}T$. This produces internal motions in the body which are associated with an energy dissipation P (erg/sec). The energy is drawn from the spin energy of the body, which means that the spin is braked. As no change is produced in the total angular momentum of the system consisting of the spinning body and the tide-producing body, spin angular momentum is transferred to the orbital angular momentum.

The value of P depends on the physical state of the body, and on the amplitude of the tides.

Suppose that the tidal bulge is displaced at an angle ε in relation to the tide-producing body (see Figure 5.1). A quantity Q, defined by $Q^{-1} = \text{tg} 2\varepsilon$ (in analogy with what is customary in treating losses in electric circuits) is then often used. There is no objection to this formalism but it should be remembered that Q depends not only on the frequency but also on the amplitude. The amplitude dependence of the tidal braking is in general very large (see Jeffreys, 1962). It is not correct to assign a certain Q value to each celestial body because this value is likely to be much smaller for high tidal amplitudes than for low amplitudes. As shown by Jeffreys (1962), the relation between the solar tides and lunar tides on the Earth is very complicated and the Q-value of the Earth is different for these two tides. This is still more the case if the tidal amplitudes are very different.

5.3.1. *Fluid body*

Seen from the coordinate system of the spinning body the tidal deformation corresponds to a progressing wave. The fluid motion which in a non-structured body is associated with this wave is of the order

$$v = 2\omega\alpha R, \tag{5.6}$$

with $\omega = 2\pi/T$. For $T = 10\,h = 3.6 \times 10^4$ sec, $\alpha = 10^{-7}$, $R = 0.5 \times 10^{10}$ (applicable to the tides produced on one of the giant planets by a satellite) we have

$$v \approx 0.1 \text{ cm/sec}.$$

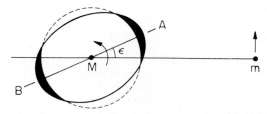

Fig. 5.1.   Classical but inadequate model of momentum transfer due to tides. The force of attraction between the satellite m and the nearest tidal bulge A exceeds that between m and B; a component of the net torque retards the rotation of the planet m and accelerates the satellite in its orbit .The real situation in the case of the Earth is seen from Figures 5.2 and 5.3. In the case of Mars, Jupiter, Saturn, and Uranus the angle ε is probably negligible.

It seems highly unlikely that such low velocities can produce any appreciable dissipation of energy. (The order of magnitude of the power dissipation with laminar flow is $P = c(v/R)^2 \cdot R^3 = cv^2 R$ where $c \approx 10^{-2}$ poise is the viscosity. With $R = 10^9$ we obtain $P = 10^5$ erg/sec.)

If instead we use values applicable to a satellite ($T = 10$ h, $\alpha = 10^{-3}$, $R = 5 \times 10^8$ cm) we find

$$v \approx 20. \, \text{cm/sec}.$$

5.3.2. Solid body

In a small solid body (asteroid-sized) only elastic deformations are produced with a minimum of energy dissipation.

In satellites which are so large that their rigidity does not prevent deformations (lunar-sized bodies) these may often be non-elastic, and hence associated with big energy losses.

As far as is known all satellites have synchronous rotation.

A planet which constitutes a fairly homogeneous solid body, probably has a negligible tidal braking. The deformations are of the order $\alpha \approx 10^{-7}$ and may be purely elastic. We are far below the yield limit of most materials.

5.3.3. Structured bodies

The most difficult case occurs when the body has a complicated structure, involving fluid layers of different densities. This is what characterizes the Earth, and in spite of all the investigations we still are far from a complete understanding of the tidal braking. Most of the dissipation of energy takes place in shallow seas, at beaches and regions near the shores. Hence a knowledge of the detailed structure of a planet is necessary in order to make any conclusion about the tidal retardation.

5.4. SATELLITE BRAKING OF PLANETARY SPINS

The Earth-Moon system is the only system where we can be sure that a significant tidal braking has taken place, and is taking place. According to the elementary theory, the Moon should produce tidal bulges in the oceans (according to Figure 5.1) and when the Earth rotates these remain stationary. Due to the viscosity of the water, the relative motion produces an energy release P_1 which brakes the spin of the Earth. At the same time, the tidal bulge is displaced a phase angle ε in relation to the vector radius to the Moon. This produces a force which tends to accelerate the Moon, but the net effect is that the lunar distance increases.

This theory, which is presented in all text books, has very little to do with reality. The observed tides do not behave at all as they should according to this theory. Instead, the tidal waves one observes have the character of standing waves excited in the different oceans and seas which act as resonance cavities. See Figures 5.2a and 5.2b.

Even if the tidal pattern on the Earth is very far from what the simple theory predicts, there is no doubt that a momentum transfer takes place between the Earth

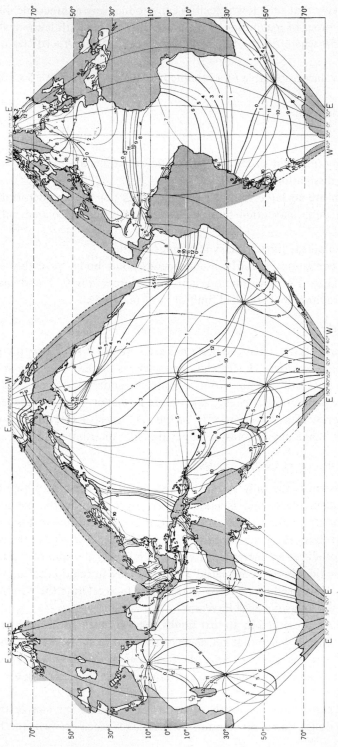

Fig. 5.2(a).   Phase relations of tides in the world ocean. The map shows the cotidal lines of the semidiurnal tide referred to the culmination of the Moon in Greenwich. The tidal amplitude approaches zero where the cotidal lines run parallel (such as between Japan and New Guinea). Much of the tidal motion thus has the character of rotary waves; among the exceptions is the south and equatorial Atlantic Ocean where the tide mainly takes the form of north-south oscillation on east-west nodal lines. This complex reality should be compared to the simple concept which is basic for existing calculations of the lunar orbital evolution and which pictures the tide as a sinusoidal wave progressing around the Earth in the easterly direction (dot and dashed curve in Figure 5.2.(b).

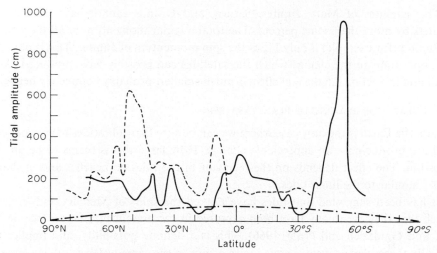

Fig. 5.2.(b). Tidal amplitude on the Atlantic coasts as an example of the actual amplitude distribution in comparison with the simple Laplacian tide concept. The curves show the average range at spring tide of the semidiurnal tide as a function of the latitude. In the comparison with the (much less known) open ocean amplitudes, the coastal amplitudes are increased by co-oscillation with the oceanic regions over the continental shelves. The distribution illustrates further the fact that tidal dissipation is governed by a series of complex local phenomena depending on the configuration of continents, shelves, and ocean basins and that the theoretical Laplacian tide obviously cannot serve even as a first-order approximation. In the figure the solid curve represents the tide on the western side of the Atlantic Ocean; dashed curve, the eastern side of the Atlantic Ocean; dot and dashed curve, the Laplacian tide. (From Alfvén and Arrhenius, 1969).

and the Moon. The effect of this has been calculated by Gerstenkorn (1955), MacDonald (1966), and most recently and in the most detail by Singer (1970). According to these theories the Moon was originally an independent planet which was captured either in a retrograde or in a prograde orbit.

There is considerable doubt as to what extent the models are applicable to the Earth-Moon system (see Alfvén and Arrhenius, 1969). Resonance effects may invalidate the models in many details. There seems, however, to be little reason to question the main result, *viz.*, that the Moon is a captured planet, brought to its present orbit by tidal action. Whether this implied a very close approach to the Earth is so far an open question. The problem will be discussed more in detail in a later section.

The Neptune-Triton system is probably an analogy to the Earth-Moon system. The only possibility of understanding how Neptune can have a retrograde satellite with an unusually large mass seems to be that Triton was captured in an eccentric retrograde orbit, which by tidal action has shrunk and become more circular.

As Neptune has a mass and a spin period similar to Uranus, it is likely to have produced a similar satellite system. The capture of Triton and the later evolution of its orbit has made Triton pass close to the primeval small satellites, either colliding with them or throwing them out of orbit. Nereid may be an example of the latter process (see McCord, 1966).

The satellites of Mars, Jupiter, Saturn, and Uranus cannot have braked these planets by more than some percent. The total angular momentum of all the satellites of Jupiter (for example) is only 1% of the spin momentum of Jupiter. This is obviously an upper limit to the change which the satellites can possibly have produced. As we shall find in Section 5.6 the real effect is much smaller, probably completely negligible.

5.5. Solar tide braking of planetary spin

Again, the Earth is the only case where we can be sure that solar tides have produced – and are producing – an appreciable change. How large this is seems to be an open question. The effect depends on the behavior of the tides on beaches and in shallow seas – similar to the lunar tides.

It has been suggested that tides have braked the spins of Mercury and Venus so much that they eventually have been captured in the present resonances (see Section 4.9, and Goldreich and Peale, 1966). This is a definite possibility, and implies that initially these planets were accreted with a spin of the same order as other planets.

However, as we have seen in Section 4, the orbit-orbit resonances are not likely to be due to tidal capture, but must have been produced at the time when the bodies were accreting. In view of this, the question arises whether the spin-orbit resonances of Mercury and Venus as well were produced during their accretion. It seems at present impossible to decide between this possibility and the tidal alternative. The latter would be favored if there are or have ever been shallow seas on these planets, but we know nothing about this.

It seems unlikely that solar tides have braked the spins of the asteroids or of the giant planets to an appreciable extent.

5.6. Tidal evolution of satellite orbits

Goldreich and Soter (1965) have made very interesting investigations into the possible tidal evolutions of the satellite systems. They have pointed out that where pairs of satellites are captured in orbit-orbit resonances, both the satellites must change their orbits in the same proportion. They have further calculated the maximum value of the dissipation (in their terminology the minimum Q-values) which is reconcilable with the present structure of the satellite systems. There seems to be no objection to these conclusions.

They further tentatively suggested that the maximum values of dissipation are not far from the real values, and that tidal effects have been the reason why satellites have been captured in resonances. This problem has already been discussed in Section 4. The conclusion is that it is difficult to understand the small librations in some of the resonances.

Further, we observe resonances in the planetary system which certainly cannot have been produced in this way, so that it is in any case necessary to assume a hetegonic mechanism for production of some resonance captures. Finally, the structure of the Saturnian rings demonstrates that Mimas' orbit cannot have changed by more than one or two percent since the formation of the Saturnian system.

These difficult and fascinating problems need to be investigated in much greater detail. Present evidence seems to speak in favor of the view that with the exception of the Moon and Triton, no satellite orbits have been appreciably changed by tidal action.

5.7. THE ISOCHRONISM OF SPINS

Photometric registrations of asteroids show intensity variations which must be interpreted as due to rotation of a body with nonuniform albedo or non-spherical shape. Several investigators have measured the periods of axial rotation of some 30–40 asteroids and find no systematic dependence on the magnitudes of the asteroids. In fact, as is shown in Figure 5.3 and Table 5.2 almost all asteroids have periods

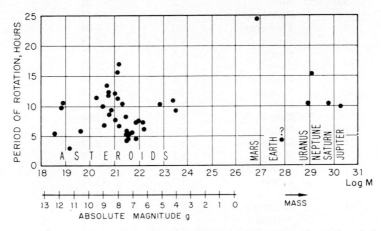

Fig. 5.3. Periods of axial rotation for the asteroids and some of the planets in relation to their masses. (From Alfvén, 1964b).

which deviate by less than 50% from an average of 8 or 9 h. It appears that this result is not due to observational selection.

Regarding the planets, we find that the giant planets as well have about the same period. It has always struck students of astronomy that the axial rotations of Jupiter, Saturn, and Uranus are almost equal. The period of Neptune is somewhat longer (15^h), but a correction for the tidal braking of its retrograde satellite reduces the period at least somewhat (see McCord, 1966). For the Earth we should use the period before the capture of the Moon, which according to Gerstenkorn was most likely five or six hours (see Alfvén, 1964).

Hence we find the very remarkable fact that the axial period is of the *same* order of magnitude for a number of bodies with *very different masses*. In fact, when the mass varies by a factor of more than 10^{11} – from less than 10^{19} g (for small asteroids) up to more than 10^{30} g (for Jupiter) – the axial period does not show any systematic variation. We may call this the *law of isochronic rotation*.

Obviously this law cannot be applied to bodies whose present rotation is regulated by tidal action or captured in resonances. (See Section 4)

TABLE 5.2
Periods of asteroids

Number	Name	Absolute magnitude (g)	Period (hours)
1	Ceres	4.0	9.08
2	Pallas	5.1	10.12
3	Juno	6.3	7.22
4	Vesta	4.2	10.68
5	Astraea	7.9	16.81
6	Hebe	6.6	7.28
7	Iris	6.7	7.13
9	Metis	7.2	5.06
11	Parthenope	7.7	10.7
15	Eunomia	6.2	6.08
16	Psyche	6.8	4.30
17	Thetis	8.6	12.27
19	Fortuna	8.2	7.45
20	Massalia	7.4	8.10
22	Kalliope	7.4	4.07
23	Thalia	8.2	12.40
24	Themis	8.1	12.00
25	Phocaea	9.0	9.95
27	Euterpe	8.5	8.50
28	Bellona	8.1	16.00
29	Amphitrite	7.2	5.00
30	Urania	8.7	13.67
39	Laetitia	7.3	5.14
40	Harmonia	8.4	9.13
43	Ariadne	9.1	11.47
44	Nysa	7.9	6.42
54	Alexandra	8.7	7.05
61	Danae	8.6	11.45
321	Florentina	11.3	2.87
349	Dembowska	7.2	4.67
433	Eros	12.3	5.27
511	Davida	7.0	5.17
624	Hektor	8.6	6.92
753	Tiflis	11.8	9.84
984	Gretia	10.6	5.76
—	PH 1931	11.7?	10.23

Excepting such bodies, the only body with a rotation known to be far from the order of 8^h is Pluto, which rotates in 6 days. Mars (period 25^h) and Icarus (2^h) each deviate by a factor of three.

5.8. Conclusions from the isochronism

Concerning the mechanism which produces the equality of spin periods to almost all the bodies, the following conclusions can be drawn:

(1) The equality of the periods cannot be produced by any factor acting today

For example, we cannot expect that the rotation of Jupiter is affected very much by any reasonable forces acting now.

(2) The equality of the periods cannot have anything to do with the rotational stability of the bodies. The giant planets, for example, are very far from rotational instability today. It is unlikely that one could find a mechanism by which the present isochronism can be connected with rotational instability during the prehistory of bodies as different as a small asteroid and a giant planet.

(3) Hence, the isochronism must be of hetegonic origin. All the bodies must have been accreted by a process which has the characteristic feature that it makes their axial period about equal, no matter how much mass is acquired. There are several two-step processes for accretion which have this property (see Section 8).

(4) The braking of the axial rotation of the bodies has not been very significant since their accretion. A braking produced by a surrounding uniform viscous medium should lengthen the period of a small body much more than the period of a larger body. The fact that asteroids as small as some ten kilometers rotate with the same periods as the largest planets indicates that even such small bodies have not been braked very much since they were formed. In this essential respect the Solar System seems to be in the same state now as when it was formed. This makes it reasonable that hetegonically important results can be obtained from detailed analysis of the present state of the Solar System.

(5) The isochronism shows further that the asteroids cannot derive from a broken-up planet. If a planet explodes (or is disrupted in some other way), we should expect an equipartition of the rotational energies of the parts. This means that the periods of axial rotation of the smallest asteroids should be much smaller than those of the larger asteroids. This is in conflict with observations. (Perhaps it is applicable to the smallest asteroids, like Icarus.)

In Section 8, we shall discuss theories which aim at an explanation of the isochronism.

6. Post-Accretional Changes in the Solar System

6.1. STABILITY OF ORBITS

Celestial mechanics applied to the motion of planets and satellites shows that of the orbital parameters the longitude π of the pericentre and the longitude θ of the ascending node vary monotonically, whereas the eccentricity e and the inclination i exhibit secular variations within certain limits. The most constant parameter is the semi-major axis a. There is a famous theorem by Lagrange and Poisson according to which there are no secular perturbations in a to the first and second approximation. Surveys of the orbital variation and the stability of the system from a mathematical point of view are given for example by Brouwer and Clemence (1961b), and by Hagihara (1961).

From a physical point of view the constancy of a is connected with the constancy of the angular momentum $C=[a(1-e^2)]^{1/2}$. It is difficult to change the orbital momentum of a body because the momentum must then be transferred either to another body or to the interplanetary medium. As the density in interplanetary space is very small the latter process is not very efficient. A transfer of angular momentum by tidal action seems to be the only important mechanism by which a considerable change can take place. Angular momentum can also be exchanged by resonance effects, but their transfer is usually small, just enough to secure a locking of the bodies in the resonance.

A possible change in the solar rotation in connection with the solar wind will be discussed in a later section.

The authors cited express – rather vaguely – the opinion that the Solar System probably is more stable than can be proved by ordinary celestial mechanics methods.

The presence of resonances seems to increase the stability of the system. Bodies locked in resonances are likely to remain in them for an indefinite time. The amplitude of the librations is a measure of the stability. If the librations increase to an amplitude of 180°, the bodies break loose from the resonance. In many cases the librations are very small (see Table 4.2) which indicates a high degree of stability.

A proportional change in the periods of all the orbiting bodies in a satellite system or in the planetary system will not alter the resonances. Such a change can be produced by an increase or decrease in the mass of the central body. Very little can consequently be said about such variations from a study of the resonance pattern. We can make more definite conclusions concerning changes in the relative positions of the orbits.

Assuming with Goldreich that if once an exact resonance is established, the bodies

TABLE 6.1

$2/5$	$T_{\saturn} = 11.783$	0.67%		3	$T_{\saturn} = 88.373$	5.2%	
	$T_{\jupiter} = 11.862$				$T_{\uranus} = 84.018$		
$1/7$	$T_{\uranus} = 12.003$	1.18%		$1/2$	$T_{\neptune} = 82.39$	2.0%	

will remain in resonance indefinitely, the existence of near-commensurabilities exclude larger changes in the orbits.

As Table 6.1 shows the period of Jupiter is intermediate between the 2:5 resonance of Saturn and the 1:7 resonance of Uranus. Similarly, the period of Uranus is intermediate between the 3:1 resonance of Saturn and the 1:2 resonance of Neptune. Hence if we assume that $T_{\uranus}$ and $T_{\saturn}$ have been constant, we can conclude that $T_{\jupiter}$ can never have been as much as 0.67% shorter because then it would have been trapped in 2:5 resonance with Saturn, nor can it have been as much as 1.18% longer, because of the 1:7 resonance with Uranus.

Similarly, if $T_{\saturn}$ and $T_{\neptune}$ have been constant, $T_{\uranus}$ cannot have been 2.0% shorter because of the 1:2 resonance with Neptune, nor 5.2% longer because of the 3:1 resonance with Saturn.

Similar arguments can be applied to the near commensurabilities in the satellite systems. The conclusion is that the orbital periods in the Solar System are likely to have varied less than a few percent since hetegonic times. The only exceptions are the Earth-Moon and the Neptune-Triton systems.

Another argument for the same view is obtained by the relations between Mimas and Cassini's division and between Janus and a dark marking in the B-ring (Alfvén, 1967, 1968). The maximum increase in Mimas' orbital distance since the formation of the rings is a few percent. For Janus the same value is less than one percent.

The conclusion from this is that tidal effects are negligible in the Jovian, Saturnian and Uranian systems (see Section 5).

6.2. CONSTANCY OF SPIN

As stated in Section 5.6 there are also good reasons to believe that for most planets and asteroids the spin has not changed very much since hetegonic times. However, for all satellites the spin has been braked greatly by tidal effects, making the spin periods equal to the orbital periods.

Much of the primeval spin of the Earth has been transferred to the Moon, and – to a smaller extent – the same is true for Neptune-Triton. The other giant planets have probably not been braked appreciably after the formation of their satellite systems. Even the transfer of angular momentum to their satellites when these were formed did not change their spin by more than a few percent. In fact, for all the giant planets the total orbital momentum is more than one order of magnitude smaller than the spin of the primary (Table 2.1).

The law of isochronism (see Sections 5.7 and 5.8) holds for bodies as different as small asteroids (mass $\sim 10^{18}$ g) and the giant planets (mass $\sim 10^{30}$ g). The conclusions from this is that the spin of most of the asteroids has not been changed very much since their formation.

To what extent the spins of the terrestial planets have been braked is uncertain. The very slow rotations of Mercury and Venus may be due to a braking produced by solar tides (in combination with resonance effects). The spin of Mars is unexpectedly slow. This cannot be due to tidal effects from its satellites because they are too small to take up an appreciable momentum. The large solar distance does not make it likely that solar tides could be very efficient but such an effect cannot be ruled out. Pluto is reported to have a very slow rotation (6^d). We know too little about this planet to speculate about this fact.

6.3. ON THE POSSIBILITY OF RECONSTRUCTING THE HETEGONIC PROCESSES

We have reasons to believe that a series of dramatic events four or five billion years ago produced the Solar System. In order to reconstruct these events it is necessary to find how the system has changed since its formation. Unless we are able to correct for this, we have little chance of understanding the formation. As we shall see later from a chemical point of view there is a rapidly increasing quantity of information relating to the formation of the Solar System. But also from a dynamical point of view there is a surprisingly large amount of data referring to the initial formation. With a few exceptions we find the Solar System at present in a state which is not too different from that immediately after the formation.

In the literature there are numerous suggestions about changes in the structure of the Solar System: Drastic changes in the orbits of planets and satellites are proposed. Most of these suggestions should never have been published if the authors had investigated the dynamical implications because celestial mechanics shows that they are unacceptable.

Summing up, there is no indication of any major change in the planetary orbits. Of the satellites, only the Moon and probably also Triton, have undergone large orbital changes. Probably both were initially independent planets, captured and brought to their present position by tidal effects.

There is no evidence that any other of the prograde satellites have changed their orbits appreciably.

Concerning the small bodies (asteroids, comets, meteoroids) the conclusion is different. As we have found viscous effects, including collisions, are of importance in many cases, and this implies that their orbital elements change. The retrograde satellites Jupiter 8, 9, 11, 12 and Phoebe belong to this category. Their capture into the present orbits may have taken place during the post-accretional phase, although it is perhaps more likely that it occurred during the accretion.

Suggestions have been made that the Martian satellites (and also other prograde satellites) are recently captured asteroids. This is extremely unlikely. A capture will lead to an almost parabolic orbit, and the transfer from this to the almost circular orbits of today is about as difficult as the transfer of a main belt asteroid into an Earth-crossing orbit, a case which we have discussed in Section 3.

Acknowledgements

Many of the ideas presented here are the result of discussions over several years with Dr. Nicolai Herlofson, Royal Institute of Technology, Stockholm. The paper has been prepared at the Division of Plasma Physics, Royal Institute of Technology, Stockholm and during several periods with H. Alfvén as visiting professor at the Department of Applied Physics and Information Science, University of California, San Diego, at La Jolla at the invitation of Dr. Henry Booker. We have profited much from discussions with colleagues and collaborators, before all Dr. Bo Lehnert and Mr. Lars Danielsson in Stockholm and Dr. J. Arnold, Dr. W. Thompson, Mrs. Jane Frazer and particularly Mr. J. Trulsen. We also thank Dr. Öpik for important criticism. The editing was initiated in Stockholm with the aid of Mr. J. D. Finkel and was completed in La Jolla by Mrs. Lynne Love assisted by Mrs. Lola Lindsay, Mrs. Linda Chen, and Mrs. Marjorie Sinkankas.

Support for the work in La Jolla was received from the National Aeronautics and Space Administration under grants No. NASA NGR 05-009-110 and No. NASA NGL 05-009-002, and for the work in Stockholm by Naturvetenskapliga Forskningsrådet.

References

Alfvén H.: 1954, *On the Origin of the Solar System*, Oxford Univ. Press, London.
Alfvén, H.: 1964a, 'On the Origin of the Asteroids', *Icarus* 3, 52.
Alfvén, H.: 1964b, 'On the Formation of Celestial Bodies', *Icarus* 3, 57.
Alfvén, H.: 1967, 'Partial Corotations of a Magnetized Plasma', *Icarus* 7, 387.

Alfvén, H.: 1968, 'On the Structure of the Saturnian Rings', *Icarus* **8**, 75.

Alfvén, H.: 1969, 'Asteroidal Jet Streams', *Astrophys. Space Sci.* **4**, 84.

Alfvén, H.: 1970, 'Jet Streams in Space', *Astrophys. Space Sci.* **6**, 161.

Alfvén, H. and Arrhenius, G.: 1969, 'Two Alternatives for the History of the Moon', *Science* **165**, 11.

Alfvén, H. and Fälthammar, C.-G.: 1963, *Cosmic Electro-Dynamics*, 2nd ed., Oxford Univ. Press, London.

Allan, R. R.: 1967, 'Resonance Effects due to the Longitudinal Dependence of the Gravitational Field of a Notating Body', *Planetary Space Sci.* **15**, 53.

Allen, C. W.: 1963, *Astrophysical Quantities*, Athlone Press, Univ. of London, London.

Allen, W. H., Krumm, W. J., and Randle, R. J.: 1967, 'Visual Observations of Lunar-Libration-Center Clouds', NASA SP-150, p. 91.

Anders, E.: 1964, 'Origin, Age and Composition of Meteorites', *Space Sci. Rev.* **3**, 583.

Anders, E.: 1965, 'Fragmentation History of Asteroids', *Icarus* **4**, 399.

Arnold, J. R.: 1964, 'The Origin of Meteorites as Small Bodies', in *Isotropic and Cosmic Chemistry*, North-Holland Publishing Co., Amsterdam, pp. 347–364.

Arnold, J. R.: 1969, 'Asteroid Families and Jet Streams', *Astron. J.* **74**, 1235.

Belton, M. J. S.: 1967, 'Dynamics of Interplanetary Dust Particles Near the Sun', NASA SP-150, p. 301.

Biermann, L.: 1967, 'Theoretical Considerations of Small Particles in Interplanetary Space', NASA SP-150, p. 279.

Blanco, V. N. and McCuskey, S. W.: 1961, *Basic Physics of the Solar System*, Addison-Wesley, Reading, Mass.

Brouwer, D.: 1945, 'The Motion of a Particle with Negligible Mass under the Gravitational Attraction of a Spheroid', *Astron. J.* **51**, 223.

Brouwer, D.: 1951, 'Secular Variations of the Orbital Elements of Minor Planets', *Astron. J.* **56**, 9.

Brouwer, D. and Clemence, G. M.: 1961a, *Methods of Celestial Mechanics*, Academic Press, New York.

Brouwer, D. and Clemence, G. M.: 1961b, 'Orbits and Masses of Planets and Satellites', in *The Solar System*, Vol. III, *Planets and Satellites* (ed. by B. M. Middlehurst and G. P. Kuiper), University of Chicago Press, Chicago.

Brown, E. W. and Shook, C. A.: 1964, *Planetary Theory*, Cambridge.

Brown, H., Goddard, I., and Kane, J.: 1967, 'Qualitative Aspects of Asteroid Statistics', *Astrophys. J. Suppl. Ser.* **14**, No. 125, 57.

Chebotarev, G. A., Beljaev, N. A., and Yeremenko, R. P.: 1970, 'The Evolution of the Hilda Group Planets and Planet Thule', *Biull. Inst. Teor. Astron.* **12**, 55.

Chebotarev, G. A. and Shmakova, M. Ja.: 1970, 'Structure of the Central Past of the Asteroid Belt', *Biull. Inst. Teor. Astron.* **12**, 82.

Cohen, C. J. and Hubbard, E. C.: 1964, 'Libration of the Close Approaches of Pluto to Neptune', *Astron J.* **70**, 10.

Cohen, C. J., Hubbard, E. C., and Oesterminter, C.: 1967, 'New Orbit for Pluto', *Astron. J.* **72**, 973.

Danielsson, L.: 1969, 'Statistical Arguments for Asteroidal Jet Streams', *Astrophys. Space Sci.* **5**, 53.

Dohnanyi, J. S.: 1969, 'Collisional Model of Asteroids and Their Debris', *J. Geophys. Res.* **74**, 2531.

Dollfus, A.: 1961, 'Visual and Photographic Studies of Planets at the Pic-du-Midi', in *The Solar System*, Vol. III, *Planets and Satellites* (ed. by G. P. Kuiper and B. M. Middlehurst) Chicago University Press, Chicago, p. 568.

Gault, D. E., Shoemaker, E. M., and Moore, H. J.: 1963, 'Spray Ejected from the Lunar Surface by Meteoroid Impacts', NASA TN D1767.

Gerstenkorn, H.: 1955, 'Über die Gezeitenreibung beim Zweikörperproblem', *Z. Astrophys.* **36**, 245.

Gerstenkorn, H.: 1969, 'The Earliest Past of the Earth-Moon System', *Icarus* **11**, 189.

Goldreich, P.: 1965, 'Commensurable Mean Motions in the Solar System', *Monthly Notices Roy. Astron. Soc.* **130**, 159.

Goldreich, P. and Peale, S.: 1966, 'Spin-Orbit Coupling in the Solar System', *Astron. J.* **71**, 425.

Goldreich, P. and Soter, S.: 1966, '*Q* in the Solar System', *Icarus* **5**, 375.

Hagihara, T.: 1961, 'The Stability of the Solar System', in *The Solar System*, Vol. III, *Planets and Satellites* (ed. by B. M. Middlehurst and G. P. Kuiper), University of Chicago Press, Chicago.

Hamid, S., Marsden, B. G., and Whipple, F. L.: 1968, 'Influence of a Comet Belt Beyond Neptune on the Motions of Periodic Comets', *Astron. J.* **73**, 727.

Herczeg, T.: 1968, 'Planetary Cosmogonies' in *Vistas in Astronomy*, Vol. 10, (ed. by A. Beer), Pergamon Press, London.

Hirayama, K.: 1918, 'Researches on the Distribution of the Mean Motions of the Asteroids', *Ann. Observ. Astron. Tokyo*, App. 4; *J. College Sci., Imp. Univ. Tokyo* **41**.

Jeffreys, H.: 1962, *The Earth*, Cambridge Univ. Press, Cambridge.

Kiang, T.: 1966, 'Bias-Free Statistics of Orbital Elements of Asteroids', *Icarus* **5**, 437.

Kiang, T.: 1966, 'Mass Distribution of Asteroids, Stars and Galaxies', *Z. Astrophys.* **64**, 426.

Kurth, R.: 1959, *Introduction to the Mechanics of the Solar System*, Pergamon Press, Los Angeles.

Lindblad, B. A.: 1970, 'A Computer Stream Search among Photographic Meteor Orbits', *Smiths. Contrib. Astrophys.*

Lovell, A. C. B.: 1954, *Meteor Astronomy*, Oxford, Univ. Press, London.

Lyttleton, R. A.: 1968, 'On the Distribution of Major Axes of Long-Period Comets', *Monthly Notices Roy. Astron. Soc.* **139**, 225.

McCord, T. B.: 1966, 'Dynamical Evolution of the Neptunian System', *Astron. J.* **71**, 585.

MacDonald, G. J. F.: 1966, in 'Origin of the Moon, Dynamical Considerations', *The Earth-Moon System*, (ed. by B. G. Marsden and A. G. W. Cameron), Plenum Press, New York, p. 165.

Marsden, B. G.: 1968, 'Comets and Non-gravitational Forces', *Astron. J.* **73**, 367.

Middlehurst, B. M. and Kuiper, G. P. (eds.): 1963, *The Solar System*, Vol. I–V, University of Chicago Press, Chicago.

Munk, W. and MacDonald, G. J. F.: 1960, *The Rotation of the Earth*, Cambridge.

Newton, H. A.: 1891, 'On the Capture of Comets by Planets, Especially by Jupiter', *Mem. Nat. Acad. Sci.* **6**, 7.

Nilsson, C. S.: 1967, 'Orbital Distributions of Meteors of Limiting Magnitude +6 Observed from the Southern Hemisphere', NASA SP-150, p. 201.

Oort, J. H.: 1963, 'Empirical Data on the Origin of Comets', in *The Solar System*, Vol. IV, *The Moon, Meteorites and Comets*, (ed. by B. M. Middlehurst and G. P. Kuiper), p. 665.

Öpik, E. J.: 1958, 'Perturbations of a Satellite by an Oblate Planet', *Irish Astron. J.* **5**, 79.

Öpik, E. J.: 1961, 'The Survival of Strong Bodies in the Solar System', *Ann. Acad. Scient. Fennicae, Series A* **3**, 61.

Opik, E. J.: 1966, 'The Stray Bodies in the Solar System. 2: The Cometary Origin of Meteorites', *Adv. Astron. Astrophys.* **4**, 302.

Öpik, E. J.: 1968, 'The Cometary Origin of Meteorites', *Irish Astron. J.* **8**, 185.

Piotrowski, S.: 1953, 'The Collisions of Asteroids', *Acta Astron. Series A* **5**, 115.

Porter, J. G.: 1963, 'The Statistics of Comet Orbits', in *The Solar System*, Vol. IV, in *The Moon, Meteorites and Comets* (ed. by B. M. Middlehurst and G. P. Kuiper), University of Chicago Press, Chicago, 550.

Ring, J., James, J. F., Dachler, M., and Mack, J. E.: 1964, *Nature* **202**, 167.

Roemer, E.: 1963, 'Comets: Discovery, Orbits, Astrometric Observations', in *The Solar System*, Vol. IV, (ed. by B. M. Middlehurst and G. P. Kuiper), *The Moon, Meteorites, and Comets*, University of Chicago Press, Chicago.

Roy, A. E. and Ovenden, M. W.: 1954, 'On the Occurrence of Commensurable Mean Motions in the Solar System', *Monthly Notices Roy. Astron. Soc.* **114**, 232.

Roy, A. E. and Ovenden, M. W.: 1955, *Monthly Notices Roy. Astron. Soc.* **115**, 296.

Russell, H. N.: 1920, 'On the Origin of Periodic Comets', *Astron. J.* **33**, 49.

Schmidt (Shmit), O. J.: 1946, 'On the Law of Planetary Distances', *Comptes Rendus (Doklady) de l'Academie des Sciences de l'URSS* **52**, 8.

Schubart, J.: 1968, 'Long Period Effects in the Motion of Hilda-Type Planets', *Astron. J.* **73**, 99.

Simpson, J. W.: 1967, 'Lunar-Libration-Cloud Photography', NASA SP-150, 97.

Singer, S. F.: 1970, 'Origin of the Moon by Capture and its Consequences', *Trans. Am. Geophys. Union* **51**, 637.

Southworth, R. B.: 1967, 'Space Density of Radio Meteors', NASA SP-150, 179.

Terenteva, A. K.: 1968, 'Investigation of Minor Meteor Streams', in *Physics and Dynamics of Meteors* (ed. by L. Kresák and P. M. Millman), Reidel, Dordrecht, The Netherlands, p. 408.

Vsekhsvyatsky, S. K.: 1964, *Physical Characteristics of Comets*, U.S. NASA Tech. Translation F80.

Wall, J. K.: 1967, 'The Meteoric Environment Near the Ecliptic Plane', NASA SP-150, p. 343.

Whipple, F. L.: 1968, 'Origins of Meteoritic Matter', in *Physics and Dynamics of Meteors* (ed. by L. Kresák and P. M. Millman), Reidel, Dordrecht, The Netherlands, p. 481.

PART II

ACCRETION OF CELESTIAL BODIES

Abstract. (7) *Formation of celestial bodies.* The basic concepts of the accretional process are discussed, and the inadequacy of the contractional model is pointed out. A comparison is made between the general pre-planetary state on the one hand and the present state in the asteroidal region on the other. A model for accretion of resonance-captured grains leading to the formation of resonance-captured planets and satellites is suggested.

(8) *Spin and accretion.* The relation between the accretional process and the spin of planets is analyzed.

(9) *Accretion of planets and satellites.* It is shown that jet streams are a necessary intermediate stage in the formation of celestial bodies. The time sequence of planet formation is analyzed, and it is shown that the newly accreted bodies have a characteristic internal heat structure; the cases of the Earth and the Moon are considered in detail. A region of high initial temperature is found at 0.4 of the present Earth radius, whereas the culminating temperature of the Moon is near its present surface. An accretional heat wave is found to proceed outwards, and may produce the observed differentiation features.

7. Formation of Celestial Bodies

7.1. GRAVITATIONAL CONTRACTION OF A GAS CLOUD

It is generally believed that stars are formed by gravitational contraction of vast interstellar clouds. The condition for contraction is, to within as order of magnitude,

$$\kappa M^2 / R > nkT , \tag{7.1}$$

where κ is the gravitational constant; M, the mass; and R, the radius of the cloud; $n = M/m_A$ is the number of atoms with average mass m_A; k, Boltzmann's constant; and T, the temperature. If the average atomic weight is $A = m_A/m_H$, we have

$$R < 10^{-15} (AM/T). \tag{7.2}$$

As pointed out e.g. by Spitzer (1968), there are serious difficulties in understanding the formation of stars on this model. In particular, a large rotational momentum and magnetic flux oppose the contraction. It is far from certain that the model is appropriate.

However, we shall not discuss here the problem of star formation but the formation of planets and satellites. Laplace made a suggestion, apparently not very seriously intended, that these bodies were formed from gas clouds which contracted gravitationally. This idea has been adopted by a number of subsequent workers, without realization of its inherent inadequacy.

If again for the order of magnitude we put $A = 10$ and $T = 100$ K for formation of planets and satellites we find

$$R < 10^{-16} M.$$

For the biggest planets with $M \approx 10^{30}$ g we find $R < 10^{14}$ cm, indicating that Jupiter and Saturn may have been formed by this mechanism. But already for Uranus and Neptune ($M \approx 10^{29}$ g) we run into difficulties because gravitational effects do not become important unless the clouds have been caused by some other means to contract to 10^{13} g, which is less than ten percent of the distance between the bodies. Going to the satellite systems or a hypothetical asteroid parent body we see immediately that gravitational contraction is out of the question. For a typical satellite mass, say 10^{23} g, we find $R < 10^7$ cm (which means that the gas cloud should be comparable to the present body).

Hence, we conclude that the gravitational contraction of gas clouds is inadequate as a general model for the formation of the bodies in the Solar System.

As another example which shows how negligible the gravitational attraction is in forming a satellite system, let us consider the inner part of the Saturnian satellite system, certainly one of the most regular examples of a system of secondary bodies. The masses of Mimas and Enceladus are of the order 10^{-7} of the mass of Saturn. At a point intermediate between Mimas and Enceladus the gravitation of these bodies is less than 10^{-5} of the gravitation of Saturn. Before the formation of the satellites the matter now forming them was spread out over the whole orbit, which makes the ratios still smaller by one or more orders of magnitudes.

The Laplacian approach cannot be saved by assuming that the present satellites once were much larger ('protoplanets' and 'protosatellites' in Kuiper's theory (Kuiper, 1951)). As shown above there are too many orders of magnitude to overcome for such a theory. Moreover, as shown by the isochronism of spins (Sections 5.7 and 8.3), the idea of very large protoplanets is not acceptable.

7.2. CONDENSATION AND ANGULAR MOMENTUM

The formation of planets and satellites by the gravitational contraction of a gas cloud also meets the same angular momentum difficulty as star formation. If a gas cloud with dimensions R is rotating with the period T, its angular momentum is $2\pi R^2/T$. If it contracts, this quantity is conserved. If the present mass of, say, Jupiter once filled a volume with the linear dimensions x times Jupiter's present radius, its rotational period must have been of the order $T' = Tx^2$, where T is the present spin period of Jupiter. If we describe the condensation of Jupiter in a coordinate system which takes part in the orbital motion of Jupiter and hence rotates with the orbital period of Jupiter, the order of magnitude of T' could not be less than the orbital period of Jupiter, which is about 10^4 times the spin period. Hence, we find $x < 100$, which means that the cloud which condensed to Jupiter must be less than 10^{12} cm. This is only one or two percent of the distance to the point intermediate between Jupiter and Saturn, which should be approximately the separation point between the gas forming Jupiter and the gas forming Saturn. (It is only 10% of the distance to the libration point, which could also be of importance.) Hence, in order to explain the spin of Jupiter if formed by condensation of a gas cloud, one has to invent some braking mechanism. Such a mechanism, however, must have the property of producing the same spin periods for bodies as different as Jupiter and the asteroids. No such mechanism is known at the present time.

7.3. The early stage of accretion (embryonic growth)

We have seen that the type of condensation discussed in Sections 7.1 and 7.2 is unacceptable. This directs our attention to the alternative, namely a gradual accretion of solid bodies (embryos) from disperse matter (grains and gas). This concept can be traced back to the nineteenth century and has recently been elaborated particularly by Levin and Safronov (1960) – for complete references see Herczeg (1968). Planets and satellites are assumed to have grown from such bodies by a rain of embryos and grains onto their surface, continuing until the bodies have reached their present size.

A number of direct observations support this concept. The saturation of the surfaces of the Moon and Mars with craters testifies to the importance of accretion by impact, at least in the terminal stages of growth of these bodies. Although now largely obliterated by geological processes, impact craters may have been a common feature also of the Earth's primeval surface.

Secondly, the isochronism of spins (Sections 5.7 and 5.8) can be understood, at least qualitatively, as a result of embryonic accretion. The isochronism requires that the same process acts over the entire observed mass range of planets and asteroids, covering twelve orders of magnitude. Consequently, all seriously considered theories of planetary spin (Marcus, 1967; Giuli, 1968; Safranov and Zvjagina, 1969) are based on the embryonic (planetesimal) growth concept.

Finally, direct evidence has recently been obtained tending to confirm that chondrules and individual achondrite crystallites in the early phase of accretion were grains, freely suspended in space for considerable lengths of time, before they accreted into meter-size embryos. Some of these grains were already known to carry a large dose of surface implanted gas and consequently to have been exposed to soft corpuscular radiation. This was first interpreted to mean that they were debris of igneous rocks, reduced to rubble and irradiated by 'solar wind' on the surface of a body large enough to be capable of producing and extruding silicate melts. Lal and Rajan (1969); Pellas *et al.* (1969); and Wilkening *et al.* (1969) demonstrated, however, by sensitive track etching techniques that the grains have been isotropically exposed in such a way as to make it necessary to assume that they have been suspended in space for periods of time ranging to the equivalent of 10^4 yr in terms of the present flux. The irradiation phenomenon was also found to be more widespread than thought earlier, and to comprise not only achondrite crystallites, but also chondrules. Analysis of grains exposed to space radiation on the lunar surface (Arrhenius *et al.*, 1970; Lal *et al.*, 1970) further confirmed the difference in track distribution geometry resulting from exposure on the surface of a large body.

All of this observational evidence lends support to the concept that aggregation of freely orbiting grains into larger embryos constituted the first stage in the hetegonic accretion process.

7.4. OBJECTIONS TO THE EMBRYONIC ACCRETION MECHANISM

In the past the major obstacle to understanding the incipient accretion process has been the difficulty in visualizing how collision can result in net accretion rather than in disruption. These difficulties have largely been eliminated by the firsthand data on collision processes in space obtained from the lunar surface, by the study of the record in meteorites, and by the consideration of the grain velocity distribution in jet streams.

In fact, as pointed out by many authors (e.g., Whipple, 1968), the relative velocities between particles considered typical, such as colliding asteroids, are of the order 0.5×10^6 cm sec^{-1}, and collisions would be expected largely to result in fragmentation of the colliding bodies. According to Whipple, at such velocities a small body colliding with a larger body will eject fragments with a total mass of several thousand times the mass of the small body. The probability of accretion would under these circumstances appear to be much smaller than the probability of fragmentation.

This is the apparent difficulty in all theories based on the embryonic accretion concept. Indeed, as will be shown in Section 7.7, such accretion requires that the orbits of the grains have eccentricities of at least $e=0.1$, in some cases above $e=0.3$. The relative velocity at collision between grains in such orbits is of the order of

$$v_i \approx v_0 e, \tag{7.3}$$

where v_0 is the orbital velocity. Since v_0 averages 10–20 km/sec, v_i necessarily often exceeds 10^5 cm/sec, so that the collisions fall in the hypervelocity range.

The solution to this problem lies in the permutation of orbits which occurs as a result of repeated collisions between grains. This process is analysed in detail in Sections 3 and 9. The net result is focusing in velocity space of the orbits and equipartition of energy between participating grains leading to relative velocities continuously approaching zero and contraction of the particle populations into jet streams.

Other important information for the understanding of the embryonic accretion comes from observation of impact effects on the lunar surface. These demonstrate (Asunmaa *et al.*, 1970) that hypervelocity impact of microscopic silicate particles results in cratering and surficial melting of the target and in melting and evaporation of the projectile. With decreasing impact velocity the melting assumes increasing proportions, and an increasing fraction of the projectile welds onto the target, adding to its mass.

At still lower velocities involving energies in excess of the cohesive energy of the solids, but too low to lead to melting, impact may result in fracture.

In the lowest range of impact velocities net accretion results from the surface electrostatic bond energies when these exceed the kinetic energies of the colliding particles. The tenacious adhesion of fine dust to larger particles in the lunar soil is an example of this type of aggregation. Bonding by recondensed impact vapor is another active process in space contributing to cementation of particles into aggregate structures and illustrated in the lunar surface material.

There is evidence from the meteorites concerning the proportional roles of various processes of disruption and accretion. The decisive importance of loosely coherent powder aggregates in absorbing impact energy is indicated by the high proportion of fine grained material in chondrites, which form by far the largest group of meteorites. The low original packing density of this material is also suggested by evidence from meteors. Such fluffy aggregates probably represent the state of matter in the jet streams at the stage when a substantial portion of the collision debris of the original grains together with recondensed impact vapor have been brought into low relative velocities by inelastic collisions and adhere electrostatically.

The embryonic stage of accretion can be considered to be over when an aggregate reaches such a mass that gravitational acceleration begins to control the terminal impact velocities. The catastrophic growth process that follows, and leads to the accretion of planets and satellites, is discussed in Section 9.

7.5. Accretion of resonance captured grains

There are some regions in our solar system where planetesimal accretion may be in progress at the present at some of the resonance points. We know three different regions where several bodies are gravitationally captured in permanent resonances. These cases are:

(1) and (2) The two libration points before and after Jupiter where the Trojans are moving.

(3) The Hilda asteroids (20 asteroids) which are in 2:3 resonance with Jupiter.

In each of these three groups the bodies are confined to move in a certain region of space. There is very likely to be a large number of smaller bodies belonging to each of the groups. Energy is pumped into these groups of bodies because the gravitational field is perturbed, in part due to the non-circular orbital motion of Jupiter, in part due to perturbations from other planets. Furthermore, other asteroids (and comets and meteoroids) pass the region and may collide with the members of the group, thereby feeding energy into it.

However, these sources of energy input are probably not very important and consequently we neglect them in the following idealized model. Hence the only change in the energy of the group of bodies is due to mutual collisions if such occur. If these collisions take place at hypervelocities they lead to fragmentation. The number of bodies increases, but as the collisions are at least partially inelastic the total internal kinetic energy of the assembly decreases. According to our assumptions there are no effects increasing the internal energy, so the result will be that the relative velocities go down, until collisions occur only in the range when accretion predominates. The result will be a net accretion. We shall expect that all the matter in each of the groups eventually will end up in one body.

Hence, if we treat the case where initially a large number of small grains, e.g. the result of primordial condensation, were injected in the velocity space of one of our idealized groups, we could expect to follow in detail the accretional process from grains to planets.

In the actual Hilda group most of the mass is found in one object (Hilda itself). From this we may conclude that the accretional process is already far advanced.

There are a number of other resonances where only one small body is found to be captured by a large body. Such cases are Thule (3:4 resonance with Jupiter), Pluto (4:3 resonance with Neptune) and Hyperion (4:3 resonance with Titan). These cases may represent a still more advanced state than that in the Trojan groups and the Hilda group with all of the observable mass gathered into one body. (There may be small still unknown companions.) We may also consider Mimas captured by Dione, and Enceladus, captured by Tethys, as similar cases.

It should be remembered that the libration amplitudes in some of the cited cases are small or very small. As we have found in Section 7.4, this is difficult to reconcile with the tidal theory of resonance capture. Our model of planetesimal accretion, on the other hand, provides a mechanism for energy loss through mutual collisions between the accreting bodies, which may give a small libration. In fact, in the accretional state we have a number of bodies librating with different phase and amplitude. Their mutual collisions will bring down the libration of the finally accreted body.

A detailed analysis of the proposed model is desirable in order to demonstrate its applicability to real cases.

7.6. Properties Required of the Accretional Process

We shall now discuss the more general case of accretion. We start from the assumption that the condensation of a plasma, distributed in different regions around a central body, has led to the formation of a large number of grains. We require that the accretion of these grains finally shall lead to the formation of the celestial bodies we observe. From this requirement we can draw certain conclusions about the chemical properties of the initially condensed grains and about their dynamical state. We shall in this section confine ourselves to a discussion of the latter requirement.

We find that the celestial mechanical data which should be explained by a theory of accretion are:

(a) *The orbital elements of the bodies.* The total angular momentum C_M of a celestial body should be the sum of the orbital momenta of all the grains which have contributed to the formation of the body. The eccentricity e and the inclination i of the orbital elements of the accreting grains and the mechanism of accretion are elements as well.

(b) *The spacing of the bodies.* The spacing quotient q is given in Tables 2.1 and 2.2. In the different groups it usually varies between about 1.2 and 2. A theory of accretion should explain the values of q. Of special interest is that with the exception of the group of Jupiter 6, 10, and 7 there are no q-values very close to one. In principle, the matter accumulated in, for example, the region of the Galilean satellite group might accrete to, say 100 satellites with spacings $q = 1.01$ or 1.02. If such a state is established it would be just as stable as the present state with only four bodies. Hence, the lumping together of the matter into a small number of bodies is an important fact which the accretional process should account for.

(c) *The spin of the bodies.* The accretional mechanism should leave the bodies with the spins they had before the tidal braking. Because all satellites and a few planets have been braked very much, the observational data we can use for checking a theory consists of the spin values of asteroids and a number of planets. In particular, we have to explain the isochronism of the spin periods.

7.7. THE ECCENTRICITY OF THE GRAIN ORBITS

From (b) we can derive an interesting property of the orbits of the grains which form the raw material for the accretional process.

Suppose that the plasma condensation has resulted in a large number of grains all moving around the central body in exactly circular orbits in the equatorial plane. Two spherical grains with radii R_1 and R_2 moving in orbits a_1 and a_2 can collide only if

$$\Delta a = a_2 - a_1 < R_1 + R_2. \tag{7.4}$$

Since in the Solar System R_1 and R_2 are usually very small compared to the size of the orbits, we find $\Delta a \ll a_2$, which means that we can have a large number of grains in consecutive circular orbits. At least in such systems (e.g., the Uranian system) where the total mass of the satellites is very small compared to the mass of the central body, such a system would be perfectly stable from a celestial mechanics point of view. Such a state would resemble the Saturnian rings and is conceivable even outside the Roche limit.

Hence, the fact that in the different groups of bodies (see Table 2.4) there are only a small number (3–6) of bodies, shows that *the grains out of which the bodies were formed cannot have orbited originally in circles in the equatorial plane.*

Suppose next that we allow the original grains to orbit in circles with certain inclinations i. Then grains with the same angular momenta C but with different values of i will collide. In case the collisions are perfectly inelastic, they will result in grains with the C-values unchanged but all with the same i-values. Such a state is again dynamically stable but irreconcilable with the present state of the Solar System.

Hence, we find that *the original grains must necessarily move in eccentric orbits.* (Originally circular orbits with different i-values would result in eccentric orbits in the case where the collisions are not perfectly inelastic. This case is probably not important.)

An estimate of the minimum eccentricity is possible, but not without certain assumptions. We assume that a satellite or planet accretes by direct capture of the grains. If two adjacent embryos during the late stages of the accretional process move in circles with radii a_1 and a_2 and the spacing ratio is $q = a_2/a$, all grains must have orbits which intersect either a_1 or a_2. If not, there would be grains which are captured neither by a_1 nor by a_2, and these would finally accrete to a body intermediate between a_1 and a_2, against our assumptions. As the ratio between the apocentre and the pericentre is $(1 + e)/(1 - e)$ we find

$$(1 + e)/(1 - e) > q, \tag{7.5}$$

or

$$e > (q - 1)/(q + 1). \tag{7.6}$$

Since in some cases (e.g., in the giant planet group) $q = 2.0$, we find that at least for some groups $e \geqslant 0.33$. For smaller q-values such as $q = 1.2$ (in the inner Saturnian satellite group) we obtain $e > 0.09$. These results are not necessarily correct for a more complicated model of the accretion. However, as we shall find in Section 9, it is essentially valid for the two-step accretional process considered there.

7.8. Comparison with the Present State in the Asteroidal Belt

We have found that the embryonic approach requires a state characterized by a number of bodies moving in eccentric Kepler orbits. This state has a striking resemblance to the present state in the asteroidal belt. In fact, as shown in Figure 3.6, the eccentricities of the asteroidal orbits vary up to about 0.30 or 0.35. There are only very few asteroids with higher eccentricities. Hence, from this point of view it is tempting to identify the present state in the asteroidal region with the intermediate step in the embryonic accretion.

This is contrary to the common view that the asteroids are fragments of one or several planets, exploded by collisions. The latter is based on several contributing arguments.

(a) There is no doubt that collisions occur between asteroids. Arguments have also been developed, particularly by Anders (1965) that the resulting fragmentation contributes to the observed size distribution of asteroids. However Anders also points out that only the small-size part of the distribution is explained by fragmentation and that the large-size asteroids show another distribution which he attributes to 'initial accretion' but which probably could be explained equally well as concurrent with the fragmentation.

(b) As discussed in Section 7.3 it has been believed that collision of small objects could not lead to accretion; in this situation it appeared necessary first to postulate the formation of one or several large parent bodies by some *ad hoc* process, and then to decompose these to generate the wide size range of objects now observed.

Obviously this approach does not solve the problem of accretion, which is either ignored or relegated to the realm of hypotheses such as the one discussed in Sections 7.1 and 7.2.

(c) It was long thought that the parent planet concept was lent credence by evidence from meteorites. These were often believed to derive from the asteroid belt, and cosmic irradiation effects strongly suggest that they are fragments of bodies with an original size of at least tens or hundreds of meters. It is also well established that the material components of most types of meteorites have been exposed to temperatures in the range of at least 700–1000 K. Furthermore some types contain microcrystalline diamond which used to be taken as safe evidence of high pressure. The combined evidence of high temperature and high pressure was interpreted as a definite indication of parent bodies larger than the present asteroids and thus as support for the exploded planet theory.

However, in recent years diamond synthesis in plasma-gas systems below atmospheric pressure has been proven by several independent workers (Angus *et al.*, 1968; Eversole, 1958; similar results are also recently reported from the U.S.S.R.). This removes the necessity of assuming high sustained pressure in meteorite parent bodies. The accretion theory also results in a high temperature history as discussed in a later section. Finally the structure of the lunar igneous rocks suggests that impact generated melts on small bodies in space account for the type of igneous meteorites previously believed to indicate melting in the interior of a large body.

In summary, the conceptual need for large bodies as predecessors of asteroids has in recent years vanished and is actually counterindicated. It is consequently possible that the asteroids are generated by competing disruption and accretion processes resulting in a net growth.

If the asteroidal region represents a state of proceeding planetesimal accretion, one must ask why the accretion has not yet led to the formation of one or more planets in this region. The answer to this question is simple and straightforward.

The evolution from a planetesimal state to accreted planets, which represent the end result, involves mutual collisions. Other things being equal the frequency of collisions is proportional to the density. As seen from Figure 2.6 and Table 2.1 the density in the asteroidal region is four or five orders of magnitude smaller than that in the regions of the terrestrial planets or the giant planets. Hence, we should expect the time scale of the accretion to be four or five orders of magnitude larger in the asteroidal belt than in the planetary groups. This means that even if we put the time of planetary accretion as low as 10^6 or 10^7 yr we will reach 10^{10}–10^{12} yr in the asteroidal belt, values which exceed the age of the Solar System. Hence, the asteroidal belt should still be in the era of accretion.

It is also possible that the density is so low that planetary accretion will never be completed.

8. Spin and Accretion

8.1. GRAIN IMPACT AND SPIN

When an embryo grows by accreting grains, the spin of the embryo is determined by the angular momentum (in relation to the center of gravity of the embryo) which the grains transfer to the embryo. Suppose that a spherical embryo has a radius R, an average density θ, and a moment of inertia I, and that it is spinning with a period T. Then its angular momentum $\mathbf{C}$ is

$$\mathbf{C} = I\omega, \tag{8.1}$$

with $\omega = 2\pi/T$.

The vector $\mathbf{C}$ is parallel to the axis of rotation. The mass of the embryo is $M=$

$= (4\pi\theta/3) R^3$. If we put

$$I = MR_i^2 = MR^2\alpha^2 = \frac{4\pi\theta}{3} \alpha^2 R^5 ,$$

(8.2)

R_i is called the radius of gyration, and α the normalized radius of gyration. If the density of the body is uniform we have

$$\alpha^2 = 0.4 .$$

(8.3)

For celestial bodies with central mass concentration, α^2 is smaller. See Table 2.1.

Suppose that a grain with mass Δm impinges with the velocity v on the embryo at an angle β with the vertical. Its angular momentum at impact is

$$\Delta\mathbf{C} = \mathbf{R} \times \mathbf{v} \cdot \Delta m ,$$

(8.4)

where $\mathbf{R}$ is the vector from the center of the embryo to the impact point. The absolute value of ΔC is

$$\Delta C = \Delta m R v \sin\beta .$$

(8.5)

Depending on the angle between $\mathbf{C}$ and $\Delta\mathbf{C}$, the impact may increase or decrease the spin of the embryo.

We shall discuss the simple case when $\Delta\mathbf{C}$ is parallel to $\mathbf{C}$ and $\Delta C \ll C$. Then we have, from 8.1),

$$\Delta C = I \Delta\omega + \omega \Delta I .$$

(8.6)

Assuming that after the impact the accreted mass Δm will be uniformly distributed over the surface of the embryo (so that it keeps its spherical shape) we have

$$\Delta m = 4\pi\theta R^2 \Delta R = 3M \frac{\Delta R}{R}$$

(8.7)

and

$$\Delta I = \frac{4\pi\theta}{3} \alpha^2 \times 5R^4 \Delta R = 5I \frac{\Delta R}{R} = \frac{5}{3} \frac{I}{M} \Delta m .$$

(8.8)

From (8.7), (8.6), (8.5), and (8.2) we find

$$\Delta C = R v \sin\beta \left(3M \frac{\Delta R}{R} \right) = MR^2\alpha^2 \Delta\omega + 5I \frac{\Delta R}{R} \omega ,$$

(8.9)

or

$$\frac{\Delta\omega}{\Delta R} = \frac{3v \sin\beta}{\alpha^2 R^2} - \frac{5\omega}{R} .$$

(8.10)

8.2. ACCRETION FROM CIRCULAR ORBITS BY A NON-GRAVITATIONAL EMBRYO

The general problem of finding the spin of an accreting body is a very complicated three-body problem which is far from solved. Important progress has been made in the treatment of the problem under the assumption that all the accreting grains are confined to move in the embryo's orbital plane. A three-dimensional treatment (where the accreting grains move in orbits out of the embryo orbit plane) gives the same qualitative results as the two-dimensional treatment. However, both treatments make certain assumptions which have not yet been tested by calculation. The conclusions we draw in the following are made with this reservation.

We shall start by treating the simple but unrealistic case where an assembly of grains moves in circular Kepler orbits (in an exact Coulomb field or in the invariant plane of a perturbed field). The space density of the grains is assumed to be $\varrho = dm/dr$. A small embryo is orbiting in the circle r_0 with a velocity $v_0 = r_0 \omega_0$. The radius of the embryo is R_0, its density $= \theta$ (assumed to be uniform). We suppose that the accreted mass is immediately uniformly distributed over the surface of the embryo. (It should be observed that we assume the embryo to be a sphere but that the distribution of the grains is two-dimensional.)

As $\omega^2 r^3 = \text{const}$, a grain at the distance $r_0 + x$ will have the angular velocity $\omega_0 + \Delta\omega$. If $x \ll r$ we have

$$\Delta\omega = -\frac{3}{2}\frac{\omega}{r_0} x. \tag{8.11}$$

It will hit the embryo at the distance Δr from the axis with the relative velocity

$$v = r\,\Delta\omega = -\tfrac{3}{2}\omega x. \tag{8.12}$$

If the mass of the grain is Δm, it will give the embryo the angular momentum

$$\Delta C = \Delta mvx = -\tfrac{3}{2}\omega x^2\,\Delta m. \tag{8.13}$$

As $\Delta m = \varrho\Delta x$ we find that when all the matter in the ring $r_0 - R_0$ to $r_0 + R_0$ is accreted, the embryo has the angular momentum

$$C = -\tfrac{3}{2}\omega_0\varrho \int_{-R_0}^{+R_0} x^2\,dx = -\omega_0\varrho R_0^3, \tag{8.14}$$

and, hence, the spin velocity

$$\omega = \frac{C}{I} = -\frac{\omega_0\varrho R_0^3}{MR_0^2\alpha^2}. \tag{8.15}$$

As the accreted mass is $M = 2\varrho R_0$, we find

$$\omega = -\omega_0/2\alpha^2. \tag{8.16}$$

Hence, the non-gravitational accretion from circular grain orbits gives a slow retrograde rotation.

One would think that the case we have treated would be applicable at least to the accretion by very small bodies. This is not the case for the following reason. It is possible to neglect the effect of gravitation on the accretion if the velocity v_∞ of a grain at large distance is

$$v_\infty \gg v_e,$$ (8.17)

where

$$v_e = (2M\kappa/R_0)^{1/2} = \left(\frac{8\pi}{3}\theta\kappa\right)^{1/2} R_0.$$ (8.18)

Comparing this with (8.12) we can write (8.17) as

$$\tfrac{3}{2}\omega_0 x \gg \left(\frac{8\pi}{3}\kappa\theta\right)^{1/2} x,$$ (8.19)

or

$$\theta \ll \frac{27}{32\pi\kappa}\omega_0^2 = 4 \times 10^6 \, \omega_0^2.$$ (8.20)

For a body at the Earth's distance we have $\omega_0 = 2 \times 10^{-7}$ sec^{-1} which gives $\theta \ll 2 \times 10^{-7}$ g cm^{-3}. For a body at Amalthea's distance from Jupiter we have $\omega_0 = 10^{-4}$ and $\theta \ll 0.04$. As the density of an embryo is likely to be $\theta > 1$, we conclude that the case we have treated is not applicable to any problem in the solar system. This is satisfactory because the spin value we derived does not agree with any observational value (except perhaps for Venus).

8.3. Gravitational accretion*

In case the gravitation of the embryo is of decisive importance to the accretion, the velocity v with which a grain hits the embryo must satisfy

$$v \gtrsim v_e = (8\pi\kappa\theta/3)^{1/2} R.$$ (8.21)

In case the space velocity is negligible, we have

$$v \approx v_e = (8\pi\kappa\theta/3)^{1/2} R.$$ (8.22)

This is an important typical case which we shall discuss.

Formula (8.10) shows how the spin of a spherical embryo changes at accretion. If $\Delta\omega = 0$ we have $\omega = \tfrac{3}{2}(v \sin\beta/R) = \omega_e \tfrac{3}{2} \sin\beta$ where we have introduced (8.22) and put

$$\omega_e = \left(\frac{8\pi\kappa\theta}{3}\right)^{1/2}.$$ (8.23)

Hence, if the accretion occurs in such a way that $\sin\beta$ – or rather the weighted mean of it – remains constant and if we put

$$C' = \tfrac{3}{2} \overline{\sin\beta},$$ (8.24)

where

$$\overline{\sin\beta} = \frac{\int dm \sin\beta}{\int dm},$$ (8.25)

---

* The term gravitational accretion should not be confused with the gravitational instability of a gas cloud which has been discussed in Section 7.1.

then the spin will tend towards the value

$$\omega = C'\omega_e.\tag{8.26}$$

This value is independent of R.

Hence we see that this accretional model has a very important property: *The spin of a body produced by planetesimal accretion is independent of the size of the body*, in case β is constant. A model with this property explains, at least in a qualitative way, the *isochronism* (see Section 5.7), i.e., the remarkable fact that the spin of bodies ranging from 10^{18}–10^{30} g does not show any systematic dependence on the size of the body.

This may be considered as a very important support of the planetesimal accretion of the type we discuss. It is also a strong argument against the idea of 'protoplanets' with properties very different from the present planets. However there are some difficulties with the application of this simple model. We shall therefore discuss two other models which also have the same property.

8.4. Giuli's theory of accretion

In order to find the numerical value of ω we must calculate C'. As stated above, we confine ourselves to a two-dimensional model. The problem is a 'three-body problem' and can only be solved by using computers. This has been done by Giuli. He starts from the general 'planetesimal' picture of accretion, and assumes that the embryo of a planet (e.g., the Earth) orbits in a circle around the Sun. At the same time there is a uniform distribution of grains which when at large distance from the Earth move in Kepler orbits around the Sun. When a grain comes into the neighborhood of the embryo, it is attracted gravitationally. If it hits the embryo, it is assumed to stick. The mass of the embryo will increase, and at the same time the grain transfers angular momentum to the embryo. The ratio between angular momentum and mass determines its spin.

Dole (1962) has demonstrated that in order to hit an embryo moving in a circular orbit around the Sun the grains must be moving within certain 'bands', the orbital elements of which he calculates for the case when the grains before approaching the Earth move in circular orbits around the Sun (see Figure 8.1). Giuli has made similar calculations including also grains moving in eccentric orbits. (Like Dole, he restricts his calculations to the case when the particles move in the orbital plane of the embryo.) Further, he has calculated the spin which a growing planet attains when it accumulates mass in this way.

He finds that a planet capturing exclusively grains moving in circular orbits will acquire a *retrograde* rotation. However, if accretion takes place also from eccentric orbits, the rotation will be *prograde* (assuming equal grain density in the different orbits). This result is essentially due to a sort of resonance effect which makes the accretion from certain eccentric orbits very efficient. Such orbits are ellipses with $a > 1$ (a = semi-major axis, with the orbital radius at the Earth taken as unity) which at perihelion graze the planet's orbit in such a way that the grain moves with almost

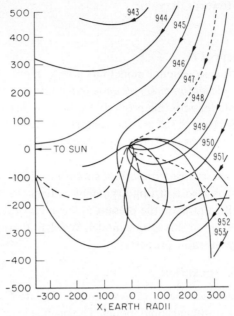

Fig. 8.1.   Particle orbits in a rotating coordinate system (according to Dole, 1962). Small bodies ('grains') which originally move in circular orbits around the Sun with orbital radii greater than 1 AU will gradually be overtaken by the Earth. In a rotating coordinate system which fixes the Earth at the origin and the Sun on the abscissa to the left at a distance of 1 AU (thus assuming the Earth has a circular orbit), the particles will approach the Earth and will move in the complicated trajectories depicted in the figure. If their original heliocentric orbital radii fall within seven ranges of values ('bands'), they will hit the Earth. Otherwise, they will depart from the neighborhood of the Earth and return to heliocentric (but non-circular) orbits. Seven similar bands exist for particles with initial orbital radii less than 1 AU.

the same velocity as the Earth. There is also a class of orbits with $a < 1$, the aphelion of which gives a similar effect. In both cases a sort of focusing occurs in such a way that the embryo receives a strong prograde spin.

Consider a coordinate system (xy) which has its origin at the center of the Earth. The Sun is very far in the $-x$ direction. The coordinate system rotates with the period of one year. Taking the Sun-Earth distance as length unit and one year/2π as time unit, the equations of motion close to the Earth can be written approximately as

$$\mathrm{d}^2 y/\mathrm{d}t^2 = -(Mx/r^3) + X,\tag{8.27}$$
$$\mathrm{d}^2 x/\mathrm{d}t^2 = -(My/r^3) + Y,\tag{8.28}$$

where M is the Earth-Sun mass-ratio, and

$$X = 2(\mathrm{d}y/\mathrm{d}t) + 3x,\tag{8.29}$$
$$Y = -2(\mathrm{d}x/\mathrm{d}t).\tag{8.30}$$

The rotation of the coordinate system introduces the Coriolis force $(2(\mathrm{d}y/\mathrm{d}t), 2(\mathrm{d}x/\mathrm{d}t))$ and the inhomogeneity of the solar gravitation the force $(3x, 0)$. These forces together disturb the ordinary Kepler motion around the planet. The capture is most efficient for particles moving through space with approximately the same speed as the Earth. These particles will hit the Earth at approximately the escape velocity v_e. We can discuss their orbits under the combined gravitation of the Earth and the Sun in the following qualitative way (see Figure 8.2).

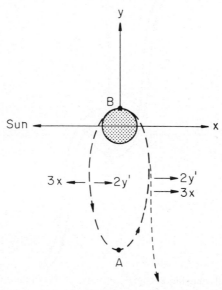

Fig. 8.2. Particles shot out tangentially with approximately the escape velocity from the point B at the Earth's equator (at 0600 local time) will move in an ellipse with apogee at A. The motion is disturbed by the Coriolis force and by the tidal effect from the Sun.

Let us reverse time and shoot out particles from the Earth. In case a particle is shot out from the 6^h point of the Earth ($x=0$, $y=r$) in the eastward direction with slightly less than the escape velocity, it will move in an ellipse out in the $-y$ direction towards its aphelion A. The Coriolis force $2(\mathrm{d}y/\mathrm{d}t)$ and the solar gravitation gradient $3x$ will act in opposite directions so as to minimize the net disturbance. On the other hand, on a particle shot out in the westward direction from the 6^h point the two forces will add in such a way as to deflect it from the ellipse far out from the Earth's gravitational field, where it will continue with a very low velocity.

Reversing the direction of motion we find that particles from outside can penetrate into the Earth's field in such a way that they hit the 6^h point of the Earth's equator from the west direction but not from the east direction. Hence the particles form a sort of a jet which gives a prograde spin. Similarly, particles moving inside the Earth's orbit can hit the 18^h point only from the west direction and they also give a prograde momentum.

Thus we have an efficient capture mechanism for two 'jets' both giving prograde rotations (see Figure 8.3). They derive from particles moving in the solar field with about $a=1.04$ and $a=0.96$ and an eccentricity of 0.03. Most other particles hit in such a way that on the average they give a retrograde momentum.

Applied to the Earth, the net effect of the process is (according to Giuli) a prograde spin with a period of 15 h – a value which is of the correct order of magnitude, but

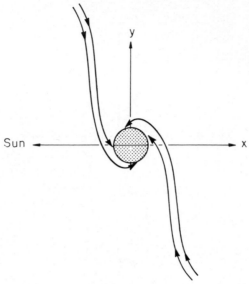

Fig. 8.3.   Particles originally moving in slightly eccentric Kepler ellipses in the solar field may hit the Earth in two jets, both giving prograde rotation.

larger by a factor of two or three than the Earth's period before the capture of the Moon (5 or 6 h). Giuli finds that a body with the radius 0.1 $R_\oplus$ and the same density will get the same period. It is likely (although not proved mathematically) that the period is proportional to $\theta^{-1/2}$ ($\theta=$ density of the body, assumed to be homogeneous). The value of $T\sqrt{\theta}$ which is obtained in this way is

$$T\sqrt{\theta} = 35\,hg^{1/2}\ \text{cm}^{-3/2}.\tag{8.31}$$

This value is larger by a factor of two than the average for all planets – including asteroids – which are not affected by tidal braking.

Giuli's calculations are based on the simplest possible planetesimal model, *viz.* that an embryo grows by accretion of those grains which hit it. Hence, he has neglected collisions between the grains. It is highly satisfactory that this simple model gives the correct order of magnitude for the spin. It is reasonable to interpret this agreement as a strong support for the theory of planetary accretion.

It should be mentioned that, if for some reason a planet accretes mainly from grains moving in orbits with small eccentricities, it should have a retrograde rotation. This means that if one could find some reason why Venus has accreted in this way, its retrograde rotation might be explained.

Three-dimensional calculations (where the accreting particles move on orbits out of the ecliptic plane) give the same qualitative results. Quantitatively, the results vary approximately by a factor of 2 from the two-dimensional calculations.

8.5. STATISTICAL THEORY OF THE ACCRETION

In Section 8.4 it has been assumed (1) that each grain accreted by an embryo has a mass which is infinitely small compared to the mass of the embryo, and (2) that the grains hit randomly.

There is no reason to doubt the second assumption, but it is doubtful whether the first one is correct. When the grains are coalescing their size distribution will no doubt be a continuous function, probably of the kind we find among the asteroids. The body which we call the 'embryo' is not fundamentally different from the other grains: it is *primus inter pares*. Hence, the largest grains it is accreting are – although by definition smaller – not necessarily very much smaller than the embryo. If a grain with a mass $m_1 = \alpha m_e$, where m_e is the mass of the embryo, hits the embryo, one single grain with a reasonably large α can change the state of rotation drastically. Take as an extreme case that the grain hits the embryo tangentially with the escape velocity. In fact it will give the embryo an additional angular velocity

$$\omega_s = C_s \omega_e, \qquad C_s = \frac{m_1}{m_e} \left(\frac{R_1}{R_i}\right)^2 \tag{8.32}$$

where $(R_i/R_1)^2$ is the relative moment of inertia of the embryo, which typically $=0.33$ (see Table 2.1). The Giuli process gives for the order of magnitude $C'=0.1$. In order to make C_s comparable, we need only have $\alpha = m_1/m_e = 3\%$. Hence, already one grain as large as some percent of the mass of the embryo mass under favorable conditions can completely change the rate of rotation.

Levin and Safronov (1960); Safronov (1958, 1960); and Safronov and Zvjagina (1969) on one hand, and Marcus (1967) on the other, have considered the question of the relative sizes of bodies which collide randomly. Their results are not in quantitative agreement with each other, but both results provide the possibility that the growing embryos were impacted by grains sufficiently large to give the resultant planet an anomalous rotation axis (Uranus) or a slow rotation rate (Mars).

Whereas Giuli's mechanism gives spin axis perpendicular to the orbital plane, the statistical mechanism gives a random distribution of spin axis. It is possible that the statistical mechanism is applicable to the spin of asteroids. However, for the small asteroids the escape velocity is very small and our models may meet difficulties because the approach velocities must be correspondingly small. It is possible that this is reconcilable with a jet stream accretion, but the problem no doubt needs further clarification.

9. On the Accretion of Planets and Satellites

9.1. EMBRYONIC ACCRETION

According to the embryonic model of accretion, all planets and satellites have been formed by accretion of smaller bodies. The craters of the Moon and Mars give clear evidence that an accretion of smaller bodies has been of major importance at least during the last phase of their formation. Theories of the spin of planets (see Section 9.8), indicate that the planetesimal picture is useful for the explanation of the rotation of planets. The isochronism of spin periods (Section 5) indicates that both planets and asteroids are likely to have been formed in this way.

The planetesimal accretion theory encounters difficulties in three respects.

(a) If planets are accreting by capturing grains moving in elliptic orbits in their neighborhood, one can calculate how long a time is needed before most of the grains are accreted to a planet or satellite. As shown by Safronov (1960), the time which Neptune and Pluto require to capture most of the grains in their environment is several times the age of the solar system. Safronov concludes from this that, e.g. Neptune has captured only a small fraction of the matter accumulated in its neighborhood and the rest is assumed to be dispersed. This is not very likely. Although the matter in the asteroidal region has not accreted to form a large planet, it is not dispersed. Similarly, if Neptune had not yet captured all the mass in its environment one should expect the rest to be found as asteroid-like bodies. According to Safronov the 'missing mass' must be some orders of magnitude larger than Neptune's mass. So much mass could not possibly be stored as asteroids because it should produce detectable perturbations of the orbits of the outer planets.

(b) The second difficulty is that according to practically all pictures of the embryonic state, it must have resembled the present state in the asteriodal region. In fact, if any embryo should be growing by accretion it is necessary that a large number of asteroid-size bodies move in Kepler orbits in its surrounding. But the relative velocities between two colliding asteroids is of the order of 5 km/sec. It is known that collisions at such 'hypervelocities' usually lead to disruption or erosion so that larger bodies are fragmented into smaller bodies. Collisions are not likely to lead to an accretion of smaller bodies to larger bodies unless the relative velocity is below a value v_l of about 0.5 km/sec (see Gault *et al.*, 1963).

(c) A third difficulty is associated with the gaseous structure of the giant planets. Even if one can work out a planetesimal mechanism by which solid grains accrete, this does not help us to understand the accretion of largely gaseous bodies like the giant planets. Öpik (1962) has suggested that Jupiter was accreted from snow-flakes of solid hydrogen, but the required temperature (~ 2 K) is unreasonably low.

9.2. JET STREAM AS AN INTERMEDIATE STEP IN THE FORMATION OF PLANETS AND SATELLITES

The jet stream concept discussed in Sections 3 and 7 seems to resolve these difficulties. We shall devote this chapter to a study of this possibility.

There is strong indication (although perhaps not a rigid proof) that a large number of grains in Kepler orbits constitute an unstable state. Even if the mutal gravitation between them is negligible, mutual collisions tend to make the orbits of the colliding grains similar. Hence the 'viscosity' of an assembly of grains in Kepler orbits tends to focus the grains into a number of jet streams.

The general structure of the jet streams we are considering should resemble the jet streams found in the asteroidal region (see Section 3.6). There is also a similarity with the meteor streams, although their eccentricity is usually very large. This does not necessarily imply that the asteroidal jet streams or the meteor streams are focussed by the mechanism we consider. In both cases we know too little about the density of small bodies in the streams to be able to make such a conclusion. Also the theory of jet streams is not yet very well developed.

In the simplest case a jet stream is a toroid with a large radius r_0 (equal to the orbital radius of a grain moving in a circular orbit around a central body) and a small radius $x = \beta r_0$. The jet stream consists of a large number of grains, moving in Kepler orbits with a semi-major axis close to r_0 and with eccentricities e and inclinations i of the order of β or less. If a particle moving in the circle r_0 has an orbital velocity v_0 other particles in the jet stream have this velocity superimposed by the 'internal velocity' of the stream, which is characterized by a value v_i. This is the vector sum of differential velocities of the order $v_0 e$, $v_0 i$, and $(v_0/2)\,(\Delta a/a)$ produced by the eccentricity, inclination and difference in a of the individual orbits.

In our qualitative model we put

$$v_i = \beta v_0 \tag{9.1}$$

and assume β to be constant. Hence,

$$v_i/v_0 = x/r_0 . \tag{9.2}$$

The 'characteristic volume' U of the jet stream is

$$U = 2\pi^2 r_0 x^2 = 2\pi^2 r_0^3 \beta^2 = \tfrac{1}{2}\mu T^2 \beta^2 = \tfrac{1}{2} r_0 T^2 v_i^2 \tag{9.3}$$

or

$$U = \frac{2\pi^2}{\mu}\, v_i^2 r_0^4 , \tag{9.4}$$

where

$$\mu = \kappa M_c = 4\pi^2 r_0^3/T^2 = r_0 v_0^2 \tag{9.5}$$

(κ = gravitational constant, M_c = mass of central body) and $T = 2\pi r_0/v_0$ is the orbital period.

9.3. ACCRETION OF AN EMBRYO

The accretion of large bodies may take place in two steps. The grains condensed from a partially corotating plasma will be treated in a later section (see also Alfvén, 1967). The process results in grains in elliptic orbits. The precession of the ellipses will sooner or later bring them to collide with a jet stream in the region where they move. This will eventually lead to capture of the grains by the jet stream. Before a capture has taken place, or in connection with the capture, a grain may make a hypervelocity collision with another grain and hence be vaporized, melted, or fragmented. This is not very important because the ultimate result will be that the mass of the grain is added to the jet stream. The collisions will reduce the relative velocity of the grain or its fragments until it reaches the internal velocity v_i of the jet stream.

The result of a collision may either be fragmentation-erosion, leading to a decrease in the size of at least the largest of the colliding bodies, or an accretion to larger bodies. These phenomena have not been studied very much, especially not for the type of bodies we are concerned with. The processes depend very much on the impact velocity, the chemical composition, the size and the physical properties, whether the bodies are brittle or fluffy. We know from the studies of Gault *et al.* (1963) that for high velocities fragmentation and erosion dominates, but that these effects decrease very rapidly with decreasing impact velocities. Extrapolating Gault's curves we find that the fragmentation and erosion should got to zero for a limiting velocity v_l of about 0.5 km/sec. It is uncertain whether this is true for different types of bodies.

Accepting that for $v > v_l$ fragmentation dominates, we can not be sure what happens for $v < v_l$. At very low velocities we have good reason to suppose that accretion dominates. This is certainly true for $v < v_e$ (the escape velocity of the larger body). Investigations of microscopic surface features in the lunar material have revealed a number of impact phenomena of obvious importance to accretion in space (see Asunmaa *et al.*, 1970; and Section 7.3).

Although admittedly our knowledge of the accretional processes is rudimentary it seems reasonable to assume that accretion dominates at collisions below a certain limiting velocity v_l. In our model we shall assume $v_l = 0.5$ km/sec and that in general collisions with $v > v_l$ produce an increase in the number of grains but for $v < v_l$ the number decreases.

Hence, if $v_i < v_l$ grains inside the jet stream will accrete. Their size will be statistically distributed. In our model we choose the biggest grain, which we call the 'embryo', and study how it accretes by capturing smaller grains. We assume it to be spherical with radius R.

In case such an embryo is immersed in a stream of infinitely small particles, which has the pre-accelerated velocity v in relation to the embryo, the capture cross-section is

$$\sigma = \pi R^2 (1 + v_e^2/v^2), \tag{9.6}$$

where v_e is the escape velocity. If we put

$$v_e = R/t_e, \tag{9.7}$$

we find that the 'time of escape'

$$t_e = \left(\frac{8\pi}{3} \kappa \theta\right)^{-1/2} = 1340 \ \theta^{-1/2} \ \text{sec g}^{1/2} \ \text{cm}^{-3/2} \tag{9.8}$$

is independent of R. Hence, for $v_e \gg V$, the capture cross-section is proportional to R^4.

We cannot be sure that formula (9.6) holds for the case where the embryo is moving in a Kepler orbit in a gravitational field. As shown by Giuli (1968), an embryo moving in a circular orbit will accrete grains under certain conditions. His calculations are confined to the two-dimensional case when all grains move in the same orbital plane as the embryo. As shown by Dole (1962), if the grains also move in circles (far away from the embryo), there are fourteen different 'bands' of orbits which lead to capture. Of these only four are broad enough to be of importance. Hence, formula (9.6) can at best be approximately true. Unfortunately, Giuli has not yet solved the three-dimensional case, which means that a comparison between formula (9.6) and his exact calculations is not possible. A comparison which necessarily is qualitative seems to indicate that (9.6) gives the right order of magnitude for the capture cross-section. We shall therefore use it until a better one is available.

We denote by ϱ the space density of condensable substances. The jet stream may also contain volatile substances which are not condensing to grains. The plasma condensation takes place essentially outside the jet streams, and the grains resulting from it are captured by the jet streams. However, the non-condensable substances may also partly be brought into the jet streams. Fortunately it is not necessary to make any specific assumption about the amount of volatile substances. (Indirectly they may contribute to the damping of the internal velocities and help to dissipate the kinetic energy.)

The mass increase of the embryo is

$$\frac{dM_e}{dt} = v_i \varrho \pi R^2 \left[1 + v_e^2/v_i^2\right] = \frac{R_i}{t_e} \varrho \pi R^2 \left[1 + R^2/R_i^2\right], \tag{9.9}$$

with

$$R_i = v_i t_e. \tag{9.10}$$

As

$$dM_e/dt = 4\pi R^2 \theta \, dR/dt, \tag{9.11}$$

(θ = embryo density) we can rewrite (9) as

$$\frac{dR}{R_i(1 + R^2/R_i^2)} = \frac{v_i \varrho}{4\theta R_i} \, dt = \frac{\pi}{2\tau_a} \, dt, \tag{9.12}$$

where we have introduced the *accretion time* τ_a which is given by

$$\tau_a = 2\pi t_e \frac{\theta}{\varrho} = 8.4 \times 10^3 \, \theta^{1/2} \varrho^{-1} \text{ sec.} \tag{9.13}$$

9.4. Accretion in a Medium of Constant Density

Under the assumption that ϱ and v_i, and, hence, τ_a and R_i, are constant, we can integrate (9.12). We obtain $\tan^{-1}(R/R_i) = \pi/2\tau_a(t - t_0)$ or

$$\frac{R}{R_i} = \tan \frac{\pi}{2\tau_a} (t - t_0). \tag{9.14}$$

Hence, in a medium of constant density and constant v_i, an embryo increases from zero to infinity in the finite time τ_a. Half of this time is needed for reaching R_i, the size where the gravitation of the embryo becomes important. When $t - t_0$ approaches τ_a, dR/dt approaches infinity, and the increase becomes catastrophic.

It should be noted that this collapse occurs according to a mechanism which has very little resemblance to the gravitational contraction of a gas cloud (see Section 7.1).

9.5. MASS BALANCE OF THE JET STREAM

Let us assume that in a certain region there is a constant injection of plasma during a time τ_i resulting in a production of grains, which are all captured in a jet stream. In the jet stream an embryo is accreting, so that finally all the injected mass is accumulated to one body, a planet if the region we consider is interplanetary space, or a satellite if it is space around a planet. The final mass of the accreted body is denoted by M_p (mass of final planet or satellite). Hence the rate of mass injection into the jet stream is M_p/τ_i. We assume that this mass is uniformly distributed over the volume U of the jet stream. This jet stream loses mass to the embryo which is accreting according to (9.12). Hence, we have

$$U \frac{d\varrho}{dt} = \frac{M_p}{\tau_i} - \frac{d M_e}{dt} = \frac{M_p}{\tau_i} - 4\pi R^2 \theta \frac{dR}{dt}. \tag{9.15}$$

Introducing (9.12) we find

$$U \frac{d\varrho}{dR} = \frac{\tau_a}{\tau_i} \frac{2R_i M_p}{\pi(R_i^2 + R^2)} - 4\pi\theta R^2. \tag{9.16}$$

We put

$$M_p = \frac{4\pi\theta}{3} R_p^3, \tag{9.17}$$

which means that R_p is the radius of the eventually accreted body. Further we introduce

$$x = R/R_p; \quad x_i = R_i/R_p; \quad y = \varrho/\varrho_0; \quad \varrho_0 = M_p/U; \quad \tau_a^0 = \tau_a y. \tag{9.18}$$

Hence, we find from (9.16) that

$$\frac{dy}{dx} = \frac{B}{y(x_i^2 + x^2)} - 3x^2, \tag{9.19}$$

with

$$B = \frac{2}{\pi} \frac{\tau_a^0}{\tau_i} x_i = \frac{4t_e\theta}{\tau_i\varrho_0} x_i.$$

9.6. ENERGY BALANCE IN A JET STREAM

The jet stream we consider is fed by an injection of condensed grains, each having a relative velocity v in relation to the jet stream. Hence the energy input is $\frac{1}{2}M_p v^2/\tau_i$. On the other hand the jet stream loses energy by internal collisions. In a qualitative model we assume that the mass is distributed in N equal spherical grains, each with radius R_g, a cross-section $\sigma_g = \pi R_g^2$, and a mass $m_g = \frac{4}{3}\pi\theta R_g^3$. Their space density is given by

$$n_g = \frac{N}{U} = \frac{2N}{r_0 T^2 v_i^2}. \tag{9.20}$$

They collide mutually with the frequency $v_g = n_g v_i \sigma_g$, where v_i is the internal velocity of the jet stream. We assume that at each collision a fraction α of the kinetic energy $W_g = \frac{1}{2}m_g v_i^2$ is lost. Hence, the loss rate per grain is

$$dW_g/dt = -\alpha v_g W_g, \tag{9.21}$$

which gives

$$dv_i/dt = -\alpha v_g v_i/2 = -\alpha\pi R_g^2 N/r_0 T^2 \tag{9.22}$$

or

$$dv_i/dt = -3\alpha M_j/4\theta r_0 T^2 R_g, \tag{9.23}$$

where M_j is the total mass of the jet.

According to our assumption in Section 9.3 there is a limiting velocity v_l such that if $v_i > v_l$ a rapid decrease in the size R_g will occur leading to an increase in the internal loss. If, on the other hand, $v_i < v_l$, an accretion takes place so that the average value of R_g increases and the loss will decrease.

The conclusion is that, within wide limits, a jet stream will adjust itself in such a way that it keeps balance with the injected energy. The process will tend to make $v_i = v_l$. Hence, the volume U of the jet stream is likely to remain constant, and the energy balance is produced by a change in the size of the grains in the stream.

When the injection stops, there is no energy input into the jet stream. Collisions will decrease the internal velocity. As $v_i = \beta v_0 = 2\pi r_0 \beta/T$ we have

$$\frac{d\beta}{dt} = -\frac{3\alpha}{8\pi\theta} \frac{M_j}{r_0^2 T R_g}. \tag{9.24}$$

Putting

$$M_j = \frac{4\pi}{3}\theta R_F^3 \tag{9.25}$$

we find

$$\frac{d\beta}{dt} = -\frac{\alpha}{2} \frac{R_F^3}{R_g r_0^2 T}. \tag{9.26}$$

Eventually all the mass in the jet stream is accreted to one spherical body, the radius of which is R_F according to (9.25). If R_g is constant, the thickness of the stream will decrease linearly and reach zero after a time

$$\tau_j = T \frac{2}{\alpha} \frac{r_0^2}{R_F^3} R_g. \tag{9.27}$$

9.7. Accretion when the injection in the jet stream is constant

We have found reasons for putting $v_i=$ constant and, hence, $U=$ constant. If the injection starts at $t=0$, and we neglect the mass accreted by the embryo, we have

$$\varrho = M_j t / \tau_i U . \tag{9.28}$$

Introducing this into (9.12) we obtain

$$\frac{dR}{R_i(1 + R^2/R_i^2)} = \frac{M_j}{4\theta U} \frac{t \, dt}{t_e \tau_i}; \tag{9.29}$$

or

$$\tan^{-1}(R/R_i) = \frac{M_j}{8\theta U} \frac{t^2}{t_e \tau_i}, \tag{9.30}$$

$$\frac{R}{R_i} = \tan \frac{\pi}{2} \frac{t^2}{2\tau_i \tau_a}, \tag{9.31}$$

with

$$\tau_a = 2\pi t_e (\theta U / M_j). \tag{9.32}$$

These formulae are valid only for $t \leqslant \tau_i$. We obtain as an approximate value for the time τ_c after which there is a catastrophic increase of the embryo

$$\tau_c = (2\tau_i \tau_a)^{1/2} . \tag{9.33}$$

We distinguish two typical cases:

(a) $\tau_c \ll \tau_i$. The density in the jet stream increases in the beginning linearly, and the radius of the embryo increases as t^2, its mass as t^6. The linear increase in jet stream density continues until rather suddenly the embryo consumes most of its mass. The catastrophic growth of the embryo stops even more rapidly than it has started, and for $t > \tau_c$ the embryo accretes mass at about the rate it is injected ($dM_e/dt \approx M_p/\tau_i$).

(b) $\tau_c \gg \tau_i$. The injection stops before any appreciable accretion has taken place. The jet stream begins to contract because no more energy is fed into it to compensate the loss due to collisions. When it has contracted so much that its density is large enough, the accretion sets in. This is also catastrophic.

9.8. Numerical values

Table 9.1 shows the numerical values for ϱ and τ_a. As both v_i and the geometrical factor in U are uncertain, the values relative to the Earth are first calculated and given in the column $\theta^{1/2}\varrho_\oplus/\varrho_0$. The values of τ_a are calculated from

$$\tau_a = 2\pi t_e \theta \frac{U}{M} = 8.4 \times 10^3 \text{ sec cm}^{3/2}/\text{g}^{1/2} \, \theta^{1/2} \frac{U}{M},$$

with

$$U = \frac{2\pi^2}{\mu} v_i^2 R^4,$$

TABLE 9.1

	$a = R/R_\oplus$	$U/U_\oplus$	$b = M/M_\oplus$	θ g/cm³	$\varrho_0/\varrho_\oplus = b/a^4$	$\theta^{1/2}\,\varrho_\oplus/\varrho_0$	τ_a yr	τ_c (if $\tau_i = 3 \times 10^8$ yr) 10^6 yr
Mercury	0.387	0.026	0.056	6.03	2.18	1.24	1.1×10^6	26×10^6
Venus	0.72	0.27	0.82	5.11	3.03	0.75	0.64×10^6	20
Earth	1.00	1.00	1.00	5.52	1.00	2.36	2.0×10^6	35
Moon	1.00		0.012	3.5	0.012	1.54×10^2	130×10^6	280
Mars	1.52	5.40	0.11	4.16	0.020	1.02×10^2	87×10^6	230
Jupiter	5.20	730	316	1.34	0.43	2.69	2.28×10^6	37
Saturn	9.58	8400	95	0.68	0.011	0.75×10^2	84×10^6	220
Uranus	19.14	134000	14.5	1.55	1.1×10^{-4}	1.23×10^4	10.5×10^9	> 300
Neptune	30.20	850000	17.2	2.23	2.1×10^{-5}	7.4×10^4	64×10^9	> 300

$$\mu = \kappa M_c = 6.7 \times 10^{-8} \times 2 \times 10^{33} \, \text{dyn cm}^2 \, \text{g}^{-1} = 1.34 \times 10^{26};$$
$$v_i = 5 \times 10^4 \, \text{cm sec}^{-1}; \quad U_\oplus = 1.9 \times 10^{37} \, \text{cm}^3;$$
$$\tau_{a\oplus} = 8.4 \times 10^3 \times \theta_\oplus^{1/2} \times U_\oplus / M_\oplus = 8.4 \times 10^3 \times 5.52^{1/2}$$
$$\times 1.9 \times 10^{37} / 6.0 \times 10^{27} = 6.3 \times 10^{13} \, \text{sec} = 2 \times 10^6 \, \text{yr}.$$

The table shows that the values of τ_a fall into three groups. Mercury, Venus, Earth, and Jupiter have all values around 10^6 yr, which must be much shorter than τ_i. Uranus and Neptune have values which are larger than the age of the solar system. Hence, $\tau_a > \tau_i$. There is an intermediate group, consisting of the Moon, Mars, and Saturn with $\tau_a \approx 10^8$ yr. This is probably of the same order of magnitude as τ_i. In any case we cannot be sure whether τ_i or τ_a is the larger quantity. Our conclusion is that for Mercury, Venus, Earth and Jupiter the catastrophic growth of the embryo took place at an early time of the injection, but for the Moon, Mars and Saturn near the end of the injection. Uranus and Neptune cannot have accreted until, after the end of the injection, the jet stream had contracted, so that β was considerably smaller than its original value.

9.9. Temperature of Accreted Body

In order to study the temperature of the accreting body we calculate the power w per unit surface from

$$4\pi R^2 w = \frac{v^2}{2} \frac{dM_e}{dt}, \tag{9.34}$$

where

$$v^2 = v_i^2 + v_e^2. \tag{9.35}$$

We find from (9.7) and (9.10) that

$$w = \frac{1 + R_i^2 / R^2}{8\pi t_e^2} \frac{dM_e}{dt}; \tag{9.36}$$

or, from (9.9),

$$w = \frac{\varrho}{8} \frac{(v_i^2 + v_e^2)^2}{v_i}, \tag{9.37}$$

which shows that for $R \gg R_i$ the heat delivered per cm^2 and sec is proportional to the mass increase of the *whole* embryo. The function w has a maximum at the time of the catastrophic increase, i.e. for $t \approx \tau_c$. If w is compensated by radiation from the surface of the accreting body, its surface temperature should vary similarly to w. This means that the maximum temperature is reached when a fraction φ of the mass is accumulated, given by

$$\varphi = \tau_c / \tau_i = (2\tau_a / \tau_i)^{1/2} \quad (\tau_a < \tau_i). \tag{9.38}$$

Hence, in an accreted body the region at a radial distance $r = Rx$ (where R is the final radius) has received most heat. We have

$$x = \varphi^{1/3} = (2\tau_a / \tau_i)^{1/6} \quad (\tau_a < \tau_i). \tag{9.39}$$

For the Earth $\tau_a = 2 \times 10^6$ yr. If as above we tentatively put $\tau_i = 3 \times 10^8$ yr, we have

$$x = (4/300)^{1/6} = 0.5 \,. \tag{9.40}$$

Hence, the layers were accreted with different temperatures (see Figure 9.1). The innermost part was cold, the layers somewhat below $x=0.5$ were hot and again the outer parts were cold. The value $x=0.5$ depends on a guess for τ_i, but is very insensitive to the value. If we choose $\tau_i = 10^8$ or 10^9 yr, x is changed to $x=0.58$ or $x=0.40$.

We know neither the chemical composition nor the heat conductivity of the Earth's interior very well. Also the content of radioactive substances, which should heat the interior, is unknown. We are not in conflict with any facts or plausible conclusions if we assume that neither the radioactive heating nor the thermal conductivity has changed the temperature structure in a drastic way. Hence our results may give a simple explanation for the fact that only an intermediate part of the Earth is melted, whereas both the core and the mantle are solid. According to our result, the intermediate layers were heated most intensely whereas both the central region and the outer layers were formed cool.

As the heat per unit surface is proportional to dM/dt, the average formation temperature of a celestial body is proportional to M_p/τ_i. If we assume τ_i to be similar for the different bodies, the formation temperature (under the condition of similar accretion processes) is proportional to their present masses.

9.10. CONCLUSIONS ABOUT THE TEMPERATURE STRUCTURE OF PLANETS

From this we can draw the following general conclusions about the internal temperatures of the planets:

(1) The giant planets were formed with at least one hot region. The heat structures

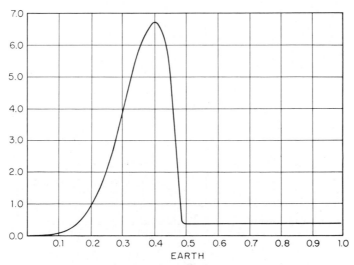

Fig. 9.1.   Temperature structure of the primeval Earth. Abscissa is size of growing Earth with 1.0 being present radius. Ordinate is thermal power delivered to surface in arbitrary units.

of these planets differ in the respect that while the heat maximum of Jupiter occurs at about half of the radius, this maximum for Saturn occurs somewhat further out, and for both Uranus and Neptune most of the heat was delivered to the outermost layers.

If the primeval heat structure of these planets is conserved, it may be an essential factor affecting their average density.

(2) Venus should have about the same heat structure as the Earth, but with the melted region closer to the center. This could be checked by measurements with soft-landed seismographs.

(3) The heating power on Mars should have been one order of magnitude less than for the Earth. The temperature maximum should therefore be rather close to the surface (perhaps at 0.9 of its radius), where – if anywhere – a liquid region may be found. The existence of this molten region is not very probable; it is more likely that the whole interior of Mars is solid.

(4) Both Mercury and the Moon are likely to have been solid and cool immediately after their formation. For Mercury the hottest region should have been at 0.9 of its radius, for the Moon at 0.8 also (see Figure 9.2).

9.11. The accretional heat front

Our conclusions about the cold origin of some celestial bodies should not be interpreted as meaning that the matter they consist of has never been melted. On the contrary, for large celestial bodies every part, with the exception of the central cores, has been heated above the melting point repeatedly. One can attribute these processes to an 'accretional heat front' which sweeps through the body outwards.

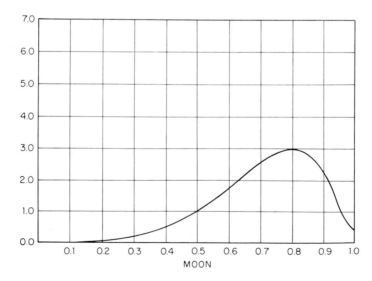

Fig. 9.2. Temperature structure of the primeval Moon. Abscissa is size of growing Moon with 1.0 being the present radius. Ordinate is thermal power delivered to surface in arbitrary units.

Suppose that an energy U_m is needed to melt a mass M of a certain substance. We define a velocity v_m by the condition

$$U_m = M_1 (v_m^2/2).\qquad(9.41)$$

As soon as a body of this substance has a velocity $v > v_m$ its kinetic energy suffices to melt it if converted into heat. For most substances v_m is of the order of 10^5 cm/sec.

If a body with mass M_1 and velocity v hits a target of the same composition its kinetic energy suffices to melt a mass

$$M_2 = \alpha M_1 ,\qquad(9.42)$$
where

$$\alpha = v^2/v_m^2 .\qquad(9.43)$$

A fraction of its energy will be used for the production of shock waves, the ejection of fragments from the place of collision and for the emission of radiation, but (9.42) and (9.43) give the correct order of magnitude.

We see that it is doubtful whether in a body as small as the Moon α has become much larger than unity. For planets like the Earth it may be $10 - 100$ for the last phase of accretion.

When the matter melted by an impacting grain has cooled down, it may be remelted many times by the impact of another grain in its close neighborhood. The impacting matter will increase the radius of the embryo, and finally the lump of matter we are considering is buried so deeply that no new impact will be able to melt it. Before this is achieved *it is likely to be molten α times* (because all impacting matter melts α times its own mass).

In retrospect we can picture this as an 'accretional heat front', starting at the center of all accreting bodies and proceeding outwards. All matter is heated α times before the front has passed. The factor α increases proportional to r^2. The front is able to melt all material as soon as $\alpha \gg 1$ which probably occurs at about 10^8 cm from the center. As long as the impact frequency is low the impacts produce locally heated regions which radiate their heat and cool down again. The accretional heat front will leave a cool region behind it. This is what is likely to have been the case in the Earth's central core and in the mantle, and also in the entire Moon.

If, on the other hand, the impact frequency is large, the heated regions have no time to cool. The accretional heat front will leave a hot melted region behind it. According to the interpretation in Section 9.9 this is what happened in the melted layers in the Earth's outer core.

9.12. SEGREGATION EFFECT OF THE ACCRETIONAL HEAT FRONT

In a volume of matter melted by meteorite impact a chemical segregation will take place when the heavy components sink and the light components float up. By iteration of this effect the accretional heat front will hence produce a bulk separation. When the front proceeds upwards it will bring with it the light components. In this way a limited amount of light material may be brought the entire distance from the interior of the body to the surface. The lower limit where this effect starts is given by $\alpha \approx 1$, which for the Earth possibly is at 10^8 cm from the center.

The heavy components will sink down in the locally heated regions, but if the heat front leaves a nonliquified region behind it, the heavy component cannot sink more than the thickness of the heated region. This thickness will depend on the size of the impinging meteorite. Typically it may seldom exceed a few kilometers.

Hence the accretional heat front may bring light components from the interior to the surface, but it will *not* bring heavy components downwards more than a very small distance. In this respect it differs from what occurs in an extended melted region.

The result is that if the accreted grains all have the same composition the light material is depleted from the radius where α passes unity out to near the surface where it becomes concentrated. The change in the percentage of heavy material is mainly a secondary effect to the displacement of the light material.

This may answer the question why the upper layers of both the Earth and the Moon contain unusually large amounts of radioactive material. It is well known that the interior of the bodies must have a much lower content of such elements, because otherwise the total heating of the bodies would be very large. In a melted region the radioactive elements are retained in the liquid phase which because of its progressively decreasing density floats on top of the heavier crystallates. Hence they may be carried by the heat front from the interior and finally be deposited in the surface layers. This could take place also in a body which is cool, and has been cool the whole time since its formation with the exception of the passage of the heat front.

References

Alfvén, H.: 1967, 'Partial Corotations of a Magnetized Plasma', *Icarus* **7**, 387.

Allen, C. W.: 1963, *Astrophysical Quantities*, 2nd ed., Athlone Press, Univ. of London, London.

Anders, E.: 1964, 'Origin, Age, and Composition of Meteorites', *Space Sci. Rev.* **3**, 583.

Anders, E.: 1965, 'Fragmentation History of Asteroids', *Icarus* **4**, 399.

Angus, J. C., Will, H. A., and Stanko, W. S.: 1968, 'Growth of Diamond Seed Crystals by Vapor Deposition', *J. Appl. Phys.* **39**, 2915–2922.

Arrhenius, G., Asunmaa, S., Drever, J. I., Everson, J., Fitzgerald, R. W., Frazer, J. Z., Fujita, H., Hanor, J. S., Lal, D., Liang, S. D., Macdougall, D., Reid, A. M., Sinkankas, J., and Wilkening, L.: 1970, 'Phase Chemistry, Structure and Radiation Effects in Lunar Samples', *Science* **167**, 659–661.

Asunmaa, S. K., Liang, S. S., and Arrhenius, G.: 1970, 'Primordial Accretion; Inferences from the Lunar Surface', Proc. Apollo 11 Lunar Science Conf., Houston, Texas, January 1970, Volume 3, Pergamon Press, p. 1975.

Dole, S. H.: 1962, 'The Gravitational Concentration of Particles in Space near the Earth', *Planetary Space Sci.* **9**, 541.

Eversole, W. G.: 1958, 'Synthesis of Diamond', U.S. 3.030.187 (Cl. 23–209.4) April 17, 1962, Appl. July 23, 1965, 5 pp.

Gault, D. E., Shoemaker, E. M., and Moore, H. J.: 1963, 'Spray Ejected from the Lunar Surface by Meteoroid Impacts', NASA TN D-1767.

Giuli, R. T.: 1968, 'Gravitational Accretion of Small Masses Attracted from Large Distances as a Mechanism for Planetary Rotation', *Icarus* **9**, 186.

Giuli, R. T.: 1968, 'On the Rotation of the Earth Produced by Gravitational Accretion of Particles', *Icarus* **8**, 301.

Herczeg, T.: 1968, 'Planetary Cosmogonics', in *Vistas in Astronomy* (ed. by A. Beer), Pergamon Press, London, **10**, 175–206.

Kuiper, G. P.: 1951, 'On the Origin of the Solar System', in *Astrophysics* (ed. by J. A. Hynek), McGraw-Hill, New York, p. 404.

Lal, D. and Rajan, R. S.: 1969, 'Observations on Space Irradiation of Individual Crystals of Gas-rich Meteorites', *Nature* **223**, 269–271.

Lal, D., Macdougall, D., Wilkening, L., and Arrhenius, G.: 1970, 'Mixing of the Lunar Regolith and Cosmic Ray Spectra: Evidence from Particle Track Studies', Proc. Apollo 11 Lunar Science Conf., Houston, Texas, January 1970, Volume 3, Pergamon Press, p. 2295.

Levin, B. J. and Safronov, V. S.: 1960, 'Some Statistical Problems Concerning the Accumulation of Planets', *Theor. Probab. Appl.* **5**, 220.

Marcus, A. H.: 1967, 'Formation of the Planets by the Accretion of Planetesimals: Some Statistical Problems', *Icarus* **7**, 283.

Öpik, E. J.: 1962, 'Jupiter: Chemical Composition, Structure and Origin of a Giant Planet', *Icarus* **1**, 200.

Pellas, P., Poupeau, G., Lorin, J. C., Reeves, H., and Audouze, J.: 1969, 'Primitive Low-energy Particle Irradiation of Meteoritic Crystals', *Nature* **223**, 272–274.

Safronov, V. S.: 1958, 'The Growth of Terrestrial Planets', *Vopr. Kosmog.* **6**, 63.

Safronov, V. S.: 1960, 'Accumulation of Terrestrial Planets', *Vopr. Kosmog.* **7**, 59.

Safronov, V. S.: 1966, 'Sizes of the Largest Bodies Falling onto the Planets during their Formation', *Sov. Astron. A.J.* **9**, 987.

Safronov, V. S. and Zvjagina, E. V.: 1969, 'Relative Sizes of the Largest Bodies during the Accumulation of Planets', *Icarus* **10**, 109.

Spitzer, L.: 1968, 'Diffuse Matter in Space', Interscience, New York.

Whipple, F. L.: 1968, 'Origins of Meteoritic Matter', in *Physics and Dynamics of Meteors* (ed. by L. Kresák and P. Millman), D. Reidel Publishing Company, Dordrecht, The Netherlands.

Wilkening, L., Lal, D., and Reid, A. M.: 1969, 'Compositions of Irradiated Pyroxenes and Feldspars in the Kapoeta Howardite', Meteoritical Society, Houston, Texas, Oct. 1969.

PART III

THE PLASMA PHASE

Abstract. Parts I and II of our analysis of the evolution of the solar system were devoted mainly to the mechanical processes. The present part (Part III) deals primarily with the plasma processes and the hydromagnetic aspects.

Much confusion in the cosmogonic field is due to the treatment of the early phases of the evolution of a circumstellar medium by pre-hydromagnetic methods, or by erroneous application of magneto-hydrodynamics. In order to reduce the speculative element as far as possible the present analysis tries to connect the cosmogonic processes as directly as possible to laboratory plasma physics and to space phenomena actually observed today (Section 10).

Models of the Laplacian type have been made obsolete by magnetohydrodynamics. Furthermore they are in conflict with observations. A new model is suggested (Section 11).

A plasma surrounding a rotating central body may attain a state of partial corotation which is determined by the balance between gravitation and the centrifugal force acting on a plasma in a dipole field. Condensation from a partially corotating plasma results in grains orbiting in ellipses with $e = \frac{1}{3}$ and finally accreting to bodies at $\frac{2}{3}$ of the central distance of the point of condensation (Section 12).

An application of the theory to the Saturnian rings and to the asteroidal belt shows that the fall-down ratio $\frac{2}{3}$ (derived from the geometry of a dipole field) is essential for the understanding of their structure. The structure of the groups of planets and satellites is also discussed but only in a preliminary way. The behavior of volatile substances is a major problem which still awaits an appropriate treatment (Section 13).

10. Plasma Physics and Hetegony

10.1. Summary of Parts i and ii and plan for Parts iii and iv

In the preceding two parts (Alfvén and Arrhenius, 1970a, b) of this discourse we have treated the later phases in the formation of planets and satellites. In doing so, we have adopted the *actualistic principle*. Starting from the present properties of planets and satellites, we have traced their history back in time in an attempt to find how these bodies have accreted from smaller bodies. The formation of *jet streams* is an essential intermediate stage in this sequence of hetegonic events. Our whole approach is of the planetesimal type (Moulton, 1905; Chamberlin, 1905; Alfvén, 1942–1945; Schmidt, 1944–1959).

Our preceding treatment has shown that the essential features of the present structure of the solar system can be understood if an original population of grains with certain properties is postulated. In a general way we can say that

(a) The grains should have such *dynamic* properties that after accretion they form celestial bodies with the orbits and spins that we observe today (with the exception of instances where post-accretional events such as tidal interaction have played a part).

(b) The grains should have such *chemical* and *structural* properties as to explain these properties in the present celestial bodies.

It is the purpose of this and the following chapters to investigate by what processes a population of grains with these required properties could have originated. We make the assumption that the grains have condensed from a partially ionized gas – a plasma. The reason for this assumption is that in a circumstellar region, partial ionization is necessarily impressed on any dilute gas not only by electromagnetic radiation from the star but also by electron collision caused by currents associated with rotation of the magnetized central bodies. In the latter process, optical opacity due to dust does not impede the ionization. We choose to use the convenient term 'plasma' instead of 'partially ionized gas' to semantically emphasize the necessity of taking magnetohydrodynamic effects into account and to stress the generality of thermal disequilibrium between grains and gas. The degree of ionization in hetegonic plasmas and in cosmic plasmas in general may vary over a very wide range depending on the specific process considered and is of importance down to very low values. In a plasma of solar composition with a degree of ionization as low as 10^{-4}, for example, the major part of the condensable components are still ionized.

This implies that the model we are trying to construct is essentially a model of a plasma which produces grains with the dynamic and chemical properties mentioned above. (The primeval plasma may also contain pre-existing grains – the limits on the possible amounts of such grains are discussed in Section 14.)

More specifically, in the present Section we analyze the general requirements of a hetegonic model for the production of the grains, whereas in Section 11 and 12 we suggest a specific model derived essentially on the basis of dynamics and properties of cosmic plasmas. In Sections 14–15, the chemical information pertaining to our discussion is reviewed and in Sections 16 and 17, a synthesis between the chemical and physical evidence within the framework of our model is attempted.

In comparison to Parts I and II, the treatment in this and following parts is necessarily more hypothetical and speculative. There are two reasons for this:

(a) We go further back in time.

(b) Plasma physics – which is essential to any realistic discussion of processes in space and hence also to the discussion of the formation of grains – is a much more complicated and less well-developed field than celestial mechanics which was the basis of Parts I and II.

One of our problems then is how we should proceed in order to reduce the hypothetical character as far as possible. This requires that we clarify the general state of the physics of cosmic plasmas. This will be done in Sections 10.2 and 10.3.

10.2. RELATION BETWEEN EXPERIMENTAL AND THEORETICAL PLASMA PHYSICS

Because plasma physics is essential to the understanding of the early phase of evolution of the solar system, we give here a brief survey of its present state. Plasma physics started along two parallel lines – one mainly empirical and one mainly theoretical. The investigations in what was called 'electrical discharges in gases', now more than a hundred years old, was to a high degree experimental and phenomenological. Only very slowly did it reach some degree of theoretical sophistication. Most theoretical physicists looked down on this complicated and awkward field in which plasma exhibited striations and double-layers, the electron distribution was non-Maxwellian, and there were many kinds of oscillations and instabilities. In short, it was a field which was not at all suited for mathematically elegant theories.

On the other hand, it was thought that with a limited amount of work, the highly developed field of kinetic theory of ordinary gases could be extended to include also ionized gases. The theories that thus emerged were mathematically elegant and claimed to derive all the properties of a plasma from first principles. The proponents of these theories had very little contact with experimental plasma physics and all the poorly understood phenomena which had been observed in the study of discharges in gases were simply neglected.

In cosmic plasma physics, the experimental approach was initiated by Birkeland (1908) who was the first to try to synthesize what are now known as laboratory plasma physics and cosmic plasma physics. Birkeland observed aurorae and magnetic storms in nature and tried to understand them through his famous terrella experiment. He found that when his terrella was immersed in a plasma, luminous rings were produced around the poles (under certain conditions). Birkeland identified these rings with the auroral zones. As we know today, this was essentially correct. Further, he constructed a model of the polar magnetic storms supposing that the auroral electrojet was closed through vertical currents (along the magnetic field lines). This idea also is essentially correct. Hence, although Birkeland could not know very much about the complicated structure of the magnetosphere, research today follows essentially Birkeland's lines, of course supplemented by space measurements; see Dessler (1968), Boström (1968) and Cloutier (1971).

Unfortunately, the progress along these lines did not proceed uninterrupted. Theories about plasmas – at that time called ionized gases – were developed without any contact with the laboratory plasma work. In spite of this, the belief in such theories was so strong that they were applied directly to space. One of the results was the Chapman-Ferraro theory which soon became generally accepted to such an extent that Birkeland's approach was almost completely forgotten, and for thirty or forty years, was seldom even mentioned in textbooks and surveys. All attempts to revive and develop it were neglected. Similarly, the Chapman-Vestine current system according to which magnetic storms were produced by currents flowing exclusively in the ionosphere, took the place of Birkeland's three-dimensional system.

The victory of the theoretical approach over the experimental lasted as long as a confrontation with reality could be avoided. However, from the theoretical approach it was concluded that plasmas could easily be confined in magnetic fields and heated to such temperatures as to make thermonuclear release of energy possible. When attempts were made to construct thermonuclear reactors, a confrontation between the theories and reality was unavoidable. The result was catastrophic. Although the theories were 'generally accepted' the plasma itself refused to behave accordingly. Instead, the plasma showed a large number of important effects which were not included in the theory. It was slowly realized that one had to construct new theories but this time in close contact with experiments.

The 'thermonuclear crisis' did not affect cosmic plasma physics very much. The development of theories could continue since they dealt largely with phenomena in regions of space where no real verification was possible. The fact that the basis of several of the theories had been proven to be false in the laboratory had very little effect: it was considered that this did not necessarily prove that they must be false in the cosmos!

The second confrontation came when space missions made the magnetosphere and interplanetary space accessible to physical instruments. The first results were interpreted in terms of the generally accepted theories or new theories were built up on the same basis. However, when the observational technique became more advanced it was obvious that these theories were not applicable. The plasma in space was just as complicated as laboratory plasmas. Today, very little is left of the Chapman-Ferraro theory and nothing of the Chapman-Vestine current system. Many theories which have been built on a similar basis may have to share their fate.

10.3. THE FIRST AND SECOND APPROACH TO COSMIC PLASMA PHYSICS

10.3.1. *General considerations*

The result is that the 'first approach' has been proven to lead into a dead-end street and we have to make a 'second approach' (Alfvén, 1968a). The characteristics of these approaches are shown in Table 10.1.

10.3.2. *Ionized gas vs plasma*

The basic difference between the first and the second approach is to some extent

TABLE 10.1

Cosmic Electrodynamics

First approach	Second approach
Homogeneous models	Space plasmas often have a complicated inhomogeneous structure
Conductivity $\sigma = \infty$	σ depends on current and often suddenly vanishes
Electric field $E_\parallel$ along magnetic field $= 0$	$E_\parallel$ often $\neq 0$
Magnetic field lines are 'frozen-in' and 'move' with the plasma	Frozen-in picture is often completely misleading
Electrostatic double layers are neglected	Electrostatic double layers are of decisive importance in low-density plasma
Instabilities are neglected	Many plasma configurations are unrealistic because they are unstable
Electromagnetic conditions are illustrated by magnetic field line pictures	It is equally important to draw the current lines and discuss the electric circuit
Filamentary structures and current sheets are neglected or treated inadequately	Currents produce filaments or flow in thin sheets
Maxwellian velocity distribution	Non-Maxwellian effects are often decisive
Theories are mathematically elegant and very 'well-developed'	Theories are not very well developed and are partly phenomenological

illustrated by the terms *ionized gas* and *plasma* which, although in reality synonymous, convey different general notions. The first term gives an impression of a medium which is basically similar to a gas, especially the atmospheric gas we are most familiar with. In contrast to this, a plasma, particularly a fully ionized magnetized plasma, is a medium with basically different properties: typically it is strongly inhomogeneous and consists of a network of filaments produced by line currents and surfaces of discontinuity. These are sometimes due to current sheaths, sometimes to electrostatic double layers.

If we observe an aurora in the night sky we get a conspicuous and spectacular demonstration of the difference between *gas* and *plasma* behavior. Faint aurorae are often diffuse and spread over large areas. They fit reasonably well into the picture of an *ionized gas*. The degree of ionization is so low that the medium still has some of the physical properties of a gas which is homogeneous over large volumes. However, in certain other cases, e.g. when the auroral intensity increase, the aurora becomes highly inhomogeneous, consisting of a multitude of rays, thin arcs, and draperies – a conspicuous illustration of the basic properties of all magnetized plasmas.

In the solar atmosphere the border between the photosphere and the chromosphere marks a transition similar to that between the two auroral states. The photosphere can be approximated as a homogeneous medium – at least to some extent – but in the chromosphere and upwards we have a typical plasma, a basic property of which is inhomogeneity manifest in filaments, streamers, flares, etc. To describe it by means of homogeneous models and according to the 'first approach' ideas is a fundamental mistake which may lead to conclusions and conjectures that are totally divorced from reality.

10.3.3. *Some laboratory results relevant to cosmic physics*

Following Birkeland, the first laboratory experiments with reference to cosmic physics had the character of *scale model experiments* (see Malmfors, 1945; Block, 1955, 1956, 1967; Danielsson and Lindberg, 1964, 1965; Schindler, 1969; and Podgorny and Sagdeev, 1970). It was soon realized, however, that no real scaling of cosmic phenomena down to laboratory size is possible, partly because of the large number of parameters involved which obey different scaling laws. Hence, laboratory experiments should aim at clarifying a number of basic phenomena of importance in cosmic physics rather than trying to reproduce a scaled-down version of the cosmic example.

The laboratory experiments have already demonstrated the existence of a number of such basic phenomena which had been neglected earlier, particularly the following:

(a) Quite generally a magnetized plasma exhibits a large number of *instabilities*. Lehnert (1967) lists 32 different types but there may still be a few more.

(b) A plasma has a tendency to produce *electrostatic double layers* in which there are strong localized electric fields. Such layers may be stable but they often produce oscillations. The phenomenon is basically independent of magnetic fields. If a magnetic field is present, the double layer *cuts* the *frozen-in* field lines. A survey of the laboratory results and their application to cosmic phenomena (especially in the ionosphere) has been given by Block (1972).

(c) If a current flows through an electrostatic double layer (which is often produced by the current itself) *the layer may cut off the current*. This means that the voltage over the double layer may reach any value necessary to break the circuit (in the laboratory say 10^5 or 10^6 V – in solar flares even 10^{10} V). The plasma 'explodes', and a high vacuum region is produced (see Carlqvist, 1969, Babic *et al.*, 1971; Torvén, 1972; see also Figure 10.1).

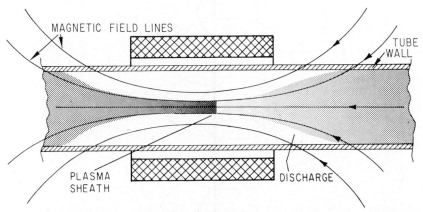

Fig. 10.1. Electrostatic double layers or sheaths are often produced in a plasma. The figure shows a discontinuity produced spontaneously by the plasma in a discharge tube where the plasma is confined by a longitudinal magnetic field, thus ensuring the absence of wall effects. Over the double layer a voltage drop is produced which sometimes suddenly becomes large ($\sim 10^5$V) and may disrupt the discharge.

(d) Currents parallel to a magnetic field (or even in absence of magnetic fields) have a tendency to *pinch*, i.e. to *concentrate into filaments* and not flow homogeneously (Alfvén and Fälthammar, 1963, p. 193; see also Figure 10.2). This is one of the reasons why cosmic plasmas so often exhibit filamentary structures. The beautiful space experiments by Lüst and his group (see Haerendel and Lüst, 1970) are important in this connection (although not fully interpreted yet).

(e) The inevitable conclusion from phenomena (a)–(d) above is that *homogeneous models are often inapplicable*. Striation in the positive column of a glow discharge and filamentary structures (arc and flash of lightning at atmospheric pressure, auroral rays, coronal streamers, prominences, etc.) are typical examples of inhomogeneities. Nature does not always have a *horror vacui* but sometimes instead a *horror homogeneitatis* resulting perhaps in an *amor vacui*. For instance, a magnetized plasma has a tendency to separate into high density regions such as prominences, coronal streamers and low density 'vacuum' regions, e.g. the surrounding corona.

(f) If the relative velocity between a magnetized plasma and a nonionized gas surpasses a certain critical velocity, v_{crit}, obtained by equating the kinetic energy $\frac{1}{2}mv_{crit}^2$ to the ionization energy eV_{ion} (V_{ion} = ionization voltage, m = atomic mass), so that

$$v_{crit} = (2 \, eV_{ion}/m)^{1/2}, \tag{10.1}$$

Fig. 10.2.   Simple model of a filamentary current structure in a low density plasma. Currents flow parallel to the magnetic field. The lines in the figure represent both current paths and magnetic field lines.

The magnetic field derives partly from an external axial field and partly from the toroidal field produced by the current itself (see Alfvén and Fälthammar, 1963). The current is strongest at the axis and becomes weaker further away from the axis as shown by the decreasing thickness of the lines.

the interaction becomes very strong and leads to a quick ionization of the neutral gas (see Section 17). The phenomenon is of importance in many thermonuclear experiments, in the problem of the origin of the solar system and probably in several other cosmic problems.

(g) The *transition* between a fully ionized plasma and a partially ionized plasma, and vice versa, is often *discontinuous* (Lehnert, 1970). When the input energy to the plasma increases gradually, the degree of ionization jumps suddenly from a fraction of one percent to full ionization. Under certain conditions, the border between a fully ionized and a weakly ionized plasma is very sharp.

(h) *Flux amplification*: If the toroidal magnetization in an experimentally produced plasma ring exceeds the poloidal magnetization, an instability is produced by which the poloidal magnetization increases at the expense of toroidal magnetization (Lindberg *et al.*, 1960; Lindberg and Jacobsen, 1964). This phenomenon may be of basic importance for the understanding of how cosmic magnetic fields are produced (Alfvén, 1961; see also Figure 10.3).

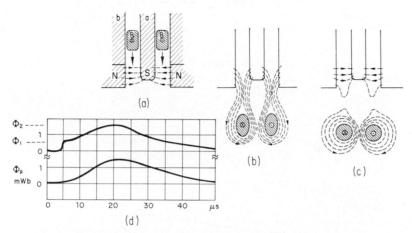

Fig. 10.3.   Geometry of the Lindberg plasma ring experiment. (a) Before leaving the gun, the plasma has a toroidal magnetization *B*. It is shot through the radial field N–S. (b) On leaving the gun, the plasma ring pulls out the lines of force of the static magnetic field. (c) Plasma ring with captured poloidal field. If the toroidal magnetic energy is too large, a part of it is transferred to poloidal magnetic energy (through kink instability of the current). (d) The poloidal magnetic flux Φ_p during the above experiment. The upper curve shows how the ring, when shot out from the gun, first acquires a poloidal flux Φ_1. An instability of the ring later transforms toroidal energy into poloidal energy thus increasing the flux from Φ_1 to Φ_2. The upper and lower curves represent the flux measured by two loops at 15 and 30 cm distance from the gun respectively.

   (i) When a plasma moving parallel to a magnetic field reaches a point where the field lines bend, a laboratory plasma may *deviate in the opposite direction* to the bend of the field lines (Lindberg and Kristoferson, 1971; see also Figure 10.4), contrary to what would be natural to assume in most astrophysical theories.

   (j) *Shock* and *turbulence* phenomena in low pressure plasmas have to be studied in the laboratory before it is possible to clarify the cosmic phenomena (Podgorny and Sagdeev, 1970).

   (k) Further physical experiments of importance include studies of *magnetic conditions at neutral points* (Bratenahl and Yeates, 1970).

   Condensation of solid matter from plasma differs decisively from condensation of a saturated or supersaturated gas at low temperature. This is partly because of the pronounced thermal disequilibrium which develops between radiation cooled solid grains and a surrounding, optically thin, hot gas. Important effects also arise because of the marked chemical differences between neutral and ionized components of mixed plasmas.

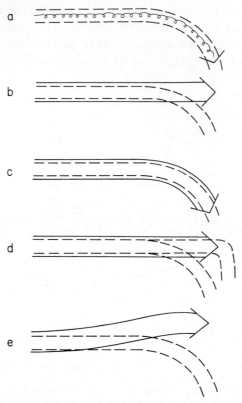

Fig. 10.4.   (a) In a magnetic field which has a downward bend, charged particles shot parallel to the field will follow the bend. (b) If instead, a plasma beam is shot, one would expect either that it (c) follows the bend (like a), *or* (d) continues to move straightforward bringing the 'frozen-in' field lines with it *or* gets electrically polarized and moves straight forward without bringing the field lines with it. (e) In reality, it does neither. In the experiment quoted, the plasma beam bends instead in the opposite direction to that of the magnetic field. In hindsight, this is easily understandable ad due to an electric field transmitted backward by fast electrons (Lindberg and Kristoferson, 1971).

Cosmic plasmas contain at least twenty elements controlling the structural and major chemical properties of the solid materials that form from them. With this degree of complexity, condensation experiments in partially ionized media are a necessary complement to theoretical considerations if we wish to understand the chemical record in primordial solid materials. Such experiments are discussed by Arrhenius and Alfvén, 1971; Meyer, 1969, 1971; and by Brecher, 1972b.

Thus we find that laboratory investigations begin to demonstrate many basic properties of a plasma previously unknown or neglected. These properties differ drastically from those assumed in many astrophysical theories. The difference between the laboratory plasma and the plasma of these theories may in some cases be due to the dissimilarity between laboratory and space but more often it reflects the confrontation between the first and the second approach. In other words, it is the difference between a hypothetical medium and one that has physical reality. The treatment of the former leads to speculative theories of little interest except as intellectual exercise. The latter medium is basic to the understanding of the world we live in.

10.4. The strategy of the analysis of hetegonic plasmas

What has been said in the preceding Section makes it evident that it is essential to work in close contact with laboratory plasma physics and chemistry. Furthermore, the study of present-day cosmic phenomena is essential. We cannot hope to construct a reasonable model of the hetegonic plasma processes by abstract reasoning alone but it is conceivable that we can extrapolate from present situations to the hetegonic conditions. Hence our strategy should be the following:

(a) Fundamental principle: Pre-magnetohydrodynamic models (Laplace, Berlage, Cameron, etc.) and 'first approach' theories (Hoyle) are of limited interest. We should follow the 'second approach' as defined above. This implies that we should rely to a large extent on laboratory and space experiments – especially those which clarify hetegonic problems.

(b) Extrapolation from magnetospheric physics: The transfer of angular momentum from a rotating magnetized central body to a surrounding plasma has some similarity to the present situation in the terrestrial magnetosphere. The hetegonic situation differs from this in two respects: (1) The plasma density must have been much higher. (2) The present solar wind effects (magnetic storms, etc.) may not necessarily be very important.

An extrapolation of our knowledge of the magnetosphere encounters difficulty because this field is not yet very advanced. Space research has certainly supplied us with a wealth of observations but the theories are not yet well developed. Most theories are of the 'first approach' type and hence of limited interest. The first systematic attempts to transfer laboratory plasma knowledge to the magnetosphere (according to the 'second approach' principle) have recently been made by Lindberg, Block and Danielsson. The works of these authors have been referred to elsewhere in our discussion.

(c) Extrapolation from solar physics: In some respects the hetegonic phenomena can be extrapolated from the sunspot-prominence phenomena. As a vertical magnetic field $\mathbf{B}$ and a rotational motion $\mathbf{v}$ often exist in sunspots, a sunspot is in principle somewhat similar to a rotating magnetized body. A voltage difference

$$V = \int_a^b (\mathbf{v} \times \mathbf{B})\, d\mathbf{r} \qquad (10.2)$$

is produced between a point a in the sunspot and a point b in the environment. It may give rise to an electric discharge along a magnetic field line from a to the point c in the environment where the field line through a intersects the solar surface (Figure 10.5; also see Stenflo, 1969). The current circuit is closed by currents in the photosphere.

The current along the magnetic field line is the basic phenomenon in prominences. A *filamentary current* of this type has the property of drawing ionized matter from the environment into itself. This phenomenon is somewhat similar to the 'pinch effect'

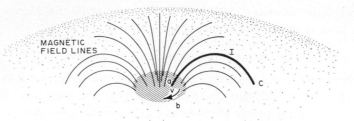

Fig. 10.5. The rotational motion v and the magnetic field in a sunspot may give rise to a voltage between the points a in the sunspot and b outside the sunspot, causing a discharge current I to flow along the magnetic field line from a to c. The circuit is closed through currents below the photosphere from c to b (and back to a).

(Alfvén and Fälthammar, 1963). As a result, the density in the prominence is orders of magnitude larger than in the surrounding corona. At the same time the temperature is orders of magnitude lower ($\sim 10^4$ K in the prominence compared to 10^6 K in the corona).

A typical value of the current in a prominence is 10^{11} A (Carlqvist, 1969). As the currents in the magnetosphere are typically of the order 10^5 A and the linear dimensions are not very different – in both cases of the order of 10^{10} cm – the solar situation merely represents a high-current and high-density version of the magnetospheric situation. As we shall see in 11.3, the hetegonic situation generally implies very high currents. Hence to some extent the hetegonic magnetosphere is similar to the present-day solar corona. In some hetegonic planetary magnetospheres, the linear dimensions of the filamentary structures are comparable to the present-day solar prominences whereas in the hetegonic solar corona (the supercorona), the dimensions should be 3 or 4 orders of magnitude larger.

Unfortunately, most of the theoretical solar physics is still in the state of the 'first approach' and hence of limited use for our purpose. Still, the analogy between the solar prominences and the hetegonic filamentary structures is important because it reduces the hypothetical ingredients of the model. Hence, a reasonable model is to consider the rotating magnetized central body to be surrounded by a network of prominence-like structures joining the surface of the central body with a surrounding plasma.

It is interesting to note that Chamberlin (1905) and Moulton (1905) connected their 'planetesimal' theories with solar prominences although in a different way.

(d) Extrapolation from circumstellar dust envelopes: During the era of formation of planets and satellites, the amounts of gas flowing into the circumsolar region from inter-stellar space gave rise to coronal-type concentrations in a volume comparable to the size of the solar system. The production of solid particle condensates in the filamentary structures extending through this medium must have been high enough to produce at least the total mass of companion bodies (10^{30} g) in a time period of the order of 10^8 yr.

Situations more or less similar to this may be represented by stars with optically thin envelopes of silicate dust. The relatively common occurrence of these attests to the substantial duration of the phenomenon. The fact that the central stars in such systems are of widely varying types ranging from early to late type stars (Neugebauer *et al.*, 1971; Stein, 1972) suggests that, in general, the circumstellar matter is gathered by the star from outside rather than being ejected from the star itself.

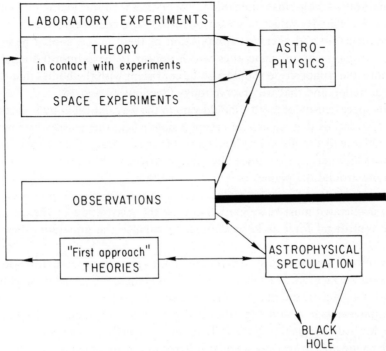

Fig. 10.6.   An illustration of the present strategic situation in astrophysics. Before we are allowed to combine them with observations, the 'first approach' theories must be processed through the laboratory where many of their ingredients will no doubt be filtered away. This is the only way of building up astrophysics with a minimum of speculation. Again we have to learn that science without contact with experiments is a very dangerous enterprise which runs a big risk of going astray.

At the present time it is very uncertain how close the parallelism may be between the circumstellar envelopes and our solar system in its formative state. Again, because of the importance of relying on actually observed rather than imaginary phenomena, the continued, refined investigation of these objects is of great interest from a hetegonic point of view.

Figure 10.6 is an attempt to illustrate the general scheme along which astrophysical theories should be developed in order to be realistic and consistent with observations.

10.5. REQUIRED PROPERTIES OF A MODEL

In the last Section of Part II, we studied the accretion of grains to asteroids, planets and satellites. We shall try here to find a model of the process which provides a suitable original population of grains for the accretional process. As we have found, the requirements for such a model are essentially:

(a) In the environment of a central body, a large number of grains should be produced which move in Kepler orbits in the same sense as the spin of the central body. This implies a transfer of angular momentum from the central body to the medium surrounding it (see Sections 11 and 12).

(b) The orbits of the grains should initially be ellipses with considerable eccentricities. In Section 7.7 (Part II) values of $e > 0.1$ or $e > 0.3$ have been suggested.

(c) The structure and chemical composition of these grains should be consistent with those components of meteorites which appear to be primordial condensates; furthermore, the composition should also be consistent with the bodies that they later form by accretion and that we observe today (Sections 14 and 15).

(d) The space density of matter should vary in the way indicated by Figures 2.6–2.9 (Part I). This means that we cannot accept a state where the density distribution has any resemblance to a uniform Laplacian disc. On the contrary, there should be certain regions with high density surrounded by (or interspaced with) regions with much lower density both around the planets and around the Sun.

(e) As the transfer of angular momentum is necessarily a slow process, the medium to be accelerated must be supported against the gravitation of the central body until the centrifugal force is large enough to balance the gravitation (see Section 11.4).

(f) The orbital axis of each system is close to the spin axis of its central body. Thus, regardless of the fact that the spin axis of Uranus is tilted $97°$, all of its satellites lie in the equatorial plane of the planet, not in the ecliptic plane.

The requirements a, e, and f specifically suggest that the model we are looking for must employ hydromagnetic effects. Indeed, there are well-known hydromagnetic processes which are able to transfer angular momentum from a magnetized rotating central body to a surrounding plasma (Alfvén, 1943; Alfvén and Fälthammar, 1963, p. 109). Furthermore, the magnetic fields may support a plasma against gravitation (*ibid.* p. 111), at least for a certain length of time (until instabilities develop). However, the mass which can be suspended with reasonable values of the magnetic field is orders of magnitude smaller than the distributed mass in the Laplacian theory as will be shown in Section 11. This implies that the mass density existing in the cloud at any particular time during the hetegonic age must be orders of magnitude lower than the distributed mass density. This is possible if *plasma is continually added to the cloud from outside and is simultaneously removed from the cloud by conversion of plasma to particulate matter or grains by a condensation process taking place in the plasma.* This state will be discussed in Section 11.5.

In the following Sections we shall show that the questions of angular momentum transfer, support of the cloud, and condensation of grains with the specific properties observed in meteoric material can all be resolved by considering suitable hydromagnetic processes.

10.6. Some existing theories

As was stated at the outset, we are abstaining in this work from consideration of such theories as do not offer an explanation of the basic structural similarities within the four well developed hetegonic systems within our solar system. At this point is seems worthwhile, however, to mention briefly some of the existing theories on the origin of the planetary system alone that have received attention in the literature over the past two decades. A somewhat arbitrarily chosen list includes the work of von Weizsäcker (1944), Berlage (1930–1948), Kuiper (1951), Cameron (1962, 1963), Hoyle (1960, 1963), McCrea (1960), Schmidt (1944–1956) and ter Haar (1948). Detailed reviews of the work of some of these authors may be found in ter Haar (1967).

All these theories start by postulating certain properties of the primeval sun and assuming that it was formed from an interstellar cloud by a certain series of processes. There is very little – if any – observational evidence in support of these assumptions. Hence the basic assumptions of these theories are highly speculative.

Berlage's theory is based essentially on the concept of the Laplacian disc which condenses to form a central body leaving behind concentric rings which form the planets. This, like most other theories, does not account for the density distribution in the satellite systems.

Cameron's theories envision an initial nebula of very large dimension ($\sim 10^5$ AU) with a mass of 1–2 solar masses which contracts gravitationally, and through the intermediate formation of a Laplacian disc, produces the solar system. The gravitational energy released is used up in ionizing the gases in the nebula. The planets are thought to retain their own Laplacian discs which form the satellites. However, in a recent version of his theory (Cameron, 1972), Cameron has changed his approach considerably. This model visualizes a turbulent interstellar cloud containing grains which during the collapse of the cloud, accumulate to form planets and satellites in the course of a few thousand years.

McCrea's theory approaches the problem of angular momentum by breaking up a very large initial cloud into a number of 'floccules', some of which later coalesce to form the Sun while others form planets.

Recently Hattori et al. (1969) and Kumar (1972) have also shown that even the largest planet, Jupiter, could not have formed by the collapse of a gaseous cloud.

The importance of electromagnetic processes in the primordial solar cloud is recognized by ter Haar (1949) and Hoyle (1960, 1963), although Hoyle introduces these processes in a highly implausible way. His theory of hydromagnetic angular momentum transfer from the Sun is based on the concept of 'frozen-in' field lines – a concept which is applicable only under exceptional circumstances in the cosmos. In this theory, a highly spiralized magnetic field is essential, implying that a large magnetic energy is stored in a toroidal magnetic field (Figure 10.7). Such a configuration is, however, unstable, as shown by Lundquist (1951). Consequently, it is not surprising that this phenomenon has never been observed in space or in the laboratory. The process that precludes it has been demonstrated experimentally by Lindberg et al. (1960, 1964; for

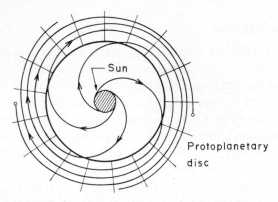

Fig. 10.7. Schematic representation of Hoyle's theory. According to Hoyle (1960, 1963), the rotation of the primeval Sun produced highly spiralized magnetic field lines both in the Sun and in the surroundings. The figure shows only five turns but Hoyle's theory requires 100000 turns storing an energy of 5×10^{45} ergs. Hoyle claims that this magnetic energy caused the protoplanetary disc to expand and form the planetary system.

    If Hoyle's mechanism were physically reasonable, it would have had important technological applications. But as shown both theoretically (Lundquist, 1951) and experimentally (Lindberg *et al.*, 1960; see also Figure 10.3), such a configuration is unstable and can never be achieved.

details see Section 10.3.3 and Figure 10.3) who showed that if the toroidal component of a magnetic field becomes too large compared to the poloidal component, an instability occurs which transfers energy from the toroidal to the poloidal field. (In the solar wind the toroidal field is likely to be larger than the poloidal field at large solar distances but this does not necessarily produce a similar instability because the magnetic energy is much smaller than the kinetic energy.)

    If we forget for a moment the question of hydromagnetic processes, a theory which has some elements of special interest is that of Schmidt (1944–1959). This is essentially a planetesimal accretion theory and treats what has been covered in the previous sections along similar lines in some respects. It starts with the assumption that the Sun captured swarms of small particles and bodies from interstellar clouds but makes no serious effort to explain why the grains have the required properties. Schmidt's theory, further developed by B. J. Levin, E. Ruskol and V. Safronov, has attracted considerable interest as a theory for the formation of satellites and particularly for the formation of the Moon.

    In Section 5.4 (Part I) of this work, and in Alfvén and Arrhenius (1972), it is shown that the Moon is not relevant in discussing the formation of satellites around the planets. But it is not immediately obvious that Schmidt's theory cannot be applied to the regular satellite systems. According to this theory, the matter now forming the satellites was injected into the neighborhood of the central body in parabolic or hyperbolic orbits which through viscous effects (mutual collisions) were transformed into the present nearly circular orbits. Hence the picture is similar to the conditions we have discussed (Alfvén, 1942–1945) before the formation of jet streams. However, a main difference is that the grains we have treated acquire their angular momenta from the plasma from which they are condensing. In Schmidt's theory, the angular momentum is due to an asymmetric injection of dust grains from 'outside'.

A number of objections can be raised against this process:

(a) The asymmetric injection is an ad-hoc assumption. It has not been shown that any reasonable dynamical distribution of grains in interplanetary space can lead to such an injection.

(b) The grains collected by a system should, on the one hand, give the central body a certain angular momentum per unit mass, and at the same time, give the satellites angular momenta per unit mass which are two or three orders of magnitude larger. (See Figures 2.2–2.5, Part I.) It is difficult to see how this could be achieved by the mechanism invoked.

(c) The spin axis of Uranus is tilted by 98 degrees – being almost in the orbital plane of the planet. (In Section 8.5 of Part II, we ascribe this to the statistical mechanism which gives the planets their spins.) The Uranian satellite system is perhaps the most regular and undisturbed of all systems, with the remarkable property that the satellites move in circular orbits *in the equatorial plane of Uranus* with negligible eccentricities and inclinations. The angular momentum transferred by the Schmidt mechanism should produce satellites moving in the orbital plane of the planet.

(d) The cloud of dust which captures the injected dust must extend far beyond the present orbits of the satellites. Suppose that a cloud with radius R captures grains from 'infinity'. We know from Giuli's work (1968a, b) that the value of C' in Equation (24) of Section 8.3 (Part II) is not likely to be more than about 10%. This means that the momentum which the cloud gains does not suffice to support the final orbits at a distance larger than $C'^2 R = R/100$. For the outermost Saturian satellite Iapetus, $R_{\text{Iapetus}} = 3.56 \times 10^{11}$ cm. This means that the cloud must be situated at a distance of 3.56×10^{13} cm. This is far outside the libration point which can be taken as the outer limit of the gravitational control of Saturn.

It seems unlikely that these objections to Schmidt's theory of satellite formation can be resolved without introducing too many ad-hoc assumptions. On the other hand, the jet streams in which the satellites are formed according to our model must necessarily capture some of the grains of the planetary jet stream in which the central body is accreting. Hence Schmidt's mechanism deserves further attention. For example, a satellite may receive a considerable number of impacts from grains which should have accreted on the planet if they had not been captured by the satellite.

11. Model of the Hetegonic Plasma

11.1. MAGNETIZED CENTRAL BODY

The simplest assumption we could make about the nature of the magnetic field in the hetegonic cloud is that the field derives from a magnetized central body. This implies that the formation of satellites around a planet and the formation of planets around a star cannot take place unless the central body is magnetized. We know that the Sun and Jupiter are magnetized. The magnetic states of Saturn and Uranus, which are also surrounded by secondary bodies, are not known. However, for our study, it is not essential that the central bodies be magnetized at present if they only possessed sufficiently strong magnetic fields in the hetegonic era (see Section 11.3 and Table 11.1). This must necessarily be introduced as an ad-hoc assumption. However, this assumption can be checked experimentally by analysis of remanent magnetization in preserved primordial ferromagnetic crystals in the way it has been done for crystals that now make up meteorites (Brecher, 1971, 1972b, Part I).

A considerable amount of work has been done on theories of the magnetization of celestial bodies, but none of the theories is in such a state that it is possible to calculate the strength of the magnetic field. However, the theories give qualitative support to our assumption that the central bodies were magnetized at hetegonic times. It should also be noted that certain stars are known to possess magnetic fields of the order of several thousand gauss, and one (HD 215 441) even as high as 35 000 G (Gollnow, 1962).

In order to make a model of the state of the plasma surrounding such a body, we assume that the central body is uniformly magnetized parallel or antiparallel to the axis of rotation. In case there are no external currents, this is equivalent to assuming that *the magnetic field outside the body is a dipole field with the dipole located at the center of the body and directed parallel or antiparallel to the spin axis.*

As we shall find later, neither the strength nor the sign of the dipole appears explicitly in our treatment. The only requirement is that the strength of the magnetic field be sufficient to control the dynamics of the plasma. We shall also see later that only moderate field strengths of the planets are required to produce the necessary effect. The dipole moment of the sun must have been much larger than it is now (see Table 11.1) but this does not necessarily mean that the surface field was correspondingly large since the latter would depend on the solar radius and we know very little about the actual size of the Sun in the hetegonic era.

11.2. ANGULAR MOMENTUM

For understanding the evolutionary history of the solar system, it is important to examine the distribution of the angular momentum in the system. Figures 2.2–2.5 in Part I show that the specific angular momenta of the respective secondary bodies exceed that of the spinning central body by one to three orders of magnitude.

This fact constitutes one of the main difficulties of all Laplacian type theories; these theories claim that the secondary bodies as well as the central body derive from an initial 'nebula' which during its contraction has left behind a series of rings which later form the secondary bodies. Each of these rings must have had essentially the same angular momentum as the orbital momentum of the secondary body formed from it whereas the central body should have a specific angular momentum which is much less. No reasonable mechanism has been found by which such a distribution of angular momentum can be achieved during the contraction. The only possibility one could think of is that the central body has lost most of its angular momentum after it had separated from the rings.

In the case of the Sun, such a loss could perhaps be produced by the solar wind.

Using the present conditions in the solar wind, an *e*-folding time for solar rotational braking turns out to be in the range $3-6 \times 10^9$ yr (Brandt, 1970). The currently accepted age of the Sun is about 5×10^9 yr. Thus, allowing for the errors in the estimate, it is not unlikely that the solar wind may have been an efficient process for the loss of solar angular momentum. However, the above value is very uncertain since there is as yet no way of deciding whether the solar wind has had its present properties at all times in the past. The emission of the solar wind is connected with the heating of the corona, which is produced by hydromagnetic turbulence within the Sun. It is possible that one or more links in this complicated causality chain has varied in such a way as to change the order of magnitude of the rate of loss of angular momentum. Hence, on the basis of the solar wind braking hypothesis, it is possible that the newborn Sun had about the same angular momentum as it has now, but it may also have been larger by an order of a magnitude or more. There are speculations about an early period of intense 'solar gale'. This is mainly based on an analogy with T Tauri stars but aside from the uncertainties in interpreting the T Tauri observations the relations between such stars and the formation of planets is questionable.

This shows how difficult it is to draw any conclusions about the hetegonic process from the study of the formation of planets around the Sun. It is much safer to base our discussion on the formation of satellites around the planets.

In all the satellite systems we find that the specific angular momentum of the orbital motion of satellites is orders of magnitude higher than that of the spinning central planet. A braking of this spin by the same hypothetical process as suggested for the Sun is out of the question since this would require a mechanism which should give almost the same spin period to Jupiter, Saturn and Uranus, in spite of the fact that these planets have very different satellite systems. From the spin isochronism discussed in Section 5 (Part I), we have concluded instead that the planets could not have lost very much angular momentum. We have also found that Giuli's theory of planetary spins (Section 8.4, Part II) strongly supports the theory of planetesimal accretion which is fundamentally different from the picture of a contracting Laplacian nebula. As shown by Hattori *et al.* (1969) and by Kumar (1972), not even the largest planet Jupiter could possibly have formed as an extended gaseous object, but would then rather have formed by collision-accretion processes in a dust-gas cloud around the Sun.

11.3. The transfer of angular momentum

The transfer of angular momentum from a rotating central body is a problem which has attracted much interest over the years. It has been concluded that an astrophysically efficient transfer can only be produced by hydromagnetic effects. The hydromagnetic transfer was studied by Ferraro and led to his law of isorotation. Lüst and Schlüter (1955) demonstrated that a hydromagnetic braking of stellar rotation could be achieved.

The Ferraro isorotation law assumes that not only the central body but also the surrounding medium has infinite electrical conductivity which means that the magnetic field lines are frozen-in. However, recent studies of the conditions in the terrestrial magnetosphere indicate the presence of components of electric field parallel to the magnetic field $(E_\parallel)$ over large distances in a few cases (Mozer and Fahleson, 1970; Kelley *et al.*, 1971). Such electric fields may occur essentially in two different ways. As shown by Persson (1963, 1966), anisotropies in the velocity distribution of charged particles in the magnetosphere in combination with the magnetic field gradient will result in parallel electric fields under very general conditions. However, $E_\parallel$ may also be associated with field-aligned currents in the magnetosphere, which are observed to have densities of the order of 10^{-6}–10^{-4} A m^{-2} (Zmuda *et al.*, 1967; Cloutier *et al.*, 1970). Such currents have a tendency to produce electrostatic double layers. A recent review by Block (1972) gives both theoretical and observational evidence for the existence of such layers, preferentially in the upper ionosphere and the lower magnetosphere.

The existence of an electric field parallel to the magnetic field violates the conditions for frozen-in field lines (see Alfvén and Fälthammar, 1963, p. 191). It results in a decoupling of the plasma from the magnetic field lines. Hence the state of Ferraro isorotation is not necessarily established and the outer regions of the medium surrounding the central body may rotate with a lower velocity than the central body itself.

11.3.1. *A simplified model*

We shall study an idealized, and in certain respects (see 11.3.2) unrealistic, model of the hydromagnetic transfer of angular momentum from a central body with radius R_c, magnetic dipole moment $\mathbf{a}$, and angular velocity Ω (see Figure 11.1)

Seen from a coordinate system fixed in space, the voltage difference between two points b_1 and b_2 at latitude λ_1 and λ_2 of a central body has a value

$$V_b = \int_{b_1}^{b_2} \left[(\Omega \times \mathbf{R}_c) \times \mathbf{B} \right] ds = \frac{a\Omega}{R_c} \left(\cos^2 \lambda_2 - \cos^2 \lambda_1 \right). \qquad (11.1)$$

Similarly, if there is a conducting plasma element between the points c_1 and c_2 situated on the lines of force through b_1 and b_2, but rotating around the axis with the angular velocity ω, there will be a voltage induced between c_1 and c_2 given by

$$V_c = \int_{c_1}^{c_2} \left[(\omega \times \mathbf{r}) \times \mathbf{B} \right] d\mathbf{r} = \frac{a\omega}{R_c} \left[\cos^2 \lambda_2 - \cos^2 \lambda_1 \right]. \qquad (11.2)$$

If we have Ferraro isorotation, i.e., if the magnetic field lines are frozen into the medium, ω will be equal to Ω, and hence $V_c = V_b$. If, however, there is no isorotation, $\omega \neq \Omega$ and hence $V_c - V_b$ will be nonzero, resulting in a current flow in the circuit $b_1 c_1 c_2 b_2 b_1$. In the sectors $c_1 c_2$ and $b_1 b_2$ this current together with the magnetic field gives rise to a force $\mathbf{I} \times \mathbf{B}$ which tends to accelerate ω and retard Ω (in the case $\omega < \Omega$), thus transferring angular momentum and tending to establish isorotation. The current $\mathbf{I}$ flows outward from the central body along the magnetic field line $b_1 c_1$ and back

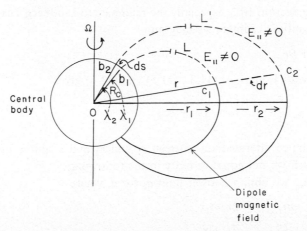

Fig. 11.1. In the absence of Ferraro isorotation, the angular velocity ω in the outer regions of the magnetosphere is different from the angular velocity Ω of the central body. This results in a current flow in the loop $b_1 b_2 c_2 c_1 b_1$ (shown by broken lines) which may result in the electrostatic double layers L and L'. Along part of the paths $b_1 c_1$ and $b_2 c_2$, the electric field has nonzero parallel components resulting in a decoupling of the plasma from the magnetic field lines.

again along the field line $b_2 c_2$. In a time dt the current between c_1 and c_2 transfers the angular momentum

$$D = dt \int_{c_1}^{c_2} [\mathbf{r} \times (\mathbf{I} \times \mathbf{B})] \, dr = I \, dt \int_{c_1}^{c_2} Br \, dr = \frac{Q\Phi}{2\pi}, \tag{11.3}$$

where $Q = I \, dt$ is the charge passing through the circuit $b_1 c_1 c_2 b_2 b_1$ in time dt and Φ is the magnetic flux enclosed between the latitude circles at λ_1 and λ_2.

Suppose that the plasma is situated in the equatorial plane between r_1 and r_2 and condenses and forms a celestial body with mass m_f and mov ng in a circular orbit of radius r Kepler period τ_K. Its orbital momentum D_{m_f} is

$$D_{m_f} = m_f r v = \frac{2\pi m_f r^2}{\tau_K} = \frac{\mu m_f}{2\pi r} \tau_K, \tag{11.4}$$

where $\mu = \kappa M_c = 4\pi^2 r^3 \tau_K^{-2}$, κ and M_c being the gravitational constant and the mass of the central body, respectively.

In an axisymmetric model with a constant current I flowing during a time τ_I we have

$$Q = I \tau_I, \tag{11.5}$$

$$\Phi = 2\pi a \left(\frac{1}{r_1} - \frac{1}{r_2} \right). \tag{11.6}$$

The current I produces a tangential magnetic field B_ϱ which at r ($r_1 < r < r_2$) is $B_\varrho = 2I/r$. This cannot become too large in comparison to B. One of the reasons for this is that if the magnetic energy of B_ϱ exceeds that of B by an order of magnitude, instabilities will develop (see Section 10.3, especially the reference to Lindberg's experiment). For an order of magnitude estimate we may put

$$I = \alpha' B r,$$
(11.7)

which together with Equations (11.3), (11.5) and (11.6) gives

$$D = \alpha' \frac{a^2}{r^2} \tau_I \left(\frac{1}{r_1} - \frac{1}{r_2} \right) = \alpha \frac{a^2}{r^3} \tau_I,$$
(11.8)

where r is a distance intermediate between r_1 and r_2 and α and α' constants of the order unity (which we put equal to unity in the following).

Putting $D = D_{m_f}$ we obtain a lower limit a_c for a, where

$$a_c^2 = \frac{\mu m_f r^2}{2\pi} \gamma = \frac{2\pi}{\tau_I} \frac{m_f r^5}{\tau_K}$$
(11.9)

with

$$\gamma = \tau_K / \tau_I.$$
(11.10)

In order to estimate the necessary magnetic field we assume that τ_I is the same as the injection time τ_i and introduce $\tau_I = 10^{16}$ s (3×10^8 yr), a value we have used earlier (Section 9) and obtain Table 11.1.

From the study of the isochronism and the planetesimal accretion (Section 8) we know that the size of the planets cannot have changed very much since their formation. As it is likely that the satellites were formed at a late phase of planet formation, it is legitimate to use the present value of the planetary radii in calculating the minimum surface magnetic field. From Table 11.1 we find that surface fields of less than 10 G are required. We have no possibility to check these values until the remanent magnetism of small satellites can be measured but with our present knowledge they seem to be acceptable.

As we know next to nothing about the state of the Sun when the planets were formed we cannot make a similar calculation for the surface field of the Sun. We can be rather confident that the solar radius was not smaller than the present one, and the formation of Mercury at a distance of 5.8×10^{12} cm places an upper limit on the solar radius. A dipole moment of 5×10^{38} G cm^3 means the following values of the surface field: (see Table 11.2).

In the absence of magnetic measurements from unmetamorphosed bodies such as, for example, asteroids, it is impossible to verify any of these values. If carbonaceous chondrites are assumed to be such samples, field strengths of the order of 0.1 to 1 G would be typical at a solar distance of 2–4 AU (Brecher, 1972a, b). If this field derived directly from the solar dipole, its value should be 10^{40}–10^{41} G cm^3, i.e. more than two orders of magnitude higher than the value in Table 11.1. However, the field may also have been strengthened locally by currents as shown by De and Arrhenius (1973) and discussed further in Section 12.

As stars are known to possess surface fields as high as 35 000 G, at least the values corresponding to $R > 3 \times 10^{11}$ cm do not seem unreasonable.

TABLE 11.1

Minimum values of magnetic fields and currents for transfer of angular momentum

Central body	Secondary body	Mass of secondary body m_f (g)	Orbital radius of secondary body r (cm)	Orbital period of secondary body τ_K (s)	$m_f r^5/\tau_K$	Dipole moment a_c (G cm³)	Equatorial surface field B (G)	Total current I (A)
Sun	Jupiter	1.9×10^{30}	0.778×10^{14}	3.74×10^8	0.15×10^{92}	0.97×10^{38}	⎱ See table 11.2	1.6×10^{11}
Sun	Neptune	1.03×10^{29}	0.45×10^{15}	5.2×10^9	3.6×10^{92}	4.8×10^{38}	⎰	0.23×10^{11}
Jupiter	Callisto	0.95×10^{26}	1.88×10^{11}	1.44×10^6	1.5×10^{76}	3.1×10^{30}	9	9×10^8
Saturn	Titan	1.37×10^{26}	1.22×10^{11}	1.38×10^6	0.27×10^{76}	1.3×10^{30}	6	9×10^8
Uranus	Oberon	2.6×10^{24}	0.586×10^{11}	1.16×10^6	1.54×10^{72}	3.1×10^{28}	2	0.9×10^8

a_c = minimum dipole moment of central body calculated from Equation (11.9).

B = minimum equatorial surface field.

I = current which transfers the momentum.

Note: If the angular momentum is transferred by filamentary currents (produced by pinch effect), the values of B and I become smaller, possibly by orders of magnitude.

TABLE 11.2

Minimum solar equatorial field for different radii of the primeval Sun

$R = 10^{11}$	3×10^{11}	10^{12}	3×10^{12} cm
$B = 5 \times 10^5$	18000	500	18 G

Table 11.1 also gives the value of the current I which transfers the angular momentum. It is calculated from

$$I = a_c r^{-2}, \tag{11.11}$$

which is obtained from Equation (11.7) by putting $\alpha' = 1$ and $B = a_c/r^3$.

11.3.2. *Discussion of the model*

The model we have treated is a steady state, homogenous model and open to the objections of Sections 10.2 and 10.3. It is likely that we can have a more efficient momentum transfer, e.g. through hydromagnetic waves or filamentary currents. This means that the magnetic dipole moments need not necessarily be as large as found here. It seems unlikely that we can decrease these values by more than one or two orders of magnitude but that can be decided only by further investigations. On the other hand, we have assumed that all the plasma condenses to grains and thus leaves the region of acceleration. This is not correct in the case when most ingredients in the plasma are non-condensable. If, for example, the plasma has a composition similar to the solar photosphere, only about 1% of its mass can form grains. As the behavior of volatile substances is not yet taken into account some modification of our model may be necessary. We may guess that if the mass of volatile substances is 1000 times the mass of condensable substances, the magnetic fields and currents may have to be increased by a factor $\sqrt{1000} \approx 33$. Hence a detailed theory may change the figures of Table 11.1 either downward or upward by one or two orders of magnitude.

11.4. THE SUPPORT OF THE PRIMORDIAL CLOUD

Closely connected with the problem of transfer of angular momentum is another basic difficulty in the Laplacian approach, namely, the support of the cloud against the gravitation of the central body. As soon as the cloud has been brought into rotation with Kepler velocity, it is supported by the centrifugal force. In fact, this is what defines the Kepler motion. But the acceleration to Kepler velocity must necessarily take a considerable amount of time, during which the cloud must be supported in some way.

Attempts have been made to avoid this difficulty by assuming that the Laplacian nebula had an initial rotation so that the Kepler velocities were established automatically. This results in an extremely high spin of the Sun, which then is supposed to be carried away by the 'solar gale'. Although this view is not in definite disagreement with present observational facts when applied to the planetary system, the proposed mechanism fails when applied to the satellite systems. One of the reasons is that it is irreconcilable with the isochronism of spins.

A plasma may be supported by a magnetic field against gravitation if a toroidal current I_ϕ is flowing in the plasma so that the force $|I_\phi \times B|$ balances the gravitational force $\mu m_B/r^2$, where m_B is the total mass of plasma magnetically suspended at any particular time. Let us assume for the sake of simplicity that the plasma to be supported is distributed over a toroidal volume with large radius r and small radius $r/2$. If n and m are the plasma density and the mean mass of a plasma particle in this volume, the condition for balance is expressed by

$$2\pi r I_\phi B = \frac{\mu m n}{r^2} 2\pi r \times \pi \frac{r^2}{4},$$ (11.12)

or

$$I_\phi = \frac{\pi}{4} \frac{\mu m n}{B}.$$

The magnetic field produced by this current is approximately homogeneous within the toroidal volume and has a value

$$B' \approx \frac{I_\phi}{r} = \frac{\pi}{4} \frac{\mu m n}{Br}.$$ (11.13)

Once again we note that if this field B' becomes too large compared to B, the dipole field will be seriously disturbed and instabilities will develop. For stability, B' must be of the same order of, or less than, B. Let us put $B' = \beta B$, with $\beta \leqslant 1$. If for B we use its equatorial value at a distance r, i.e., $B = ar^{-3}$, we obtain from Equation (11.13)

$$a^2 = \frac{\mu m_B}{2\pi\beta} r^2$$ (11.14)

which gives the value of the dipole moment a necessary for the support of the plasma. If $\beta = 1$, we get a lower limit a_g to a. Comparing a_g with a_c as given by Equation (11.9) we find that m_f and m_B in these two equations are equal if a_g is larger than a_c by a factor $\gamma^{-1/2}$. In the case of Sun-Jupiter, this is $(3 \times 10^8/10^{16})^{-1/2} \approx 5500$; for the satellite systems this factor is of the order of 10^5. Hence the magnetic fields required to suspend the whole distributed mass during the acceleration are unreasonably large. (This conclusion is not affected by the uncertainty discussed at the end of Section 11.3 which is also applicable here.)

Consequently there is no way to suspend the total mass of the plasma until it is accelerated to Kepler velocity.

11.5. THE PLASMA AS A TRANSIENT STATE

We have found that only a small fraction m_B of the final mass m_f of a planet or satellite can be supported by the magnetic field at any particular time. This means that the plasma density ϱ_B at any time can only be a small fraction γ_B of the distributed density ϱ_f (mass of final body divided by the space volume from which it derives; see Section

2.11, Part I)

$$\gamma_B = \frac{\varrho_B}{\varrho_f} = \frac{m_B}{m_f}.$$ (11.15)

This can be explained if matter is injected during a long time τ_i but resides in the plasma state only during a time $\tau_r \ll \tau_i$. This is possible if τ_r is the time needed for the plasma to condense to grains. Since during each time interval τ_r an amount of matter m_B condenses to grains, we have

$$m_f \frac{\tau_i}{\tau_r} = m_B,$$

so that

$$\gamma_B = \frac{m_B}{m_f} = \frac{\tau_r}{\tau_i}.$$ (11.16)

It is reasonable that the characteristic time for the production of grains in Kepler orbit is the Kepler period τ_K. Hence we put

$$\tau_r = \tau_K,$$

which together with Equations (11.10) and (11.16) gives

$$\gamma_B = \gamma.$$

This means that the instantaneous densities are less than the distributed densities by 10^{-7} for the giant planets to 10^{-11} for the satellite systems. Hence from Figures 2.6–2.9 (Part I) we find that the *plasma densities we should consider (compare Section 2.11) are of the same order of magnitude as the present number densities in the solar corona* (10^2–10^8 cm^{-3}).

It should be observed that these values refer to the *average* densities. Since the plasma is necessarily strongly inhomogeneous, the *local* densities are likely to be several orders of magnitude higher. Indeed the differences between the local and average densities should be of the same order as (or even larger than) the density differences between solar prominences and the solar corona in which they are embedded.

This is important because the time of condensation of a grain and its chemical and structural properties refer to the local condition. Assuming that the primordial components of meteorites were formed in the hetegonic nebula one can derive the properties of the medium from which they formed. The densities suggested in this way (Arrhenius, 1971; De and Arrhenius, 1973) are much higher than ϱ_B but still lower than ϱ_f. This will be discussed in detail in Section 15.

11.6. CONCLUSIONS ABOUT THE MODEL

Hence we can now restate the requirements of our model in the following way:
(a) Gas should be injected into the environment of the central body in such a way

as to account for the density distribution in the solar system. This is satisfied by the injection mechanism we are going to study in Section 17. In short, it means that neutral gas falling under gravitation towards the central body becomes ionized when it has reached the critical velocity for ionization. The ionization implies that a closer approach to the central body is prevented and the plasma is suspended in the magnetic field.

(b) Angular momentum is transferred from the central body to this plasma. A state of *partial corotation* is produced. This will be studied in Section 12.

(c) The condensation of the non-volatile substances of the plasma produces grains with chemical and structural properties exemplified by certain components in meteorites (Section 15; Arrhenius and Alfvén, 1971). This condensation should take place in an environment permeated by a magnetic field of the order of 0.1–1 G in the case of the planetary system (Brecher, 1972a, b, Part II).

(d) The grains should acquire such a dynamical state that they move in eccentric Kepler orbits thus satisfying the prerequisites for the planetesimal accretion. Many particle systems in this state are termed jet streams; the characteristic energy and mass balance in such systems are described in Sections 3 and 9.

The plasma state necessarily coexists with the jet streams. In fact, the grains and the plasma out of which they condense will interact mutually. As a population of orbiting grains has a 'negative diffusion coefficient' (Baxter and Thompson, 1971, 1973), the grains originally distributed through a given volume will tend to form a number of separate jet streams. Once a jet stream is formed it will collect new grains as they condense in its environment. Inside the jet streams, the grains accrete to larger bodies, and eventually to planets and satellites. A perspective of the various processes is represented by Figure 11.2. There are a number of jet streams in the equatorial plane and these are joined with the central body by plasma regions somewhat similar to the present-day solar prominences but having a very much larger dimension in case the central body is the Sun. In the following we shall refer to these regions as *super-prominences*.

11.7. The Hetegonic 'Nebula'

In Laplacian type theories, the medium surrounding the primordial Sun is called 'solar nebula' or 'circumsolar nebula', and forms the precursor for the planets. Contrary to the Laplacian theories, we are not developing a theory of the formation of planets alone but a general hetegonic theory, applicable both to the formation of planets around the Sun and the formation of satellites around planets. Since the term 'solar nebula' does not convey the need for this generalization, a term like 'hetegonic nebula' or 'hetegonic cloud' is preferable.

Moreover, the term 'solar nebula' has been used too much as a nineteenth century concept, implying a homogeneous disc of non-ionized gas with uniform chemical composition and pre-hydromagnetic dynamical properties. For a number of reasons we have discussed earlier this concept is obsolete. In terms of modern theory and observation we need instead to consider the central bodies to be surrounded by a

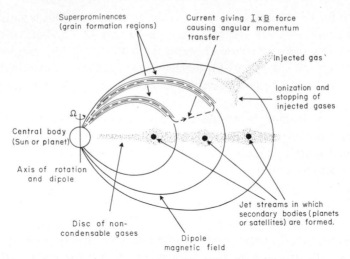

Fig. 11.2. A sketch of the series of hetegonic processes leading to formation of secondary bodies around a spinning magnetized central body. The dipole magnet is located at the center of the central body and is aligned with the spin axis. The gas injected from 'infinity' into the environment of the central body becomes ionized by collision with the magnetized plasma when its free fall velocity exceeds the critical velocity for ionization, and the ionized gas then remains suspended in the magnetic field. The rotation and magnetic field together with the conducting plasma surrounding the central body give rise to a homopolar emf. which causes a current flow in the plasma. This current $\mathbf{I}$ together with the magnetic field $\mathbf{B}$ give rise to a force $\mathbf{I} \times \mathbf{B}$ which transfers angular motion from the central body to the surrounding plasma. The current also produces prominence-like regions of gas (by pinch effect) which are denser and cooler than the surrounding regions and in these regions the condensation of grains takes place. Through viscous effects, the population of grains evolves into a number of jet streams while the non-condensable gases form a thin disc in the equatorial plane. (Not drawn to scale.)

structured medium of plasma and grains during the period of formation of the secondary bodies.

The space around the central body may be called a supercorona – characterized by a medium which is similar to the present solar corona but much larger in extent due to the flux of gas from outside into the system during the formation era. It is magnetized, primarily by the magnetic field of the central body. Its average density, to show the proper behavior, would be of the same order as that of the solar corona (10^2–10^8 cm^{-3}). This supercorona consists of four regions of widely differing properties (Figure 11.2). Note that the central body may be either the Sun *or a planet*.

(a) *Jet streams*: The theory of these is given in Section 3 (Part I). They fill up a very small part of space. The small diameter of the toroid is only a few percent of the large diameter and hence they occupy 10^{-3}–10^{-4} of the volume. They are fed by injection of grains condensed in large regions around them. The accretion of satellites or planets takes place in the jet streams (see Section 9, Part II).

(b) *Low density plasma regions*: Most of space outside the jet streams is filled with a low density plasma. This region with a density perhaps in the range 10–10^5 cm^{-3} occupies most of the volume of the supercorona. The supercorona is fed by injection of matter from a source at large distance ('infinity'). The transfer of angular momentum from the central body is achieved through processes in this plasma. This means that there is a system of strong electric currents flowing in the plasma which results in filamentary structures (superprominences).

(c) *Filamentary structures or superprominences*: The plasma structurally resembles the solar corona with embedded prominences produced by strong currents. These stretch from the surface of the central body out to the most distant regions to which angular momentum is transferred by the currents. As in the solar corona, the filaments have a density which is orders of magnitude larger and a temperature which is much lower than those of the surrounding medium. As high density and low temperature favor condensation, most of the condensation takes place in the filaments. When the condensed grains leave the filaments, they have gained a tangential velocity which determines their Kepler orbits; their interaction leads to the formation of jet streams. At the same time, plasma from the low density regions is 'pinched' into the filaments,

(d) *Non-condensable gas clouds*: As the injected matter contains a large fraction of non-condensable gases – presumably they form the main constituent – there is an increasing supply of such gases in the filaments and in the interfilamentary plasma. When partial corotation is established, this gas is accumulated close to the equatorial plane. Part of the gas is retained in the jet streams in which the apparent attraction accumulates it (Alfvén, 1971). Hence the accretion in the jet streams may take place in a cloud of non-condensable gases. When an embryo has become so large that its gravitation becomes appreciable, it may capture an atmosphere from the gas supply of the jet stream.

It is likely that the jet streams cannot keep all the gas. Some of it may diffuse away, possibly forming a thin disc of gas which may leak into the central body or transfer gas from one jet stream to another. In Figure 11.2, the gas is supposed to form tori around the jet streams which flatten out to discs. It is doubtful whether any appreciable quantity of gas can leak out to infinity because of momentum considerations.

The behavior of the non-condensable gases is necessarily the most hypothetical element in the model because we have very little and essentially indirect information about it.

The diagram in Figure 11.3 outlines the sequence of processes leading to the formation of secondary bodies around a central body. These processes will be discussed in detail in the following chapters.

11.8. IRRADIATION EFFECTS

Analysis of particle tracks and surface related gases in meteorites demonstrate that individual crystals and rock fragments become individually irradiated with accelerated particles (see Section 7.3, Part I). This irradiation evidently took place before material was permanently locked into the parent bodies of the meteorites that they are now part of. Considerable fluxes of corpuscular radiation with approximately solar photospheric composition consequently existed during that period of formation of meteorite parent bodies when individual crystals and rock fragments were free to move relative to each other, that is during the time of embryonic accretion. This process may still be going on as, for example, in the asteroidal and cometary jet streams.

With the present information it is not possible to fix the point in time when this irradiation began or decide whether it was present during or soon after the era of gas injection and condensation of primordial matter. Hence the specific irradiation phenomena are not a critical part of our treatment of these early phases. On the other hand the properties of our model are such that particle acceleration into the keV ('solar wind') and MeV ('solar flare') ranges in general is expected.

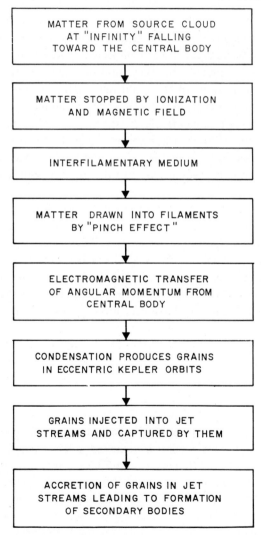

SEQUENCE OF PROCESSES LEADING
TO FORMATION OF SECONDARY BODIES

MATTER FROM SOURCE CLOUD
AT "INFINITY" FALLING
TOWARD THE CENTRAL BODY

MATTER STOPPED BY IONIZATION
AND MAGNETIC FIELD

INTERFILAMENTARY MEDIUM

MATTER DRAWN INTO FILAMENTS
BY "PINCH EFFECT"

ELECTROMAGNETIC TRANSFER
OF ANGULAR MOMENTUM FROM
CENTRAL BODY

CONDENSATION PRODUCES GRAINS
IN ECCENTRIC KEPLER ORBITS

GRAINS INJECTED INTO JET
STREAMS AND CAPTURED BY THEM

ACCRETION OF GRAINS IN JET
STREAMS LEADING TO FORMATION
OF SECONDARY BODIES

Fig. 11.3.   Sequence of processes leading to the formation of secondary bodies
around a central body.

In Section 10.4 our model is characterized as a synthesis of phenomena now observed in the Earth's magnetosphere and in the solar corona. This implies that we should expect the model to exhibit also other related properties of these regions to a certain extent. It is well known that in the magnetosphere there are processes by which particles are accelerated to keV energies (as shown by the aurora and by direct space measurements). In the van Allen belts there are also particles accelerated by magnetospheric processes to MeV energies. Furthermore, it is well known that solar activity – especially in connection with flares – produces MeV particles ('solar cosmic rays').

Our superprominences should produce similar effects in the whole region where transfer of angular momentum takes place and grains are condensing. Hence in our model grains necessarily are irradiated in various ways. Even nuclear reactions may be produced. All these effects will occur independent of whether the Sun is hot or cool or has an activity of the present type. In fact, the only properties which the central body – be it the Sun or a planet – needs to have is gravitating mass, spin and magnetization. A detailed theory of the irradiation effects is difficult and cannot be worked out until both the theory of the magnetosphere and of solar activity is much more advanced than today. When this stage is reached the irradiation effects will probably allow specific conclusions. However, we have at present no compelling reasons for attributing possible early irradiation effects ad hoc to specific solar or cosmic phenomena such as 'solar gale' or outbursts of nearby supernovae.

11.9. THE MODEL AND THE HETEGONIC PRINCIPLE

In Section 1 it was pointed out that because the general structure of the satellite systems is so similar to that of the planetary system, one should aim at a general hetegonic theory of formation of secondary bodies around a central body. This is a principle which has been pronounced repeatedly over the centuries and no one seems to have denied it explicitly.

It is an extremely powerful principle because of the severe constraints it puts on every model. In spite of this it has usually been neglected at the formulation of solar system theories. Earlier we have used the hetegonic principle for a choice between alternative explanations of the resonances in the satellite systems (Section 4, Part I). The diagram in Figure 11.4 shows how the principle is applied to the two similar series of processes leading to the formation of secondary bodies from a primeval dispersed medium. The chain of processes leading to the formation of planets around the Sun is repeated in the case of formation of satellites around the planets, but in the latter case a small part (close to the planet) of the planetary jet stream provides the primeval cloud out of which the satellites form. There is only one basic chain of processes, as summed up in Figure 11.3, which applies to the formation of both planets and satellites. This means that a complete theory of jet streams (including not only grains but also the gas component) must give the initial conditions for satellite formation.

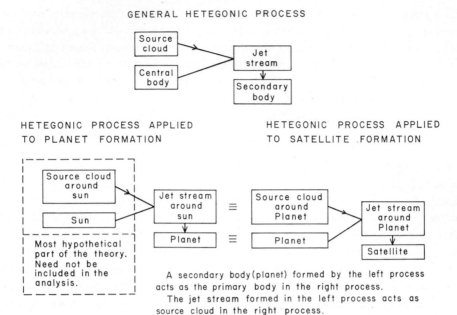

Fig. 11.4. Diagram showing how the speculative character of the theory is reduced by the hetegonic principle which implies that all theories should be applicable to both planetary and satellite systems. This eliminates the need to rely on hypotheses about the early Sun, and ties the theory closer to observations.

The consequence of this is that we can explore the hetegonic process without detailed assumptions about the properties of the early Sun. This is advantageous because these properties are poorly known. Indeed, the current theories of formation of stars are very speculative and possibly unrelated to reality. For example, the Sun may have been formed by a 'stellesimal' accretion process analogous to the planetesimal process. The planetesimal process works over a mass range from 10^{18} g (or less) up to 10^{30} g (see Section 8.3, Part II). One may ask whether to these 12 orders of magnitude one could not add three more as so to reach stellar masses (10^{33} g). Observations give no real support to any of the conventional theories of stellar formation and may agree just as well with a stellesimal accretion. As was pointed out in Section 10.3, it is now obvious that many homogeneous models are misleading and have to be replaced by inhomogeneous models. The introduction of stellesimal accretion would be in conformity with the latter type approach.

From Figure 11.4 and the discussion above we conclude that we need not concern ourselves with the question of whether the Sun has passed through a high luminosity Hayashi phase or whether the solar wind at some early time was stronger than it is now. Neither of these phenomena could have influenced the formation of satellites (e.g. around Uranus) very much. The similarity between the planetary system and satellite systems shows that such phenomena have not played a major dynamic role.

Instead of basing our theory on some hypothesis about the properties of the early Sun, we can draw conclusions about the solar evolution from the results of our theory based on observation of the four well-developed systems of orbiting bodies (the planetary system and the satellite systems of Jupiter, Saturn and Uranus).

What has been said so far stresses the importance of studying jet streams. The theoretical analysis should be expanded to include the gas (or plasma) which is trapped by the apparent attraction. One should also investigate to what extent meteor streams and asteroidal jet streams are similar to those jet streams in which planets and satellites were formed. The formation of short period comets is one of the crucial problems.

As a final remark, although the hetegonic principle is important and useful it should not be interpreted too rigidly. There are obviously certain differences between the planetary system and the satellite systems. The most conspicuous one is that the planets have transferred only a small fraction of their spin angular momentum to the satellites whereas the Sun appears to have transferred most of its spin angular momentum to the planets. The principle should preferentially be used in such a way that the theory of formation of secondary bodies is developed with the primary aim of explaining the properties of the satellite systems. We then investigate the extent to which this theory is applicable to the formation of planets. If there are reasons to introduce new effects to explain the formation of planets, we should not hesitate to do this. As we shall see in the following, there seems to be no compelling reason to assume that the general structure is different but there are local effects which may be produced by solar radiation (see Section 14).

12. Transfer of Angular Momentum and Condensation of Grains

12.1. Ferraro isorotation and partial corotation

We have shown in Section 11.3 that a difference in angular velocity between a magnetized central body and the surrounding plasma may lead to a transfer of angular momentum.

From a purely hydromagnetic point of view the final state would be a Ferraro corotation with $\omega = \Omega$. However, a transfer of angular momentum means an increase in rotational velocity of the plasma, with the result that it is centrifuged outwards. This will produce a region with low density between the central body and the plasma, and the density may decrease so much that anomalous resistance or the production of electrostatic double layers (see Section 10.3.3) impedes a further transfer of angular momentum. In this way we may reach a state where the rotational motion of an element of plasma is essentially given by the condition that the gravitation and the centrifugal force balance each other. This state is called 'partial corotation' (Alfvén, 1967).

The partial corotation can be thought of as a transient state in the process of angular momentum transfer from the central body. This state is important if the time τ_t of transfer of angular momentum from the central body to a cloud of plasma is long compared to the time τ_e it takes for the cloud of plasma to find its equilibrium position on the magnetic field line. Furthermore, the condensation of grains from a state of partial corotation becomes important if the time of condensation τ_c is small compared to τ_t and long compared to τ_e.

Partial corotation may also be a steady state. If plasma is injected at a constant rate, and the condensation products are removed at a constant rate, a state of time-independent partial corotation may be established. The condition for this is that the rate of transfer of angular momentum equals the angular momentum required to put the injected gas into rotation. The transfer of angular momentum may be regulated by the density of the plasma in the depleted region between the central body and the plasma element to be accelerated. This density determines the maximum current which transfers the momentum.

In the next section we discuss the state of equilibrium motion of an element of plasma situated in a magnetic flux tube which we have earlier called a superprominence (see Figure 11.2).

12.2. ORBITAL MOTION OF PLASMA UNDER THE ACTION OF GRAVITATION AND CENTRIFUGAL FORCE

In a spherical coordinate system $(r, \lambda, \phi$; see Figure 12.1) fixed in space, the tangential velocity of a plasma element is

$$v_\phi = \omega r \cos \lambda .\tag{12.1}$$

If the mass of this element is Δm_B, it is acted on by a gravitational force

$$\Delta m_B f_g = \frac{\kappa M_c \Delta m_B}{r^2} ,\tag{12.2}$$

and a centrifugal force

$$\Delta m_B f_c = \frac{\Delta m_B v_\phi^2}{r \cos \lambda} .\tag{12.3}$$

The gravitational force is directed toward the origin while the centrifugal force is directed perpendicular to and away from the rotation axis.

If the radius vector to the plasma element makes an angle α with the magnetic field at the element, we have, from the property of a dipole field

$$\sin \alpha = \frac{\cos \lambda}{\psi} ; \qquad \cos \alpha = \frac{2 \sin \lambda}{\psi} ,\tag{12.4}$$

where

$$\psi = (1 + 3 \sin^2 \lambda)^{1/2} .\tag{12.5}$$

The centrifugal force makes an angle $(\alpha - \lambda)$ with the magnetic field. We find from Equation (12.4)

$$\cos (\alpha - \lambda) = \tfrac{3}{2} \cos \alpha \cos \lambda .\tag{12.6}$$

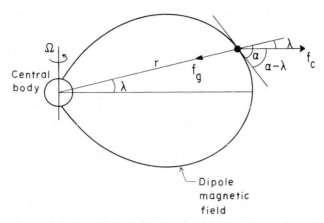

Fig. 12.1. Plasma in the magnetic dipole field of a rotating central body. A particle is acted on by the gravitation f_g which makes the angle α with the magnetic field, and by the centrifugal force f_c which makes the angle $\alpha - \lambda$ with the magnetic field.

The magnetic field line represents the central field line in a flux tube ('superprominences' and 'grain formation region' in Figure 11.2). Plasma can move freely along such flux tubes under the action of gravitation and centrifugal force.

If thermal motion of the plasma particles can be neglected, the motion of a plasma element parallel to the magnetic field will be governed by the gravitational force and the centrifugal force. An equilibrium is reached when

$$f_g \cos \alpha = f_c \cos (\alpha - \lambda). \tag{12.7}$$

This defines the state of partial corotation. In the direction perpendicular to the magnetic field in the meridian plane, the plasma is supported by the magnetic field. This implies that there is a toroidal current I_ϕ, defined by

$$\Delta m_B [f_g \sin \alpha + f_c \sin (\alpha - \lambda)] = I_\phi B. \tag{12.8}$$

We assume here that the current I_ϕ is so small that it does not modify the dipole field appreciably. In cosmic situations, this assumption is equivalent to the requirement that the gravitational energy of the plasma should be negligibly small compared to the magnetic energy.

12.3. PARTIAL COROTATION OF MAGNETIC PLASMA

We have treated an element of a medium density plasma situated at (r, λ, ϕ) in a magnetic dipole field of a rotating central body, from which it is separated by a low density plasma. We have assumed that the temperature of the plasma is low. Further, we have assumed that the mass is so small that the dipole field is not seriously disturbed. For a constant rotation of the plasma, we obtained Equation (12.7). By substituting Equations (12.2), (12.3) and (12.6) in this equation, we obtain

$$v_\phi^2 = \frac{2}{3} \frac{\kappa M_c}{r}. \tag{12.9}$$

As a comparison a circular Kepler motion with radius r is characterized by

$$v_K^2 = \frac{\kappa M_c}{r}. \tag{12.10}$$

Hence, we can state as a general theorem:

If in the magnetic dipole field of a rotating central body with the dipole located at the center of the body and aligned with the spin axis, a plasma element is in a state of a partial corotation, then the kinetic energy of it is two-thirds the kinetic energy of a circular Kepler motion at the same radial distance.

This factor $\frac{2}{3}$ derives from the geometry of a dipole field and enters because the centrifugal force makes a smaller angle with a field line than the gravitation. The plasma element is supported against gravitation in part by the centrifugal force, and in part by the current I_ϕ which according to Equation (12.8) interacts with the magnetic field to give a force. The above treatment, strictly speaking, applies only to plasma situated at nonzero latitudes. The equatorial plane represents a singularity. However, this plane will be occupied by a disc of grains and gas with a thickness of a couple of degrees – the mathematical singularity is thus physically uninteresting.

Table 12.1 compares the energy and angular momentum of a circular Kepler motion and a circular motion of a magnetized plasma.

TABLE 12.1

Comparison between Kepler motion and partial corotation

	Circular Kepler motion	Partial corotation of magnetized plasma
Gravitational energy	$-\dfrac{\kappa M_c}{r}$	$-\dfrac{\kappa M_c}{r}$
Kinetic energy	$\dfrac{1}{2}\dfrac{\kappa M_c}{r}$	$\dfrac{1}{3}\dfrac{\kappa M_c}{r}$
Total energy	$-\dfrac{1}{2}\dfrac{\kappa M_c}{r}$	$-\dfrac{2}{3}\dfrac{\kappa M_c}{r}$
Orbital angular momentum	$\sqrt{\kappa M_c r}$	$\sqrt{\dfrac{2}{3}\kappa M_c r}$

If the plasma has considerable thermal energy, diamagnetic repulsion from the dipole gives an outward force which has a component which is added to the centrifugal force. This makes the factor in Equation (12.9) smaller than $\frac{2}{3}$. It can be shown that this effect is of importance if the thermal energy $A = \delta k(T_e + T_i)$ (where δ is the degree of ionization, k is the Boltzmann's constant, and T_e and T_i the electron and ion temperatures) is comparable to the kinetic energy of a plasma particle $W_k = \frac{1}{2}mv_\phi^2$.

For a possible application to the environment close to Saturn we may put $m = 10m_H =$ $= 1.7 \times 10^{-23}$ g, $v_\phi = 2 \times 10^6$ cm s^{-1} ($=$ orbital velocity of Mimas) and $\delta = 10\%$. We find that $A/W_k = 1\%$, if $T_e = T_i = 15000$ K. This indicates that the temperature correction is probably not very important in the case we have considered.

12.4. Discussion

It is a well known observational fact that in the solar prominences matter flows down along the magnetic flux tube to the surface of the Sun presumably under the action of gravitation. This matter cannot move perpendicular to the flux tube because of electromagnetic forces. The solar prominences are, however, confined to regions close to the Sun and this state of motion is such as to make the centrifugal force unimportant. In contrast our superprominences would extend to regions very far away from the central body (see Figure 11.2), roughly to the regions where the resulting secondary bodies would be located. Here then we see a possibility that the components of centrifugal force and the gravitational attraction parallel to the flux tube on an element of plasma may balance each other, keeping the element in a state of dynamical equilibrium, i.e., the state of partial corotation. This state is analyzed in some further detail by De and Arrhenius (1973).

12.5. Condensation of the Plasma: the Two-Thirds Law

If the plasma recombines so that the current I_ϕ vanishes, the element of matter changes its motion from the type we have investigated, and under certain conditions its trajectory will be a Kepler ellipse.

Let us suppose that in the partially corotating plasma a condensation takes place so that small solid grains are produced. We shall not discuss the process of condensation in detail in this section, but refer to its product as grains. If these are very small, the electric charge they may acquire in the plasma will cause their motion to be influenced by the magnetic field. We shall confine the discussion to the simple case when the grains have grown large enough so that they are influenced neither by electromagnetic forces nor by viscosity due to the plasma. Furthermore, the condensation is assumed to be instantaneous so that the inital velocity of a grain equals the velocity of the plasma element from which it is born.

As the initial velocity of the grain is $\sqrt{\frac{2}{3}}$ of the circular Kepler velocity at its position, a grain at the initial position (r_0, λ_0, ϕ_0) will move in an ellipse with the eccentricity $e = \frac{1}{3}$ (see Figures 12.2a and b). Its apocenter A is situated at (r_0, λ_0, ϕ_0) and its pericenter P at (r_P, λ_P, ϕ_P).

$$r_P = \tfrac{1}{2}r_0, \tag{12.11a}$$

$$\lambda_P = -\lambda_0, \tag{12.11b}$$

$$\phi_P = \phi_0 + \pi. \tag{12.11c}$$

The ellipse intersects the equatorial plane $\lambda = 0$ at the nodal points $(r_n, 0, \phi_0 + \pi/2)$ and

$(r_n, 0, \phi_0 - \pi/2)$ with

$$r_n = \tfrac{2}{3} r_0 .$$ (12.12)

When the grain reaches r_n its angular velocity equals the angular velocity of a body moving in a Kepler circle in the orbital plane of the grain with radius r_n.

Suppose that grains are produced in a ring element (r_0, λ_0) of plasma. All of them cross the equatorial plane at the circle $r_n = \tfrac{2}{3} r_0$. Suppose that there is a small body (*embryo*) moving in a circular Kepler orbit in the equatorial plane with radius r_n. It will be hit by the grains, and we assume for now that all grains hitting the embryo are retained by it. Each grain has the same angular momentum per unit mass as the embryo. However, the angular momentum vector of the embryo is parallel to the rotation axis, whereas the angular momentum vector of the grain makes an angle λ_0 with the axis. In case λ_0 is so small that we can put $\cos \lambda_0 = 1$, the embryo will grow in size but not change its orbit. If $\cos \lambda_0 < 1$, the embryo will spiral inwards while growing.

Seen from the coordinate system of the embryo, the grains will arrive with their velocity vectors in the meridional plane of the embryo. These velocities have a component parallel to the rotation axis of the central body, equal to $(\tfrac{2}{3}(\kappa M_c/r_0))^{1/2} \sin \lambda_0$ and a component in the equatorial plane of and directed towards the central body, equal to $(\tfrac{1}{12}(\kappa M_c/r_0))^{1/2}$. This last component results in a slow inward motion of the embryo. If, however, the collision is perfectly inelastic, there will be no momentum transfer and the entire impact energy will be converted to heat.

The existence of an embryo in the above discussion is assumed merely to illustrate the importance of the circular orbit with radius $\tfrac{2}{3} r_0$ in the equatorial plane. All the grains which are formed at a distance r_0 from the center will cross the equatorial plane at the circumference of this circle, irrespective of what the value of λ_0 is (under the condition that we can put $\cos \lambda_0 = 1$). These grains will collide with each other and coalesce to form increasingly larger embryos until these are large enough so that they can accrete smaller grains. The large bodies thus produced will move in a circular orbit in the equatorial plane with radius $\tfrac{2}{3} r_0$.

Summarizing our results, we have found that a plasma cloud in the dipole field of a rotating central body need not necessarily attain the same angular velocity as the central body. If in the region between the plasma cloud and the central body the density is so low that the parallel electric field may differ from zero, a steady state is possible characterized by a partial corotation according to Table 12.1. If at a central distance r_0 grains condense out of such a plasma, they will move in ellipses with a semi-major axis $\tfrac{3}{4} r_0$ and an eccentricity $e = \tfrac{1}{3}$. Mutual collisions between a population of such grains will finally make the condensed matter move in a circle in the equatorial plane with the radius $\tfrac{2}{3} r_0$ (see Figures 12.2a and b).

In the more general case, when condensation takes place over a wider range of latitudes and central distances in an extended region one would expect that each grain which has condensed will ultimately be moving in a circle at a distance of $\tfrac{2}{3}$ times the distance where the condensation has taken place. This may occur under certain conditions, but is generally true because collisions between the grains are no longer res-

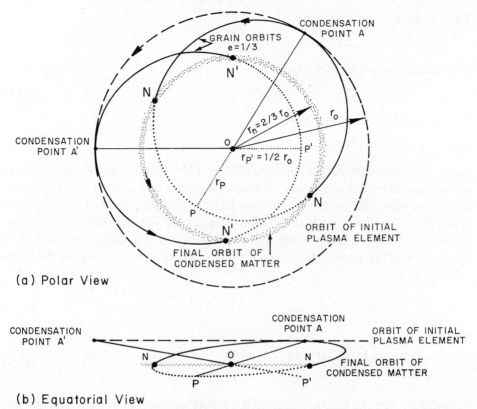

(a) Polar View

(b) Equatorial View

Fig. 12.2a and b.   The condensation process. The outer dashed line represents the circular orbit of a plasma element in the partially corotating plasma. Condensation produces small solid grains which move in Kepler ellipses with eccentricity $e = \frac{1}{3}$. Two such grain orbits are shown, one originating from condensation at A and the other, at A'. The condensation point A which hence is the apocenter of the latter orbit has the spherical coordinates (r_0, λ_0, ϕ_0). The pericenter P is at $r_P = \frac{1}{2} r_0$, $\lambda_P = -\lambda_0$, $\phi_P = \phi_0 + \pi$ and the nodal points N are at $r_n = \frac{2}{3} r_0$, $\lambda_n = 0$, and $\phi_n = \phi_0 \pm \pi/2$; Collisions between a large number of such grains result in the final (circular) orbit of condensed matter in the equatorial plane. The eccentricity $\frac{1}{3}$ of the initial grain orbit and the radius $\frac{2}{3} r_0$ of the final orbit of condensed matter are direct consequences of the plasma being in the state of partial corotation (see Sections 12.3–12.5).

tricted to the equatorial plane. There will be competitive processes through which grains agglomerate to larger embryos moving in eccentric orbits. However, the semi-major axes of these orbits are $\frac{2}{3}$ the weighted mean of the radius vector to the places of condensation (see Section 13).

12.6. ENERGY RELEASE AT THE ANGULAR MOMENTUM TRANSFER

The transfer of angular momentum from the central body to the surrounding plasma is accompanied by a conversion of kinetic energy into heat. Suppose that a central body with a moment of inertia Ξ is decelerated from the angular velocity Ω_1 to Ω_2 by accelerating a mass m' at an orbital distance r from rest to an angular velocity ω.

Then we have

$$\Xi \left(\Omega_1 - \Omega_2 \right) = m'r^2\omega.$$ (12.13)

The energy released by this process is

$$W = \tfrac{1}{2}\Xi \left(\Omega_1^2 - \Omega_2^2 \right) - \tfrac{1}{2}m'r^2\omega^2.$$ (12.14)

Putting $\Omega = \tfrac{1}{2}(\Omega_1 + \Omega_2)$, we have

$$W = m'r^2 \left(\Omega\omega - \tfrac{1}{2}\omega^2 \right).$$ (12.15)

As has been studied previously in detail (Alfvén, 1954, Chapter 4), the ionized gas will fall toward the central body along the magnetic lines of force, but at the same time its ω value is increased because of the transfer of momentum from the central body. When the velocity ωr has reached approximately the Kepler velocity, the gas will move out again and finally condense. The bodies that later are formed out of the cloud move in the Kepler orbit. Hence, the final result is that ωr equals the Kepler velocity, so that

$$m'r^2\omega^2 = \frac{\kappa M_c m'}{r}.$$ (12.16)

This gives

$$W = \frac{\kappa M_c m'}{r} \left(\frac{\Omega}{\omega} - \frac{1}{2} \right).$$ (12.17)

If to this we add the kinetic energy of the falling gas, $\kappa M_c m'/r$, we obtain the total available energy,

$$W_T = \frac{\kappa M_c m'}{r} \left(\frac{\Omega}{\omega} + \frac{1}{2} \right).$$ (12.18)

This energy is dissipated in the plasma in the form of heat. In fact, this may have been a significant source of heating for the circumsolar plasma during the hetegonic ages.

13. Accretion of the Condensation Products

13.1. SURVEY

The accretion of grains to larger bodies is one of the main problems in the theory of formation of planets and satellites. In Parts I and II we have found that this process takes place in two steps – the first one leading to the formation of jet streams and the second one – studied in Section 9 – to the formation of large bodies inside the jet streams.

The first three chapters of Part III represent an attempt to trace the plasma processes which have led to the formation of the grains. In Section 12 we have found that under certain conditions a state called partial corotation may be established, which should lead to the formation of grains moving in Kepler orbits with eccentricity equal to $\frac{1}{3}$. Whether the conditions for partial corotation were really satisfied at hetegonic times can be ascertained only by looking for evidence in the solar system today that may have resulted from such a state.

This Section shall be devoted to such evidence. More specifically, we shall study the intermediate process, namely the accretion of grains and the formation of jet streams, and compare the products of these processes with observations.

The study is facilitated by the fact that in certain parts of the solar system we find 'intermediate products' of these processes. In the asteroidal region as well as in the Saturnian ring system the accretion has not led to the formation of large bodies. In the asteroidal region the reason for this seems to be the extremely low space density of the condensed matter (see Section 7.8) whereas in the Saturnian ring system the formation of large bodies has been prevented because the region is situated inside the Roche limit.

This means that in this Section we should treat the development of a population of orbiting grains with the aim of developing three theories:

(1) A theory of formation of the Saturnian rings;

(2) A theory of the formation of the asteroid belt;

(3) A theory of the formation of jet streams as an intermediate stage in the formation of satellites and planets.

We have tried to develop the first of the above theories in Sections 13.5–13.6, the second in Sections 13.7–13.8 and the third in Section 13.10.

13.2. EVOLUTION OF ORBITS DUE TO COLLISIONS

The accretion of grains to larger bodies has been treated in Part II where observations were predominantly analyzed with the help of celestial mechanics. Here we shall treat the same problem, but as a starting point we choose the partial corotation and the state of motion of grains resulting from it. As we shall discover, it is possible to make the two pictures consonant.

In Section 12 we found that the direct result of condensation from a partially corotating plasma is an assembly of grains moving in Kepler ellipses with $e = \frac{1}{3}$ (Figure 12.2). The major axis of such an ellipse passes through the point of condensation (the apocenter) and the central body (focus) and the minor axis is parallel to the equatorial plane. The pericentric distance is $\frac{1}{2}$ and the distance of the nodes is $\frac{2}{3}$ of the apocentric distance.

We shall study the development of an assembly of such grains under the action of mechanical forces alone. Persuing the ideas of Section 3 we assume that the interaction of the grains with the plasma from which the grains have condensed is negligible, and that the charge-to-mass ratio of the grains is so small that electromagnetic forces do not influence their motion. The meaning of these assumptions has already been analyzed quantitatively in Section 3.1.

Under the idealized assumptions of a spherical homogeneous central body with a single grain orbiting around it, the orbit of this grain will remain unchanged with time. If the central body assumes an ellipsoidal shape due to rotation, this will result in a secular change (precession) in the orientation of the orbit of the grain (see Section 2). In a realistic case, one must also consider the gravitational perturbations from other neighboring celestial bodies – whether full-grown or *in statu nascendi*. Such perturbations also produce precession. (At the same time they produce long period changes in the eccentricity and inclination of the orbit, but these are of small amplitude and not very important in this connection.) If other grains are present in the same region of space the gravitation from their dispersed mass also produces secular disturbances of the same type.

However, the most important systematic change in the orbits of an assembly of grains is due to their mutual collisions, which are essentially inelastic – or at least partially so. At such a collision, kinetic energy is converted into heat but the sum of the orbital angular momenta of the two colliding grains does not change. Collisions may also result in fragmentation or in accretion.

The general result of inelastic collisions within a population of grains with inter-secting orbits is that the eccentricities of the orbits decrease with time, and so do the inclinations in relation to the invariant plane of the population (Figure 13.1). In our

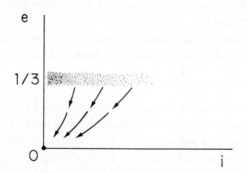

Fig. 13.1.   Development of a population of grains originally orbiting in ellipses with eccentricity $e = \frac{1}{3}$ and varying inclinations i. The final state with $e = 0$ and $i = 0$ is either a thin disc or a group of planets (or satellites) in circular orbits in an equatorial plane.

model of condensation of grains, the angular momentum of the population of grains ultimately derives from the rotation of the central body. Assuming the process of condensation to be symmetrical with respect to the equatorial plane, the invariant plane of the population will be coincident with the equatorial plane.

If gas is present in the region, the effect of viscosity on the motion of the grains may be important – particularly in the case of small grains. If we consider the gas molecules as extremely small 'grains', the presence of gas means essentially an enrichment of population in the low end of the mass spectrum of the grains. However, we must ob-serve that collisions between molecules may be perfectly elastic, whereas collisions between grains are always more or less inelastic.

Using the terminology of Sections 2.2–2.4, we can state that collisions and viscosity make both the axial oscillations and the epicycle motion decrease, eventually leaving the grains in unperturbed circular orbits.

One example of this state is the Saturnian ring system, where a large number of small bodies form an extremely thin disc (thickness ~ 2 km or less), each body moving in a circle with an angular velocity which decreases outward according to Kepler's laws. A general survey of the state of Saturnian rings may be found in Cook *et al.* (1973). Other examples are the different groups of planets or satellites (see Section 2.12 and Table 2.2.4). The bodies in a group are likely to have formed from a single population of grains that evolved through mutual collisions. Most of the planets and satellites move in almost circular orbits with small inclinations.

The asteroids represent an intermediate stage in this evolution. The present eccentricities are on the average about one half of the original value 0.33, and the present inclinations probably represent a similar decrease in relation to the unknown primeval distribution.

13.3. THE ROCHE LIMIT

Suppose that a small solid sphere of radius x and density θ moves in a circular orbit of radius r around a spherical central body of radius R_c and density θ_c, the mass of the latter body being much larger than that of the former. Consider an infinitely small test particle of mass m at the surface of the sphere x located in the direction of the central body. The particle is acted upon by the gravitation f_g of the sphere

$$f_g = \frac{4\pi}{3} \kappa \theta x ,$$ (13.1)

and by the tidal force f_t from the central body

$$f_t = \frac{4\pi}{3} \kappa \theta_c R_c^3 \left[\frac{1}{(r-x)^2} - \frac{1}{r^2} \right] \approx \frac{4\pi}{3} \kappa \theta_c \left(\frac{R_c}{r} \right)^3 2x .$$ (13.2)

The tidal force exceeds the gravitation if

$$\frac{r}{R_c} < C \left(\frac{\theta_c}{\theta} \right)^{1/3} ,$$ (13.3)

with

$$C = 2^{1/3} \approx 1.26 .$$ (13.4)

When, instead of the small solid sphere, there is a self-gravitating body consisting of a perfect fluid, the tidal force will deform it from a sphere into an ellipsoid if it is orbiting at a large distance. If the orbital radius is decreased the body will be increasingly prolonged with the long axis pointing towards the central body and at a sufficiently small distance it becomes unstable because the tidal force exceeds the self-gravitation of the body. This distance R_R is the well-known Roche limit, defined by

$$\frac{R_R}{R_c} = C\left(\frac{\theta_c}{\theta}\right)^{1/3},$$ (13.5)

with $C = 2.44$

The outer border of the Saturnian ring system is located at $r = 1.37 \times 10^8$cm which gives $r/R_h = 2.28$. As Saturn's average density is $\theta_h = 0.62$ g cm^{-3} the outer border could be identified with the Roche limit under the following conditions:

(a) the density of the grains is $\theta = 0.62(2.44/2.28)^3 = 0.76$ g cm^{-3}.

(b) the grains behave like drops of a perfect fluid.

(c) the gravitational field of all adjacent grains can be neglected.

We have no independent way of finding the density of the grains, so (a) may or may not be true.

According to some authors, the material in the rings is likely to occur in the form of more or less loosely bound particles in the form of spindle shaped aggregates with their long axes tangent to their orbits, that is, at right angles to what is supposed in the Roche theory. Other authors have proposed similar elongated aggregates but with their long axes perpendicular to the equatorial plane. So (b) is probably not satisfied.

The mass of the ring is so small that it does not perturb the Saturnian gravitational field very much. The tidal effect, however is produced by the field gradient, and adjacent grains may very well produce local perturbations of this. Hence it is doubtful whether (c) is satisfied.

The conclusion is that the identification of the outer border of the ring with the Roche limit is not very convincing from a theoretical point of view.

However, from an observational point of view there is no doubt that the outer limit of the ring marks the border between one region where matter does not accrete to larger bodies and another region where it does accrete to satellites. We shall call this limit 'modified Roche limit' (R_M). It is reasonable that this is given by the tidal disruption, but the theory of this is much more complicated and possibly rather different from the classical Roche theory.

Inside R_M matter will be much more dispersed than outside so that the mean free path between collisions will be much smaller in this region than outside R_M.

13.4. MODEL OF ORBIT DEVELOPMENT

Consider a state when the condensation of grains has proceeded for some time and a large number of grains have been produced. Inside R_M collisions between them have damped their radial and axial oscillations so that they move in circular orbits and form a thin disc in the equatorial plane. Newly condensed grains moving in orbits with non-zero inclination will pass this plane twice for every orbital revolution. Sooner or later such a new grain will collide with a disc grain knocking the latter out of the disc. The two grains will continue to oscillate about the plane of the disc, but will collide again with other disc grains. After some time the oscillations are damped out and all grains will be incorporated in the disc.

In the model we are going to develop we assume that inside R_M disturbances caused by the arrival of a newly formed grain are so small and so rapidly damped out that every new grain essentially interacts with a thin disc of grains that condensed earlier.

Outside R_M the collisions between the grains lead to accretion, first to larger aggregates or embryos and eventually to satellites. If this process is rapid enough, the mean free path between collisions may continue to be so long that the grains do not settle into the equatorial plane before new grains arrive. Hence collisions may take place also outside the equatorial plane. This may lead to a formation of jet streams. Although the theory of formation of jet streams is not yet worked out so well that it is possible to specify in detail the conditions for this, the general discussion in Sections 3.6 to 3.7 makes it reasonable that when condensation takes place outside the limit R_M a series of jet streams will be formed.

At the same time collisions will often result in production of extremely small grains by fragmentation or vapor condensation. As these small grains collide mutually they may form a thin disc (possibly with a small total mass) even outside the R_M limit, and coexisting with the jet streams.

Hence our model of orbit development should deal with two regions: one for $r < R_M$, in which the accretion leads to the formation of a thin disc, and the other for $r > R_M$, where it leads to the formation of jet streams.

13.5. ACCRETION INSIDE R_M

As discussed in Section 12.5, a grain generated at (r_0, λ_0, ϕ_0) in a coordinate system with the equatorial plane as the reference plane and the origin at the center of the central body will intersect the equatorial disc at $(\frac{2}{3}r_0, 0, \phi_0 + \pi/2)$. We center our attention on a condensation so close to the equatorial plane that we can put $\cos \lambda_0 \approx 1$ (but we exclude a very thin region close to the plane because this is a singularity. See Section 12.3). In this case the angular momentum of a new grain with reference to the axis of the coordinate system is the same as that of the disc grains at $r = \frac{2}{3}r_0$. Hence the component of its velocity at its collision with a disc grain equals the velocity of the disc grain so that seen from the disc grain, the velocity of the new grain lies in the meridional plane. Its component parallel to the axis is $(\frac{2}{3}(\mu/r_0))^{1/2} \sin \lambda_0$ and the component in the equatorial plane is $(\frac{1}{12}(\mu/r_0))^{1/2}$.

Let us first discuss the case when the collision between the new grain and the disc grain is almost perfectly inelastic by which we mean that the relative velocity between two grains after collision is small but not zero. Such a collision does not change the angular momentum of either grain, but only their velocity components in the meridional plane. After the collision, the grains will return to the equatorial plane at the point $(\frac{2}{3}r_0, 0, \phi_0 + 3\pi/2)$, where they may collide again with other disc grains. In this way new disc grains will be set in motion, but they will all reach the equatorial plane again at $(\frac{2}{3}r_0, 0, \phi_0 + \pi/2)$. A repetition of this process will result in more and more disc grains being set in motion with decreasing amplitude, so that the perturbation caused by the new grain is damped out and the grain is incorporated into the disc.

It is important to observe that this whole process *affects only the disc* grains at $r = \frac{2}{3}r_0$. The rest of the disc remains entirely unaffected (see Figure 12.2). This means that the *disc will be a kind of kinematic image of the condensing plasma* diminished in the proportion 2:3.

We shall now discuss the limitations of our idealized model.

(a) If the collision between the grains is only partially inelastic, part of the momentum contained in velocity components in the meridional plane may cause a change in the angular momentum. This will also cause a 'diffusion' of the perturbation to grains closer or more distant than $\frac{2}{3}r_0$. In a realistic case this diffusion may not be very important.

(b) If $\cos \lambda_0 < 1$, disc grains at $\frac{2}{3}r_0$ will be hit by new grains with smaller angular momenta. This will cause the grains to slowly spiral inward as they orbit around the central body.

(c) The idealized case is applicable if the disc is opaque so that the new grain collides with a disc grain at its first passage. If the disc is not opaque so that the grain is not likely to collide until after many transits, we must introduce the restriction that the collision should take place before a considerable change in the orbit of the new grain has taken place. Such a change may be due to precession, but it may also be produced by collisions outside the equatorial plane.

(d) The collision may also result in accretion or in fragmentation. In the latter case, all the fragments will move in orbits which bring them back at the point of fragmentation where they may collide again. Thus the fragments will in course of time be incorporated in the disc by the mechanism discussed above. The same is true of accretion. The entire process may be visualized as damping of oscillations around a circular orbit at $r = \frac{2}{3}r_0$ (Section 2.3).

13.6. STRUCTURE OF THE SATURNIAN RINGS

We shall apply our models of condensation and orbit evolution to the Saturian ring system. This consists of three rings: the outermost is called the A-ring and is separated by a dark region called *Cassini's division* from the B-ring which is the brightest of the rings. Inside the B-ring is the very faint C-ring also known as the crape (or crepe) ring on account of its darkness.

The photometric curve given by Dollfus (Figure 13.2) shows that near the outer edge of the A-ring there is a series of light maxima and minima. The most pronounced minimum is often referred to as *Encke's division*. A double minimum exists near the inner edge of the B-ring. In the middle of the B-ring two minima are visible. The rings lie in the equatorial plane of the planet and consist of numerous small particles that orbit around the planet with the orbital period increasing outward in accordance with Kepler's law. The thickness of the rings is about 2 km (Cook *et al.*, 1973).

13.6.1. *The resonance theory of the ring structure*

There is an old belief that the structure of the ring system is produced by resonance effects with the inner satellites. Different investigators have claimed that Cassini's division is due to a resonance with Mimas resulting in removal of particles from the dark region and that this takes place because their period is exactly $\frac{1}{2}$ of the period of Mimas. The resonance corresponding to $\frac{1}{3}$ of the period of Enceladus also falls close to Cassini's division. In a similar way the sharp change in intensity between the B-ring and the C-ring should be connected with the $\frac{1}{3}$ resonance with the period of Mimas. A list of claimed resonances has been given by Alexander (1953, 1962).

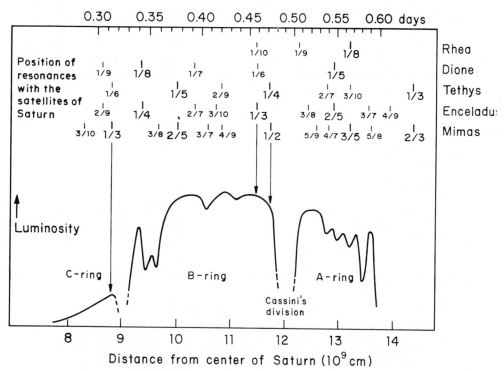

Fig. 13.2.   Photometric curve of the Saturnian rings (according to Dollfus, 1961). The abscissa gives the distance from the center of Saturn in cm calculated from the abscissa in seconds of arc given by Dollfus for Saturn placed at 10 AU from the observer.
   The top scale gives the orbital period of the particles. The periods, which are integral fractions of the periods of the inner Saturnian satellites, are marked in the upper part of the diagram. According to the resonance theory, the density minima in the ring system should be produced by resonance with these satellites. The lack of correlation between low-integer resonances and structural features show that this is not the case (see however discussion in Cook et al., 1973).

Figure 13.2 shows a plot of all resonances with denominators $\leqslant 10$. The resonances with denominators $\leqslant 5$ are marked with thick lines. A number of resonance points of Mimas and Tethys are similar because the period of Mimas is half the period of Tethys. The same is the case for the pair Enceladus-Dione. It should be remembered that the periods of Mimas, Enceladus, Tethys and Dione are approximately proportional to 2:3:4:6; see Table 4.2.

As was pointed out already in Section 4.7, a comparison between the calculated resonance points and the observed pattern of the ring system does not show any obvious connection. The $\frac{1}{2}$ resonance of Mimas falls definitely inside of Cassini's division. Half the period of Mimas differs by 1.2% from the period of the outermost particles of the B-ring and by 4% from that of the innermost particles of the A-ring. The difference between the $\frac{1}{3}$ resonance with Enceladus and Cassini's division is still larger. Nor is there any obvious connection between other markings – bright or dark – and the resonance points.

In this respect the Saturnian rings constitute a striking difference with the asteroid belt, where there are very pronounced gaps corresponding to integral fractions of Jupiter's period (the Kirkwood gaps). For example, near the resonances $\frac{1}{3}$ and $\frac{2}{3}$ of Jupiter's period there is a complete absence of observed asteroids (see Figure 13.3).

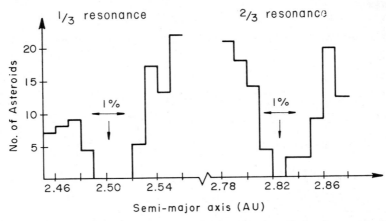

Fig. 13.3. The number of asteroids as a function of the semi-major axis showing gaps in the asteroid belt. The arrows mark the places where the period of an asteroid is $\frac{1}{3}$ or $\frac{2}{3}$ of the period of Jupiter. The reason why there are resonance gaps in the asteroid belt but not in the Saturnian rings is that the mass ratio Jupiter–Sun is 10000 times larger than the mass ratio Mimas–Saturn.

As Cassini's division has been attributed to resonances which are displaced by a few percent, it is of interest to see whether a similar asymmetry exists for the asteroids. We see from Figure 13.3 that with reference to the resonance points the asymmetry of the gaps – if any – is only a fraction of one percent. The half-width is about 1.5%. Hence with the same relative breadth any resonance gaps corresponding to $\frac{1}{2}$ Mimas' and $\frac{1}{3}$ Enceladus' periods would be altogether within the B-ring, and outside Cassini's division. Further there is not the slightest trace of a resonance gap in the B-ring corresponding to either $\frac{2}{3}$ of Mimas' period or to $\frac{1}{3}$ of Enceladus' period. Therefore, from an observational point of view there is no real similarity between the asteroid gaps on one side and the low density regions of the Saturnian rings on the other. In fact, Figure 13.2 indicates that if anything is characteristic for Cassini's division it is that not a single resonance point falls in that region.

The reason why the low density regions of the Saturnian rings show no similarity to the Kirkwood gaps is likely to be the much smaller magnitude of the perturbing force. The masses of Mimas and Enceladus are of the order of 10^{-7} of the Saturian mass, whereas the mass of Jupiter is about 10^{-3} of the solar mass. As by definition the ratio of the distances from the perturbed bodies to the central body and to the perturbing body is the same in the two cases, the relative magnitude of the perturbing force is about 10^{-4} time less in the Saturnian rings than in the asteroidal belt.

Hence it seems legitimate to doubt whether Mimas and Enceladus are large enough to produce any phenomenon similar to the asteroid gaps. In fact the sharpness of a resonance effect is generally inversely proportional to the perturbing force. Hence we should expect that the relative breadth of a Kirkwood gap in the Saturnian rings to be 10^{-4} of the breadth in the asteroid population. As the latter is of the order of 1%, the dark marking in the Saturnian rings should have a relative breadth of 10^{-4} %, which is well below the limit of observability. These objections to the resonance theory also apply to its recent development by Franklin and Colombo (1970).

Further, it should be noted that the resonance theories have so far not been able to give an acceptable explanation as to why the B-ring is brighter than the A-ring. Concerning the sharp limit between the B-ring and the C-ring it has been claimed that the $\frac{1}{3}$ resonance of Mimas should be responsible for the very large positive derivative of the light curve. However, the $\frac{1}{3}$ resonance of Enceladus is situated somewhat inside the Cassini's division in a region where the derivative of the light curve is slightly negative. There is no obvious reason why the same type of resonance with the different satellites should produce such different results.

13.6.2. Can the structure of the Saturnian rings be of hetegonic origin?

Our conclusion is that the resonance theory has not succeeded in explaining the main characteristics of the Saturnian rings. Furthermore, it is difficult to imagine that any other force acting at the present time could produce the observed structure. We therefore ask ourselves whether the structure of the rings could have been produced when the rings were formed, and preserved for four or five billion years to the present time.

Such a view implies, however, that at least some parts of the solar system have an enormous degree of dynamical stability. Many scientists object to this idea. Nevertheless, we have already found that except in the cases when tidal braking has been important (Earth, Neptune, and perhaps Mercury) planetary spins have probably not changed very much since hetegonic times (see Sections 5.7 and 5.8). Furthermore as found in Section 4 the orbit-orbit resonances must also have been produced already when the bodies were formed. The general conclusions in Sections 6.1–6.3 indicate that with a few exceptions there has been very little dynamic change in the solar system since its formation. Hence there should be no *a priori* objections to the view that the present structure of the Saturnian rings was produced when the rings were formed, and that even the details of it may have originated at the formation.

13.6.3. Hetegonic theory of the Saturnian rings

Several independent arguments, experimental as well as theoretical, suggest that the formative era of the solar system must have extended over a time period of the order 3×10^8 yr. In the case of the Saturnian region the matter which at present constitutes the satellites and the rings would consequently have been introduced around the planet during an extended period of time. This emplacement can be envisaged as a continuous injection of gas – or an injection of a series of jets – going on during a period (perhaps as long as 3×10^8 yr; see Section 9). The gas became ionized at the critical velocity, was brought into a state of partial corotation, and condensed to grains, but these processes are relatively very rapid. This means that at any given moment only a very small fraction of the total injected gas was in the plasma state. Hence the process was producing grains more or less continuously during a very long time.

In Section 13.5 we have discussed some basic processes in the formation of the rings under the assumption that they formed from a partially corotating plasma. The result was that the grains which at present are orbiting at a central distance r originally condensed out of a plasma at a distance $\frac{3}{2}r$. Therefore, if we want to find the place of origin of present grains, we should enlarge the present orbits by a factor $\frac{3}{2}$. The result is shown in Figure 13.4. We find that Cassini's division is projected into the region where Mimas moves, and the border between the B-ring and the C-ring coincides with the outer edge of the A-ring. Remembering that the grains, condensed at a certain distance, interact only with disc grains at $\frac{2}{3}$ of this distance we may interpret the figure in the following way.

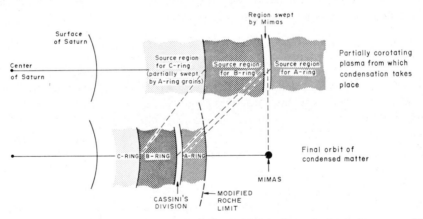

Fig. 13.4.   Condensation of grains from a partially corotating plasma in the environment of Saturn. The condensation is assumed to take place essentially from the neighborhood of the equatorial plane (but only a negligible part in the plane itself). The figure refers to a state when part of the plasma has already condensed so that Mimas (or its parent jet stream) and the rings are existing although with only a small part of their present masses. The upper part of the figure refers to the plasma which has not yet condensed.

The plasma near the orbit of Mimas condenses on this satellite (or on the jet stream in which it accretes), leaving the 'Region swept by Mimas' void of plasma. Similarly, the plasma in the region of the already existing A-ring (and B-ring) condenses directly on the grains of the ring. When the grains produced by the condensation fall down to 2/3 of their original central distances, the state depicted in the lower part of the figure is produced. Cassini's division is derived from the region swept by Mimas. The C-ring has a reduced intensity because part of the plasma has condensed on the already existing grains of the A-ring.

13.6.4. *Theory of Cassini's division*

In the region where Mimas moves, a large part of the revolving plasma will condense on Mimas (or perhaps rather on the component grains of the jet stream within which Mimas is forming). Hence in this region there will be little plasma left to form grains which later will be found at $\frac{2}{3}$ of the central distance to Mimas. In other words, we may interpret Cassini's division as what we may call the '*hetegonic shadow*' of Mimas.

The plasma outside the orbit of Mimas condenses to grains and when they have fallen to $\frac{2}{3}$ of their initial distance they form the present A-ring. However, before they reach this position they have to pass Mimas' jet stream and part of them will be captured by it. The grains condensing from plasma inside Mimas' orbit fall down to $\frac{2}{3}$ of their initial position without passing Mimas' orbit and form the B-ring. This may explain why the B-ring is brighter than the A-ring.

13.6.5. *Theory of the limit between the B and C rings*

If the limit between the B and C rings is magnified by a factor $\frac{3}{2}$, it coincides with the outer edge of the A-ring. The reason for this is that plasma injected in the region

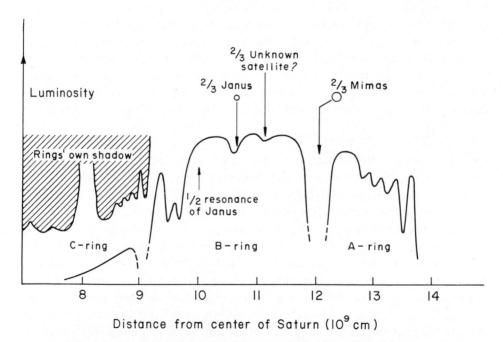

Fig. 13.5. Cosmogonic effects in the ring system. Dollfus' photometric profile compared with Mimas orbital distance reduced by a factor $\frac{2}{3}$ (or 0.65). Cassini's division may be the 'cosmogonic shadow' of Mimas. In the left corner, the photometric profile is turned upside down and reduced by the factor $\frac{2}{3}$('rings own shadow'). The rapid drop in intensity between the B-ring and C-ring coincides with the beginning of this shadow.

Effect of the recently discovered Janus: if the resonance theory were correct we should expect a gap at 1.01×10^{10} cm (marked '$\frac{1}{2}$resonance of Janus'), but this cannot be traced. On the other hand there is a light minimum close to $\frac{2}{3}$ the orbital radius of Janus as expected from the cosmogonic theory.

inside R_M will rapidly be gathered by the grains already existing there as the growing A-ring, thus depleting the plasma which gives rise to C-ring grains. In the same way as Mimas produces Cassini's division as its hetegonic shadow at $\frac{2}{3}$ of its central distance, the outer edge of the A-ring is imaged at $\frac{2}{3}$ of its distance.

The qualitative picture in Figure 13.4 can be refined and compared directly with observation: see Figure 13.5. In the upper left corner the ordinate of the light curve has been reversed and the abscissa reduced by a factor of $\frac{2}{3}$. The depletion of plasma causing the hetegonic shadow should depend on the total surface area of the matter, which is proportional to the luminosity. The figure shows that the drop in intensity from the B-ring to the C-ring occurs almost exactly where we expect the hetegonic shadow to appear. In fact Dollfus' value for the outer limit of the A-ring is 13.74×10^9 cm and for the border between the B-ring and C-ring 9.16×10^9 cm. The ratio between these values happens to be exactly $1.50 = \frac{3}{2}$.

13.6.6. *Discussion*

Considering Cassini's division as the hetegonic shadow of Mimas, we find that the fall-down ratio must be slightly higher than 1.5, namely about 1.55 ($= 1/0.65$). It is doubtful whether we should attribute very much significance to such a slight deviation (only 3%). If we look for a refinement of the theory the deviation from the value $\frac{3}{2}$ of the simple theory can be explained in two ways. It may be an indirect effect of the production of a shadow (Alfvén, 1954) or it may be due to a condensation at so large distance from the equatorial plane that $\cos \lambda_0 < 1$. In contrast, in the resonance theory of Cassini's division it is difficult to see why there should be any deviation from the theoretical resonance, which as mentioned is clearly outside Cassini's division.

13.6.7. *The discovery of Janus*

Recently a new satellite of Saturn has been discovered (Dollfus, 1967). It has been called Janus and moves in an orbit with small eccentricity at a distance of 1.60×10^{10} cm, i.e. 2.65 times the equatorial radius of Saturn. Its mass is probably about $\frac{1}{10}$ of Mimas. This discovery makes it possible to check our earlier results.

If the resonance theory of Cassini's division is correct we should expect Janus to produce a resonance gap at $(0.5)^{2/3} = 0.63$ of its central distance, i.e., at 1.01×10^{10} cm. In this region of the luminosity curve there is not the slightest trace of any gap. If on the other hand the new satellite produces a hetegonic shadow, this should be situated at $\frac{2}{3} \times 1.60 \times 10^{10} = 1.07 \times 10^{10}$ cm or at 0.65 if affected in the same way as Cassini's division. We find a clear minimum in the luminosity curve very close to the expected point. In fact the minimum in the figure is situated at 1.06×10^{10} cm, which with the orbital radius of Janus gives a fall down ratio of 1.51, a value slightly higher than the value of the simple theory and somewhat smaller than the value for Mimas-Cassini's division.

The minimum at 1.11×10^{10} cm may indicate the existence of a new satellite (Alfvén, 1968b). There is no observational confirmation of the existence of this satellite yet.

The series of minima in the outer part of the A-ring and the double minima close to the inner edge of the B-ring may be ripples produced at the accretion and associated with the edges. Whether this assumption is correct can be decided only when a detailed theory of the formation of the ring system has been worked out.

The lack of shadow around 8.00×10^9 cm due to Cassini's division may give rise to a narrow luminous ring. Furthermore, the drop in intensity at 8.94×10^9 should produce an increase in intensity at $\frac{2}{3} \times 8.94 \times 10^9 = 5.96 \times 10^9$. Recent photometric observations by Guérin seem to confirm these predictions (Guérin, 1972).

13.7. ACCRETION OUTSIDE R_M

A model of the accretion outside the modified Roche limit must necessarily include a number of hypotheses because we do not know under what conditions a collision results in fragmentation or in accretion, and to what degree it is inelastic. Also the theory of jet streams is not very well developed, and in fact, cannot be before the collision response is quantitatively clarified.

As was shown in Section 13.5 these uncertainties were not very serious for a theory of accretion inside R_M, mainly because the condensed grains were almost immediately reaching their final location. It is more serious outside R_M because the eventual formation of planets and satellites involves a long chain of processes. Our approach must necessarily be partly phenomenological, and essentially of a provisional character.

According to the model in Section 13.4 outside R_M most of the condensed grains will be captured into jet streams. This does not exclude that there may be a thin disc in the equatorial plane consisting of very small grains resulting from fragmentation and impact vapor condensation, but without substantial effect on the formation of jet streams.

Some aspects of the formation of planets and satellites may be clarified by studies of the asteroidal belt. As has been pointed out in Section 7.8 this may be considered as an intermediate 'planetesimal' state in planet formation, or in any case related to this state. The probable reason why matter has not gathered into one single body in this region is likely to be the extremely low distributed-density of matter condensed there. Indeed the distributed density is about 10^{-5} of the density in the regions of the giant planets and the terrestrial planets. This may mean that the accretion takes 10^5 times longer for completion in the asteroidal belt. Hence even if the time for complete accretion of the terrestrial planets were as short as 10^7 yr, planet formation in the asteroidal belt would require longer time than the present age of the solar system. There is also the possibility that due to the low density accretion never will proceed to this state.

Hence the study of the asteroidal region is very important because it will clarify essential features of the planetesimal state. However we need not necessarily assume that it is a close analog of an early state in, for example, the terrestrial region before the formation of the Earth. Not only the space density but also the structure and composition of the grains and the progression of the collisional processes may be different.

13.8. FORMATION OF THE ASTEROID BELT

In this paragraph we shall study whether the essential features of the asteroid belt can be interpreted as a result of a condensation from a partially corotating plasma.

There are certain similarities between the asteroidal belt and the Saturnian rings, but the structure differs in the following respects:

(1) The asteroidal belt is very far outside R_M. No tidal disruption prevents the build-up of larger bodies.

(2) The space density is very low so that collisions are rare. The time scale for development is very large.

(3) Hence whereas the reason why the grains forming the Saturn rings have not agglomerated to larger bodies is that they are moving inside the Roche limit, in the asteroid belt the bodies have not agglomerated to planets because the density is too low.

(4) Jupiter produces a large number of resonance gaps (Kirkwood gaps) of which there are no analogs in the Saturnian rings for reasons discussed in 13.6.1. The most conspicuous gap is due to the $\frac{1}{2}$ resonance.

The outer border of the main groups of asteroids is situated at a solar distance of $\frac{2}{3}$ the distance of Jupiter (Figure 13.6). On condensing, the grains move in ellipses with $e = \frac{1}{3}$; hence those grains which form outside the orbit of Jupiter repeatedly cross Jupiter's orbit and there is a high probability that either they are captured by Jupiter (or the jet stream in which Jupiter is forming) or their orbits are perturbed so that they

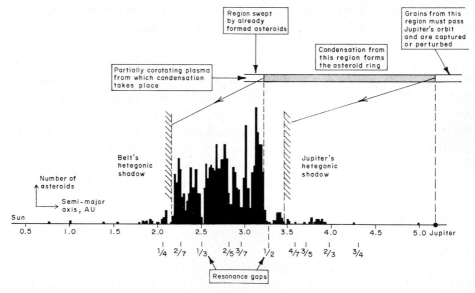

Fig. 13.6. The number of asteroids plotted as a function of their semi-major axis a. In the asteroid belt, Jupiter produces very pronounced resonance effects so that there are almost no asteroids with orbital periods equal to $\frac{1}{2}$, $\frac{1}{3}$, $\frac{2}{3}$, etc. of Jupiter's period.

The state of a partially corotating plasma is depicted in the upper part of the figure. Grains condensing outside Jupiter's orbit are captured or perturbed by Jupiter. The result is that there are very few asteroids with $a > \frac{2}{3} a_{2\downarrow}$ (called 'Jupiter's hetegonic shadow'). Plasma condensing on already existing asteroids produces a low limit cut-off ('own hetegonic shadow') of the asteroid distribution at $\frac{2}{3}$ of the upper limit.

will not ultimately be found at $\frac{2}{3}$ of their place of origin. For this reason there are very few asteroids outside $\frac{2}{3}$ of Jupiter's orbit. This means that there is no real correspondence to the A-ring of the Saturnian system. Mimas with a mass of only 10^{-7} of the Saturnian mass, has reduced the intensity of the A-ring (compared to the B-ring), but only to a limited extent.

The inner limit to the asteroid ring is given by its 'own hetegonic shadow', just as is the inner limit of the B-ring around Saturn. The very few asteroids below $a = 2.1$ AU should be an analog to the very faint C-ring.

13.9. CONCLUSIONS ABOUT PARTIAL COROTATION

We have found in Section 12.5 that condensation from a partially corotating plasma should produce bodies at a final distance of $\frac{2}{3}$ of the point of condensation. We have looked for observational confirmation of a fall-down ratio of $\frac{2}{3}$, and found several examples of this. In the Saturnian ring system the ratio $\frac{2}{3}$ is found at three (and possibly four) different places, and in the asteroidal belt in two places. We can regard this as a confirmation that partial corotation plays an important role in the condensation process.

13.10. SATELLITE AND PLANET FORMATION

The development of an asteroid-like assembly of bodies will lead to a general decrease in inclinations and eccentricities, and eventually an accretion to larger bodies. Since the space density in the present asteroidal region is very small, the time scale of development in this region is very long. In regions where the density of the primeval condensing plasma was much larger than in the asteroidal region, a more rapid development took place, leading to the formation of groups of densely populated jet streams of grains inside which satellites or planets formed.

13.10.1. *The groups*

In Section 2 we have found that the regular bodies in the solar system form several groups consisting of a number of similar bodies with regular spacings (Tables 2.1–2.2B and 2.4 of Part I). Examples of such groups are: (a) the four Galilean satellites of Jupiter, (b) the five satellites of Uranus, and (c) the giant planets possibly including Pluto.

In the Saturnian satellite system, all the inner satellites out to Rhea have orbits with spacings roughly proportional to their distances from the planet, and their sizes increase in a fairly regular way with the distance. The innermost of these satellites is connected with the outer edge of the ring system, forming an unbroken sequence of secondary bodies from the inner edge of the ring system out to Rhea. This sequence may be considered as forming a group. The distance between Rhea and the next outer satellite Titan is very large, and the disparity of masses between these two satellites is very great. It is thus possible that Titan forms an outer group with the two other outer satellites Hyperion and Iapetus, but this group is not at all as regular as the inner group. Another irregular group consists of the Jovian satellites VI, VII, and X.

We still have a few prograde satellites: one is the fifth satellite of Jupiter, Amalthea. Both its large distance to the Galilean satellites and much smaller mass makes it impossible to count this as a member of the Galilean group. If we want to classify all the satellites, Amalthea must be considered as the only known member of a separate group. Further, the highly eccentric Neptunian satellite Nereid may be a remnant of an early group of regular satellites destroyed by the capture of Triton (McCord, 1966; Alfvén and Arrhenius, 1972). Finally the very small Martian satellites may be counted as a group of regular satellites.

As grains initially move in orbits with $e = \frac{1}{3}$ after the condensation, the ratio between their apocentric and pericentric distances is 2. Hence as long as the relative spacings between bodies are smaller than 2, we can be sure that the grains will be captured – sooner or later – by one of the bodies. Inside a group the relative spacings normally do not exceed 2 (in Uranus-Saturn, it is 2.01). This means that we may have had a production of grains in the entire region of space covered by the present groups of bodies, and all this mass should now be found in the bodies.

However, the spacings *between* the groups – as we have defined them – is always greater than a factor 2. For Jupiter-Mars the ratio between their orbital radii is 3.42; for Titan-Rhea it is 2.32; for Io-Amalthea, 2.33; and for Jupiter VI-Callisto, 6.09. This means that there are regions in the gap where grains, if formed, cannot be captured by any body. From this we conclude that *there must have been regions between the groups where no appreciable condensation took place*. In other words, the different plasma clouds from which the groups have been formed were *distinctly separated by regions where the density was very low*.

One such region is found between Jupiter and Mars. From the study of the asteroids we know that the density in this region was several orders of magnitude lower than the density within the regions of the giant planets and the terrestrial planets. Similar intermediate regions where the plasma must have had an extremely low density are found between Titan and Rhea, Io and Amalthea, and Jupiter VI and Callisto. It is possible that in these regions a number of very small bodies – similar to the asteroids – may be found. The same is possible in the case between the group of Uranian satellites and the planet itself, and also outside the orbit of Oberon.

In theories of the Laplacian type it is postulated that the secondary bodies around a central body derive from a homogenous disc. We have found that the distribution of mass in both the planetary system and the satellite systems is very far from such uniformity. Mass is accumulated at certain distances where groups of bodies are formed, but between the groups there are spacings which are practically devoid of matter. The formation of groups of bodies is shown schematically in Figure 13.7.

The explanation of the low-density region between Mars and Jupiter according to Laplacian theories is that because of the large mass of Jupiter, the condensation should have been disturbed inside its orbit. This is very unlikely. The solar distance of the asteroids ($a = 2.1$–3.5 AU) is about half of the distance to Jupiter ($a = 5.2$ AU). As Jupiter's mass is 10^{-3} of the Sun's mass, the Jovian gravitation cannot be more than 0.1% of the solar gravitation in the asteroidal region. Certainly, as this is a perturba-

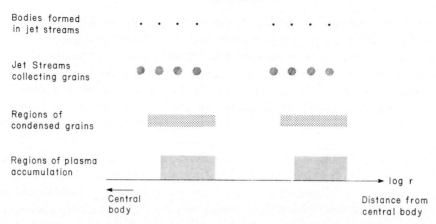

Fig. 13.7.   Diagram of the formation of a group of bodies. Injected gas is stopped and ionized at different distances from the central body. The two regions in the figure may receive plasma simultaneously or during different epoques. Condensation of the plasma is rapid during the total injection period. The grains are collected in jet streams which increase their mass during the whole injection period. Grains are stored in the jet streams (often for a long period of time) until they are finally accreted by a growing embryo. In each region of plasma injection, 3–5 bodies are formed.

tion of the Coulomb field, it produces a precession of the perihelion and the nodes of bodies orbiting in this region. Hence it may contribute to the disruption of the jet streams. However, the disruptive effect of Jupiter on jet streams active during the formation of Saturn would have been about equally large, and the same effect produced by Saturn during the accretion of Jupiter would also have been of the same order of magnitude. Hence a theory which attributes the absence of large bodies in the asteroidal region to the large size of Jupiter will easily run the risk of explaining away either Saturn or Jupiter.

13.11. THE ACCRETION OF VOLATILE SUBSTANCES

The mechanism of accretion we have considered is based on the behavior of solid grains. In the first place it is a theory of those celestial bodies which consist mainly oɪ nonvolatile elements. Such bodies are the terrestrial planets, including the Moon, and the asteroids. We know very little about the chemical composition of the satellites except that it is highly variable; existing data will be discussed in Section 14. At least the smallest of them consists entirely of materials condensable at their solar distances since their masses are not large enough to keep an atmosphere. Some of the better known satellites of Jupiter have a mean density indicating that rocky material forms a substantial fraction of their mass, others must consist largely of icy or liquid components (Lewis, 1971). The Saturnian satellite Titan with a size almost twice that of our own, is capable of retaining a thin atmosphere.

   In the case of giant planets, which mainly consist of volatile substances, our planetesimal accretionary mechanism needs supplementation. This is a general complication with all planetesimal theories, and has attracted much attention already. Öpik

(1962) has tried to solve the problem through the assumption that the accretion of Jupiter (and the other giants) took place at such an extremely low temperature as to make even hydrogen solid. This means that Jupiter should have accreted from hydrogen snow-flakes. The temperature which according to Öpik is necessary for this process is about 4 K which seems to be unreasonably low. In our mechanism, in which the gas density is lower than assumed by Öpik, the required temperature would be even lower. Hence it seems necessary to envisage the direct accretion of gas from an interplanetary medium.

In order to keep an atmosphere a body must possess a certain minimum mass. As Mars has an atmosphere but the Moon has none, we can conclude that under the conditions prevailing in this region of the inner planets the critical size should be of the order 10^{26} g. When through the accretion of solid grains an embryo has reached this mass, it is able to attract gas efficiently from the interplanetary medium and to retain it as an atmosphere.

As none of the asteroids has reached the critical size, we have no hope of studying this process by observations in the asteroidal region. Hence the accretion of gas by a growing embryo is necessarily a more hypothetical process than those for which we can find analogies in present-day phenomena. The very low density of gas in interplanetary space today together with the action of the solar wind prevents planets from gravitational gas accretion at the present time.

This process should be considered in relation to the jet stream model discussed in Sections 3.7 and 9 (Alfvén and Arrhenius, 1970a, b). The accretion of gas from an interplanetary medium may also occur in two steps. A jet stream could also have the property of drawing in gas from its environment so that it has an 'atmosphere'. This means that the gas density inside a jet stream may be much larger than in the inner-stream medium. When an embryo inside a jet stream has reached the critical state, it would then be accreting gas from a region with relatively high gas pressure.

The presence of gas as a dissipative medium in some jet streams is suggested by the state of preservation of particles in carbonaceous chondrites. In these the characteristic products of collision melting and vaporization (chondrites) are a minor component or are entirely absent. Crystals magnetized before accretion have escaped collisional heating to the Curie temperature (Brecher, 1972a, b). The highly embrittled skin of isotropically irradiated crystals in gas rich meteorites have been protected against destruction in the process of accretion of the parent body embryos (Wilkening *et al.*, 1971). Hence it is necessary to assume that the lowering of relative velocities, required for accretion, was substantially aided by viscous energy losses other than inelastic collisions.

It is believed that at least Jupiter consists mainly of hydrogen and helium which necessarily must have been accreted by direct accretion of gas from space or from the atmosphere of a jet stream. This means that the orbital characteristics should be determined by a gas accretion process and not by a solid grain accretion. The accretion mechanism that we have discussed would have to be substantially modified accordingly in the case of the giant planets, or at least for Jupiter. A detailed analysis of this problem is very important but must be left to future investigations. (In case it turns out that the hydrogen-helium model of Jupiter is not correct, such an investigation loses much of its motivation.)

Acknowledgements

The present work was carried out under grant NGR 05-009-110 from NASA's Planetology Program, and NGL 05-009-002 from NASA's Lunar and Planetary Programs Division. The generous support from these sources is gratefully acknowledged.

We wish to thank Mr Bibhas R. De for crucial help and constructive discussion. We are also indebted to Drs Nicolai Herlofson, Bo Lehnert, Carl-Gunne Fälthammar, Per Carlquist and Leon Mestel for fruitful discussions.

The competent and devoted editorial help from Mrs Marjorie Sinkankas and Mrs Dawn Rawls is also greatly appreciated, as well as assistance from the Royal Institute of Technology in Stockholm.

References

Alexander, A. F. O'd.: 1953, *J. Brit. Astron. Assoc.* **64**, 26.
Alexander, A. F. O'd.: 1962, *The Planet Saturn: A History of Observation, Theory and Discovery*, Macmillan, New York, 474 pp.
Alfvén, H.: 1942–45, *Stockholm Obs. Ann.* **14**, 5 and 9.
Alfvén, H.: 1943, *Arkiv Math. Astron. Fysik* **29A**.
Alfvén, H.: 1961, *Astrophys. J.* **133**, 1049–1054.
Alfvén, H.: 1967, *Icarus* **7**, 387–393.
Alfvén, H.: 1968a, *Ann. Geophys.* **24**, 1.
Alfvén, H.: 1968b, *Icarus* **8**, 75–81.
Alfvén, H.: 1971, *Science* **173**, 522–525.
Alfvén, H. and Arrhenius, G.: 1970a, *Astrophys. Space Sci.* **8**, 338–421.
Alfvén, H. and Arrhenius, G.: 1970b, *Astrophys. Space Sci.* **9**, 3–33.
Alfvén, H. and Arrhenius, G.: 1972, *Moon* **5**, 210–230.
Alfvén, H. and Fälthammar, C.-G.: 1963, *Cosmic. Electrodyn., Fundamental Principles*, 2nd ed., Oxford Univ. Press.
Arrhenius, G.: 1971, in A. Elvius (ed.), 'From Plasma to Planet', *Nobel Symp.* **21**, Wiley, New York.
Arrhenius, G. and Alfvén, H.: 1971, *Earth Planetary Sci. Letters* **10**, 253–267.
Babic, M., Sandahl, S., and Torvén, S.: 1971, 'The Stability of a Strongly Ionized Positive Column in a Low Pressure Mercury Arc', *Proc. Xth Internat. Conf. on Phenomena in Ionized Gases*, Oxford. p. 120.
Baxter, D. and Thompson, W. B.: 1971, in T. Gehrels (ed.), 'Physical Studies of Minor Planets', NASA SP-267.
Baxter, D. and Thompson, W. B.: 1973, 'Elastic and Inelastic Scattering in Orbital Clustering', *Astrophys. J.* **183**, 323.
Berlage, H. P.: 1930, *Koninkl. Ned. Akad. Wetenschap., Amsterdam* **33**, 719.
Berlage, H. P.: 1932, *Koninkl. Ned. Akad. Wetenschap., Amsterdam* **35**, 553.
Berlage, H. P.: 1940, *Koninkl. Ned. Akad. Wetenschap., Amsterdam* **43**, 532, 557.
Berlage, H. P.: 1948, *Koninkl. Ned. Akad. Wetenschap., Amsterdam* **51**, 796.
Berlage, H. P.: 1948, *Koninkl. Ned. Akad. Wetenschap., Amsterdam* **51**, 965.
Birkeland, K.: 1908, *The Norwegian Polaris Expedition, 1902–1903*, Aschenhong and Co., Christiania, Norway.
Block, L. P.; 1955, *Tellus* **7**, 65–86.
Block, L. P.: 1956, *Tellus* **8**, 234–238.
Block, L. P.: 1967, *Planetary Space Sci.* **15**, 1479–1487.

Block, L. P.: 1972, *Cosmic Electrodyn.* **3**, 349–376.

Boström, R.: 1968, *Ann. Géophys.* **24**, 681

Brandt, J. C.: 1970, *Introduction to the Solar Wind*, W. H. Freeman and Co., San Francisco.

Bratenahl, A. and Yeates, G. M.: 1970, *Phys. Fluids* **13**, 2696–2709.

Brecher, A.: 1971, 'On the Primordial Condensation and Accretion Environment and the Remanent Magnetization of Meteorites,' in C. L. Hemmenway, A. F. Cook, and A. M. Mullman (eds.), NASA SP-319, p. 311. *The Evolutionary and Physical Problems of Meteoroids.*

Brecher, A.: 1972a, 'Memory of Early Magnetic Fields in Carbonaceous Chondrites,' in H. Reeves (ed.), *On the Origin of the Solar System*, Nice Centre Nationale de la Recherche Scientifique, Paris, p. 260.

Brecher, A.: 1972b, 'I: Vapor Condensation of Ni-Fe Phases and Related Problems'; 'II: The Paleomagnetic Record in Carbonaceous Chondrites', *Ph. D. Thesis*, University of California, San Diego.

Cameron, A. G. W.: 1962, *Icarus* **1**, 13–69.

Cameron, A. G. W.: 1963, *Icarus* **1**, 339–342.

Cameron, A. G. W.: 1973, Accumulation Processes in the Primitive Solar Nebula', *Icarus* **18**, 407.

Carlqvist, P.: 1969, *Solar Phys.* **7**, 377–392.

Chamberlin, T. C.: 1905, in *Carnegie Institution Yearbook No. 4*, 171–185.

Cloutier, P. A., Anderson, H. R., Park, R. J., Vondrack, R. R., Spizer, R. J., and Sandel, B. R.: 1970, *J. Geophys. Res.* **75**, 2595–2600.

Cloutier, P. A.: 1971, *Rev. Geophys.* **9**, 987–996.

Cook, A. E., Franklin, F. A., and Palluconi, F. D.: 1973, *Icarus* **18**, 317.

Danielsson, L. and Lindberg, L.: 1964, *Phys. Fluids* **7**, 1878–1879.

Danielsson, L. and Lindberg, L.: 1965, *Arkiv Fysik* **28**, 1.

De, B. and Arrhenius, G.: 1973, 'Inhomogeneous Plasmas and Primordial Condensation: Inferences from Present-Day Observations (in preparation).

Dessler, A. J.: 1968, *Ann. Geophys.* **24**, 333.

Dollfus, A.: 1961, in G. P. Kuiper and B. M. Middlehurst (eds.), *The Solar System*, **3**, p. 568, Univ. of Chicago Press.

Dollfus, A.: 1967, *Compt. Rend.* **264**, 882.

Franklin, F. A. and Colombo, G.: 1970, *Icarus* **12**, 338–347.

Gollnow, H.: 1962, *Publ. Astron. Soc. Pacific* **74**, 163–164.

Guérin, P.: 1972, Personal communication.

Giuli, R. T.: 1968a, *Icarus* **8**, 301–323.

Giuli, R. T.: 1968b, *Icarus* **9**, 186–190.

Haerendel, G. and Lüst, R.: 1970, in B. M. McCormac (ed.), *Particles and Fields in the Magnetosphere*, D. Reidel Publ. Co., Dordrecht, Holland, 213–228.

Hattori, T., Nakano, T., and Hayashi, C.: 1969, *Prog. Theor. Phys.* **42**, 781–798.

Hoyle, F.: 1946, *Monthly Notices Roy. Astron. Soc.* **106**, 406.

Hoyle, F.: 1960, *Quart. J. Roy. Astron. Soc.* **1**, 28–55.

Hoyle, F.: 1963, in R. Jastrow and A. G. W. Cameron (eds.), *Origin of the Solar System*, Academic Press, New York, pp. 63–71.

Kelley, M. C., Mozer, F. S., and Fahleson, U. V : 1971, *J. Geophys. Res.* **76**, 6054–6066.

Kuiper, G. P.: 1951, in J. A. Hynek, (ed.), *Astrophysics*, McGraw-Hill, New York, p. 404.

Kumar, S. S.: 1972, *Astrophys. Space Sci.* **16**, 52–54.

Lehnert, B. 1967, *Plasma Phys.* **9**, 301–337.

Lehnert, B.: 1970, *Cosmic Electrodyn.* **1**, 397–410.

Levin, B. J.: 1962, *New Sci.* **13**, 323–325.

Lewis, J. S.: 1971, *Icarus* **15**, 174–185.

Lindberg, L., Witalis, E., and Jacobson, C. T.: 1960, *Nature* **185**, 452–453.

Lindberg, L. and Jacobsen, C. T.: 1964, *Phys. Fluids. Suppl.* **S44**, 844.

Lindberg, L. and Kristoferson, L.: 1971, *Cosmic Electrodyn.* **2**, 305–308.

Lundquist, S.: 1948, *Arkiv. Math. Astron. Fysik* **35A**.

Lundquist, S.: 1951, *Phys. Rev.* **83**, 307–311.

Lüst, R. and Schlüter, A.: 1955, *Z. Astrophys.* **38**, 190–211.

Malmfors, K. G.: 1945, *Arkiv. Math. Astron. Fysik* **32A**, No. 8.

McCord, T. B.: 1966, *Astron. J.* **71**, 585–590.

McCrea, W. H.: 1960, *Proc. Roy. Soc. London* **256**, 245–266.

McCrea, W. H.: 1963, *Contr. Phys.* **4**, 278–290.

Meyer, C., Jr.: 1969, 'Sputter Condensation of Silicates', Ph.D. thesis, Scripps Institution of Oceanography, University of California, San Diego.

Meyer, C., Jr.: 1971, *Geochim. Cosmochim. Acta* **35**, 551–566.

Moulton, R. F.: 1905, *Carnegie Institution Yearbook No. 4*, pp. 186–190.

Mozer, F. S. and Fahleson, U. V.: 1970, *Planetary Space Sci.* **18**, 1563–1571.

Neugebauer, G., Becklin, E., and Hyland, A. R.: 1971, *Ann. Rev. Astron. Astrophys.* **9**, 67–102.

Öpik, E. J.: 1962, *Icarus* **1**, 200.

Persson, H.: 1963, *Phys. Fluids* **6**, 1756–1759.

Persson, H.: 1966, *Phys. Fluids* **9**, 1090–1098.

Podgorny, I. M. and Sagdeev, R. Z.: 1970, *Soviet Phys. Usp.* **98**, 445.

Safronov, V. S.: 1958, *Vopr. Kosmog.* **6**, 63.

Safronov, V. S.: 1960, *Vopr. Kosmog.* **7**, 59.

Schindler, K.: 1969, *Rev. Geophys.* **7**, 51–75.

Schmidt, O. Yu: 1944, *Dokl. Akad. Nauk S.S.S.R.* **45**, 245–249.

Schmidt, O. Yu.: 1946, 'A New Theory on the Origin of the Earth', *Priroda*, No. 7, 6–18.

Schmidt, O. Yu.: 1947, *Trans. All-Union Geog. Soc.*, No. 3, 265–274.

Schmidt, O. Yu.: 1959, *A Theory of the Origin of the Earth; Four Lectures* (tr. by G. H. Hanna), Lawrence and Wishart, London, 139 pp.

Stein, W.: 1972, 'Circumstellar Infrared Emission – Theoretical Overview', *Pub. Astron. Soc. Pac.* **84**, 627.

Stenflo, J. O.: 1969, *Solar Phys.* **8**, 115–118.

Ter Haar, D.: 1948, *Det kgl. Danske Videnskab. Selskab. Mat.-Fys. Meddelelser, København* **25**, No. 3.

Ter Haar, D.: 1949, *Astrophys. J.* **110**, 321–328.

Ter Haar, D.: 1967, *Ann. Rev. Astron. Astrophys.* **5**, 267–278.

Torvén, S.: 1972, Personal communication.

Von Weizsäcker, C. F.: 1944, *Z. Astrophys.* **22**, 319–355.

Wilkening, L., Lal, D., and Reid, A. M.: 1971, *Earth Planetary Sci. Letters* **10**, 334–340.

Zmuda, A. J., Heuring, F. T., and Martin, J. H.: 1967, *J. Geophys. Res.* **72**, 1115–1117.

PART IV

CHEMICAL DIFFERENTIATION. THE MATRIX OF THE
GROUPS OF BODIES

Abstract. In this fourth and last part of our analysis, the first section (14) contains a study of the chemical composition of the planets and satellites. A sharp distinction is made between the large quantity of speculations about the interiors of the bodies and the rather meager *facts known with a reasonable degree of certainty*. It is shown, however, that the latter are sufficient to *disprove the old concept of a Laplacian disc* of homogeneous chemical composition. There is a *systematic variation in the chemical composition of planets* (and probably also of satellites) so that heavy elements are more abundant in the outermost and in the innermost regions of the systems.

Section 15 contains *a study of meteorites*. These have earlier been interpreted in terms of 'exploded planets' and condensation processes in thermodynamic equilibrium. It is shown that such models are irreconcilable with the laws of physics and also with the meteoritic observations. These instead are found to *provide abundant information on the processes in jet streams* and on early fractionation and condensation. Further work along these lines supplemented with other solar system materials studies may lead to a detailed reconstruction of important events in the evolution of the solar system.

Section 16 demonstrates that the location of the different groups of secondary bodies is a result of a plasma phenomenon occurring at the critical velocity limit. These have recently been studied in detail in the laboratory but have not yet been fully applied to astrophysics. *Groups of bodies* in the planetary and the satellite systems related by the critical velocity should *have the same gravitational potential*. There are large chemical differences between groups of different gravitational potential. This is reconcilable with the chemical differentiation found in Section 14.

Finally, Section 17 deals with the *structure of the different groups* of bodies and shows that the mass distribution *is a function of the spin of the central body*. Summarizing the properties and distribution of bodies in the solar system against this background, it is shown that there is *no need for 'missing planets' or to explode hypothetical large bodies. Nor* is there any justification for involving *drastic ad hoc changes in the orbits of existing bodies.* The scheme is complete in the sense that in all places where groups of bodies are expected, such bodies are actually found. All of the existing bodies are accounted for (with the exception of the small Martian satellites!).

The general conclusion is that already with the empirical material now available *it is possible to suggest a series of basic processes leading to the present structure of planet and satellite systems* in an internally consistent way. With the expected flow of data from space research the evolution of the solar system may eventually be described with about the same confidence and accuracy as the geological evolution of the Earth.

14. Chemical Compositions in the Solar System

14.1. Survey

In the theories derived from the Laplacian concept of planet formation it is usually postulated that both the Sun and the planets – satellites are often not even mentioned – derive from a solar nebula with a chemical composition assumed to be uniform and characterized by 'cosmic abundances' of elements. The Sun and the giant planets are supposed to have condensed directly from the solar nebula and are assumed to have the same composition as this nebula. The solar photosphere has been proposed as the closest available approximation to this composition (Suess and Urey, 1956). The terrestrial planets should consist of the non-volatile ingredients of the nebula condensing in the inner regions of the solar system.

We have summarized earlier a number of objections to Laplacian type theories, including the difficulty that not even bodies as large as Jupiter can condense directly from a nebula (Kumar, 1972). The only reasonable alternative was found to be the planetesimal approach. To the objections discussed earlier we should add that the composition of the solar system appears far from uniform. It is well known that densities derived from mass and size indicate substantial differences in chemical composition among the different outer planets, among the terrestrial planets and among the small bodies in the solar system. The notable variability in surface composition of asteroids supports this conclusion.

The marked differences in composition among the various groups of meteorites and comets also point at fractionation processes operating on matter in the solar system, before or during the formative stage. The observational evidence for the chemically fractionated state of the solar system will be discussed in this and the next section.

14.2. SOURCES OF INFORMATION ABOUT CHEMICAL COMPOSITION

The empirical knowledge we have about the chemical composition of the solar system may be categorized with regard to level of certainty:

14.2.1. *Surface Layers and Atmospheres*

The surface layers of the Earth and Moon have been analyzed under well defined conditions. The data refer to less than 10^{-3} of the total mass of these bodies. Fragmentary information from landed instruments has been obtained from Venus.

The surface layers of the Sun have been analyzed by remote spectroscopy; however, the error limits are generally large in comparison to elemental fractionation factors characteristic of planetary processes (Urey, 1972; Worrall and Wilson, 1972; Aller, 1967). Independent conclusions derived from chemical analysis of corpuscular radiation from the Sun, but they are also uncertain due to fractionation at the source and in the targets and to limited sensitivity of the analysis.

Emission, absorption and polarization of electromagnetic radiation by planets, satellites and asteroids give some qualitative information about the structure and chemical composition of their surface layers (of the order of a fraction of a millimeter up to a few centimeters in depth) and of their atmospheres (Dollfus, 1971; Gehrels, 1972a; Chapman, 1972a; Newburn and Gulkis, 1973).

14.2.2. *Bulk Composition*

Our knowledge of the bulk composition of the planets and satellites is extremely uncertain. Parameters that yield information on this question are:
   (1) mass and radius, from which average density can be calculated;
   (2) moment of inertia, which allows conclusions about the density distribution;

(3) seismic wave propagation, electric conductivity, heat flow, magnetic properties and free oscillations have been studied in the case of Earth and Moon. The resulting data can be inverted to model internal structure and, indirectly, composition, but generally with a wide latitude of uncertainty.

In the case of Jupiter, the observation of a net energy flux from the interior also places a limit on the internal state.

Extrapolation of bulk composition from chemical surface properties of Earth, Moon and Sun has been attempted but is necessarily uncertain.

Several hundred meteorites have been analyzed. These are of particularly great interest since they are likely to approximate the bulk composition of the bodies from which they came, and of the parent streams of particles from which these bodies accumulated. A major limitation of this material as a record of the formative processes in the solar system comes from the fact that the regions of origin and the genetic interrelation of different types of meteorites are unknown.

14.3. CHEMICAL DIFFERENTIATION BEFORE AND AFTER THE ACCRETION OF BODIES IN THE SOLAR SYSTEM

The solar system is generally considered to have formed by emplacement of gas and possibly solid dust in some specific configuration in space and time. Regardless of the details assumed with regard to this configuration and with regard to the state of the components it appears highly unlikely that such an emplacement would proceed without accompanying chemical separation effects (see Sections 16.11 and 16.12). It is also improbable that the subsequent thermal evolution of each emplaced portion of matter would take place without chemical separation of the components to some degree. Hence the solid condensates, forming in the solar system in different regions and at different times as precursor material of the subsequently accreting bodies, were probably chemically different from each other.

If we could precisely determine the chemical differences among bodies from known and widely separated regions in the solar system, planets, satellites, comets and the Sun, it should be possible to study in detail the effects of fractionation processes active in the hetegonic era. However, chemical measurements of the bulk composition of large celestial bodies do not exist since in the course of accretion and subsequent thermal evolution all such bodies must have become stratified, and since we are unable to obtain samples deeper than a thin outer layer. We know with certainty that even a body as small as the Moon has thoroughly altered the primordial material from which it accreted. Consideration of accretional heatings as a function of terminal velocity of the source particles at impact suggest that the effect of the accretional heat front would be marked for bodies larger than a few hundred kilometers (Section 9). Hence even the surfaces of the largest asteroids would not be representative of the bulk composition of these bodies. For this same reason we have no certain knowledge of the deep interior chemical composition of any planet, not even our own (Sections 14.4.1 and 14.5.1).

On the other hand bodies with sizes of tens of kilometers and smaller are likely not to have been subject to accretional and post-accretional differentiation of this kind and it should be possible to determine their bulk composition from samples of surface material. The small asteroids, comets and the matter trapped in the Lagrangian points of the larger planets (such as the Trojan asteroids of Jupiter), are possible sources for such samples. The meteorites (Section 15) in all likelihood constitute samples of such small bodies.

14.4. UNKNOWN STATES OF MATTER

As stated in Section 14.2 in many cases the measurement of the average density is our main source of information about the bulk chemical composition of a body. However, interpretation of the mean density in terms of chemical composition is often difficult because we know so little about the state of matter at high pressure. Nor do we have satisfactory information about the properties of solid bodies aggregated in low gravitational fields.

14.4.1. *Matter at High Pressure*

Static pressure experiments with satisfactory calibration extend into the range of a few hundred kilobars (Drickamer, 1965) corresponding to pressures in the upper mantle of the Earth. In transient pressure experiments using shock waves, pressures in the megabar range can be reached (see e.g., McQueen and Marsh, 1960). Although such experiments are useful in studying elastic compression effects their general applicability is more questionable in studies of materials undergoing high pressure phase transformations. The reason is that the material in the shock front is strongly heated, and the relaxation time for phase transformations may be long compared to the duration of the pressure pulse.

Under these circumstances it has been difficult to predict with certainty the structure and composition of matter in the deep interior of the planets. The interpretation of the nature of the cores of the Earth and Venus, for example, has important consequences with regard to the inferred chemical composition of these planets. Lodochnikov (1939) and Ramsey (1948, 1949) proposed that the high density of the core of the Earth and the high bulk density of Venus could be due to pressure induced transformation of magnesium-iron silicate into a high density phase. If this were the case, the Earth's core and mantle could have the same chemical composition. Although the formation of an unknown high density phase may possibly have escaped detection in transient compression experiments, it has been considered unlikely (see, e.g. Samara, 1967) that such a density change could assume the magnitude required, about 70%, at the core-mantle boundary. Recent experiments (Simakov *et al.*, 1973) suggest, however, that originally dense-packed oxide and fluoride structures undergo phase transitions of this kind at shock pressures in the megabar range.

The alternative explanation is that the Earth's core consists of material with higher mean atomic mass than that of the mantle, for example, nickel-iron with some lighter elements such as silicon or sulfur (Birch, 1964; Ringwood, 1966; Rama Murthy and Hall, 1970; Lewis, 1971a). Much of the uncertainty concerning the properties of materials in the pressure range typical of the terrestrial planets could probably be clarified in the near future due to progress in high pressure experimental studies. This does not, however, necessarily solve the problems of the state of matter in the giant planets.

14.4.2. *Grain Aggregates*

It is generally conceded that planets and satellites must have formed from smaller bodies (planetesimals) and ultimately from small condensed particles. Such particles can accumulate to form larger bodies only if they are held together by an attractive force. Since gravity is negligible in the incipient growth stages, the main initial cohesive effect is likely to have been provided by electric charge and vapor deposition, as exemplified by the lunar soil (Arrhenius and Asunmaa, 1973). Particle aggregates generated by these processes characteristically have an open network structure (Figure 14.1). High porosity and hence low bulk density may thus have been common in the initial stages of planetesimal accretion and does still today occur in bodies which have remained at a small size. A major portion of the solid matter intercepted by Earth appears to have fluffy texture with bulk densities considerably below 1.0 g cm^{-1} and becomes destroyed in the atmosphere (Verniani, 1969; McCrosky, 1970).

Although gravitational compaction would be practically absent in bodies of small size, shock compaction of the original texture would be expected as a result of collisions leading to repeated breakup and re-accumulation during the evolution of jet stream assemblages of such bodies. Evidence of a wide variety of such effects is given by textures in meteorites ranging from complete melting (compact achondrite parent rocks; e.g. Duke and Silver, 1967) and shock induced reactions and phase changes (Neuvonen *et al.*, 1972 and subsequent articles) to dense packing (10–20% porosity) without particle bonding, such as in carbonaceous meteorites; Figure 14.2, and some chondrites.

The largely unexplored fluffy state in some small bodies in the solar system could have important consequences for their response to collision and hence for the processes of disruption and accumulation and for chemical fractionation. Projectiles impacting at hypervelocity on a body partly or entirely consisting of grains with small contact area, would presumably melt and vaporize material along their trajectory through the body, but shock wave propagation would largely be limited to the vapor phase. The vapor would largely be recondensed in the voids of the material surrounding the projectile path. Hence hypervelocity impact could result in net growth of the target body and retention of volatile materials. There are, however, as yet no controlled observations of hypervelocity impact on fluffy target material, and the physical and chemical effects of this process remain largely speculative.

In contrast to the possible reaction of fluffy material, hypervelocity impact on compact objects is always found to lead to net loss of mass due to ejection of both projectile and target material from the impact region (Gault *et al.*, 1968; Neukum *et al.*, 1970, Kerridge and Vedder, 1972).

The Martian satellites are the first small objects in space studied with sufficient resolution to record discrete surface features. The two satellites are saturated with

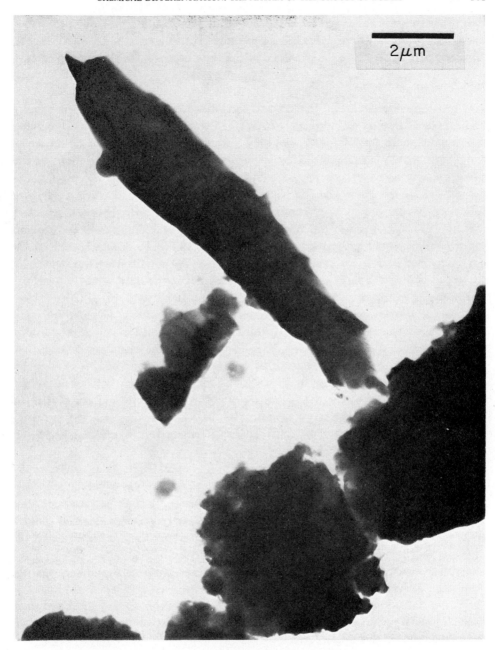

Fig. 14.1.   Head-on contacts of elongated grains are characteristic of particle clustering in lunar soil caused by electrostatic field effects. Analysis of these effects indicates that they are due to persistent internal polarization of the dielectric grains, induced by irradiation.
(From Arrhenius and Asunmaa, 1973.)

0.1 μm

Fig. 14.2.   Freely grown crystals of olivine and pyroxene forming porous aggregate material in the carbonaceous chondrite Allende. The delicate crystals are frequently twinned and have a thickness of the order of a hundred Å, thinning toward the edges. The growth of the crystals and chemical composition of the material suggest that they condensed from a vapor phase and subsequently evolved into orbits with relative velocities sufficiently low to permit accretion by electrostatic adhesion such as illustrated in Figure 14.1. (From Arrhenius and Asunmaa, 1973.)

impact craters and these have characteristics that suggest the possibility of porous target material (Figure 14.3).

14.5. THE COMPOSITION OF PLANETS AND SATELLITES

Physical data available for the planets and satellites are listed in Table 14.1, together with estimated uncertainties, and with references and footnotes indicating the accuracy and reliability of available information.

In many respects the information from the Earth is most reliable. For this reason

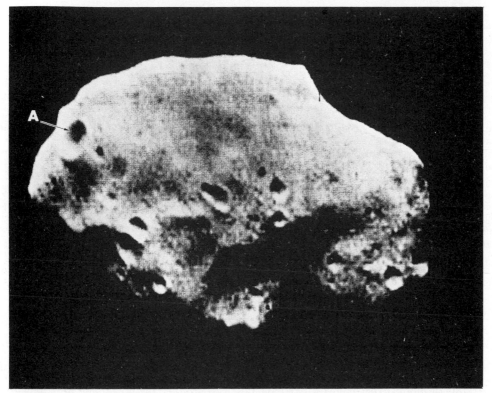

Fig. 14.3.   Impact cratering of the Martian satellite Phobos. In suitable illumination craters such as *A* above can be seen to have rims of substantial height above the surrounding terrain. Since ejecta with velocities exceeding a few meters per second will leave the satellite, the crater cones cannot be generated by fallout from the impact as is the case on Earth, Mars and the Moon. The dimensions of the cones also appear larger than the elevation of crater rims observed on the Earth as a result of shock rebound.

A possible explanation of this phenomenon is that Phobos or its outer regions consists of aggregate material with low bulk density, and that impacting projectiles dissipate their energy largely below the target surface. (NASA Photograph 71-H-1832.)

we shall begin with the data and theories relating to the composition of our own planet.

14.5.1. *Earth*

In the Earth and Moon, the only sampled terrestrial planets, the surface composition implies that oxygen is the most abundant element to considerable depth. At a few hundred kilometers depth in the Earth, the density is likely to be controlled essentially by close-packed oxygen ions.

The steep increase in density indicated at the core-mantle boundary has been interpreted in different ways:

(1) One suggestion is that the boundary represents a pressure-induced phase transformation associated with a substantial decrease in specific volume and with band gap

TABLE 14.1A

Physical data for the planets, former planets (Moon and Triton) and asteroids*

	Mean orbital radius r_{orb} from Sun 10^{12} cm	Radius 10^8 cm	Mass 10^{27} g	Average density g cm^{-3}		
				Best estimate	Upper limit	Lower limit
Mercury	5.791[a]	2.434 ± 0.002[b]	0.3299 ± 0.0029[b]	5.46[h]	5.53[h]	5.40[h]
Venus	10.821[a]	6.050 ± 0.005[b]	4.870[b]	5.23[b]		
Earth	14.960[a]	6.378[a]	5.976[a]	5.52[b]		
Moon	~ 14–20[j]	1.738[a]	0.0735[b]	3.35[b]		
Mars	22.794[a]	3.400[b]	0.6424[b]	3.92[b]		
Vesta	35.331[i]	0.285 ± 0.015[d]	0.00024 ± 0.00003[e]	2.5[h]	3.3[h]	1.9[h]
Ceres	41.402[i]	0.567 ± 0.042[d]	0.00119 ± 0.00014[e]	1.6[h]	2.2[h]	1.1[h]
Jupiter	77.837[c]	71.60[c]	1899[c]	1.31[c]		
Saturn	142.70[c]	60.00[c]	568[c]	0.70[c]		
Uranus	286.96[c]	25.40[c]	87.2[c]	~ 1.3[c]		
Neptune	449.67[c]	24.75 ± 0.06[c]	102[c]	1.66[c]		
Triton	~ 450[j]	1.88 ± 0.65[c]	0.135 ± 0.024[c]	~ 4.8[c]	20.1[h]	1.6[h]
Pluto	590.00[c]	3.20 ± 0.20[cg]	0.66 ± 0.18[k]	4.9[c]	7.4[h]	2.9[h]

* For notes to Table 14.1A see Table 14.1B.

closure resulting in metallic conductivity. The general background of this proposition has been discussed in Section 14.4.1. Objections against it are partly based on the results of model experiments which have failed to produce the postulated high density silicate phase. These results are, however, not entirely conclusive since the experiments employ transient shock rather than static pressure, hence transformations with relaxation times longer than the shock duration would not necessarily be reproduced.

(2) To avoid the assumption of a hypothetical high density silicate phase, the other current interpretation assumes that the core differs distinctly from the mantle in chemical composition and consists mostly of nickel-iron alloyed with 10–20% of light elements such as silicon or sulfur. This hypothesis requires a/mechanism to explain the heterogeneous structure of the Earth. It also implies a high concentration of iron in the source material from which the Earth was formed.

Four types of mechanisms have been suggested which would account for the proposed separation of an oxygen free metal core from a mantle consisting mainly of silicates:

(a) *A metallic core developed as a result of accretional heating.* The progression of the accretional heat front has been discussed in Section 9.12; this analysis shows that (1) the Earth's inner core should have accreted a low temperature, (2) runaway exhaustion of the source material in the terrestrial region of space would have coincided roughly with the formation of the outer core, and (3) the mantle accreted at a low mean temperature but with local heating at each impact causing light melts to migrate outward with the surface of the growing planetary embryo. Hence heavy differentiates including metal would not be able to sink further than to the bottom of locally melted

TABLE 14.1B

Physical data for the regular satellites with radii $> 10^8$ cm

	Mean orbital radius r_{orb} 10^{10} cm	Radius 10^8 cm	Mass 10^{24} g		Average density g cm^{-3}		
					Best estimate	Upper limit	Lower limit
Jovian							
Io	4.22[c]	1.82 ± 0.01[d]	72.4	± 5.6[c]	2.9[h]	3.1[h]	2.6[h]
Europa	6.71[c]	1.55 ± 0.08[d]	47.1	± 1.0[c]	3.0[h]	3.5[h]	2.6[h]
Ganymede	10.71[c]	2.64 ± 0.02[d]	155	± 2[c]	2.0[h]	2.1[h]	1.9[h]
Callisto	18.84[c]	2.50 ± 0.15[d]	96.6	± 7.4[c]	1.5[h]	1.9[h]	1.1[h]
Saturnian							
Enceladus	2.38[c]	0.27 ± 0.15[c]	0.085 ± 0.028[c]		~ 1.0[c]	16[h]	0.2[h]
Tethys	2.94[c]	0.60 ± 0.10[c]	0.648 ± 0.017[c]		0.7[c]	1.3[h]	0.4[h]
Dione	3.77[c]	0.41 ± 0.20[c]	1.05 ± 0.03[c]		~ 3.6[c]	28[h]	1.1[h]
Rhea	5.27[c]	0.72 ± 0.10[f]	1.8 ± 2.2[f]		1.2[h]	4.1[h]	0.0[h]
Titan	12.22[c]	2.42 ± 0.15[c]	137 ± 1.[c]		2.3[c]	2.8[h]	1.9[h]
Iapetus	35.62[c]	0.85 ± 0.10[f]	2.24 ± 0.74[f]		0.9[h]	1.7[h]	0.4[h]

References and notes to Tables 14.1A, B.

[a] Allen (1964).
[b] Lyttleton (1969).
[c] Newburn and Gulkis (1973).
[d] Morrison (1973).
[e] Schubart (1971).
[f] Murphy *et al.* (1972).
[g] Upper limit of radius obtained from near occultation.
[h] Average density is calculated from the mean values of mass and radius given in the table. The upper density limit is calculated by combining the lower estimated error limit for the radius and the upper estimated error limit for the mass and vice versa for the lower density limit. Spherical shape is assumed for all calculations.
[i] *Ephemerides of Minor Planets* for 1969.
[j] The distances of interest in the present discussion are those at the time of formation. Since Moon and Triton are considered to be captured planets, their original orbital radius can only be approximated.
[k] Seidelmann *et al.* (1971); these authors suggest a mass error of 16–17 %. We have here arbitrarily used a higher value (± 25 %) in order to, if anything, exaggerate the uncertainty margin of the density estimate.

pools. Large scale simultaneous melting and sinking of metal over large radial distances would be limited to the still liquid outer core, entirely melted in the runaway phase of accretion.

Complete melting of the entire planet at catastrophic accretion has recently been proposed by Hanks and Anderson (1969) as a means for gravitational separation of a metallic core. This approach, however, does not take into account a distribution of matter preceding accretion, which satisfies the boundary conditions for obtaining the present structure of the planet and satellite systems. Furthermore it meets with the same objection as any scheme involving complete melting of the Earth, further discussed in (b) below.

(b) *The Earth's core developed during or after the accretion of the planet.* This type of theory has been developed in detail by Elsasser (1963) and Birch (1965). They suggested that the Earth accreted as a homogeneous body consisting of a mixture of metal, sillicates and sulfides, similar to meteorite material. The interior of the planet heated up gradually due to radioactive decay reaching the melting point of iron (or the eutectic point in the iron-sulfide system at about 44 atomic percent S (Rama Murthy and Hall, 1970; Lewis, 1971a)) at a depth below the surface determined by the pressure effect on melting. At further heating the point would be reached where the strength of the supporting silicate material became insufficient to sustain the gravitational instability due to the higher density of the iron (or iron sulfide) liquid. At this point the liquid would drain toward the center of the Earth, releasing potential energy. The energy release would be sufficient to completely melt the entire planet.

This scheme encounters difficulties from the time constraints in the Earth's thermal evolution. On one hand the core formation process is not allowed to begin until radioactive heating has raised the initiating material to the melting point range and the supporting silicate material to its yield temperature. On the other hand, preserved segments of the crust are found which are as old as 3.6 Gy. It is questionable if these limitations would allow complete melting of the planet to occur at any time in its early history, even as early as at the time of accretion (Majeva, 1971; Levin, 1972). Such an event would also generate a heavy atmosphere containing the major fraction of the planet's accreted volatiles (Arrhenius *et al.*, 1974). This would be likely to prevent cooling to such a temperature that an ocean could form even today; nonetheless evidence for condensed water and development of life are found in the earliest preserved sediments, exceeding 3 Gy in age.

Another observation of importance in connection with the question of core formation is the consistently high content ($\sim 0.2\%$) of nickel ion in the magnesium silicates from the upper mantle. If the metallic iron, now assumed to form the core, at one time was homogeneously distributed as small particles throughout the protoplanet, such as in stone meteorites, the melting, migrating droplets of iron would be expected to reduce nickel ion in the silicate phase and to remove the resulting metallic nickel into solution in the melt (Ringwood, 1966); hence, a metallic core is generally thought of as consisting mainly of nickel-iron (e.g., see Birch, 1964). Accretional melting indeed leads to such extraction of nickel, as demonstrated by the conditions in the lunar surface rocks. These are low in metallic nickel iron, and have an order of magnitude less nickel ion in the magnesium silicates than terrestrial mantle rocks. Generation of core metal by accretional or post-accretional reduction of iron silicates with carbon (Ringwood, 1959) would doubtlessly be a still more efficient way to remove nickel from the silicate phases. Hence the presence of substantial concentrations of oxidized nickel in the Earth's mantle also speaks against melt extraction of a metallic core from an originally homogeneous planet.

(c) *The differentiation, ultimately leading to the formation of an iron core, is due to a solid grain interaction process in the Earth's jet stream.* It has been suggested that condensed nickel-iron metal particles would aggregate together at higher relative velocities, and hence at an earlier time in the evolution of the jet stream than silicate grains. This would be due to the plastic properties of the metal (Orowan, 1969) or to a high accretion cross section caused by magnetization of the grains (Harris and Tozer, 1967). Such a selective accretion of metal grains, if possible at all, could only occur when relative velocities had been brought down to the subsonic range since hypervelocity impact invariably leads to breakup and vaporization in the metal grains (Gault *et al.*, 1968; Neukum *et al.*, 1970).

Observations in meteorites do not provide support for this type of mechanism as far as preferential accretion of metal by collisional or magnetic processes is concerned. Studies of the state of metal grains in chondrites such as those by Urey and Mayeda (1959) do not indicate collision induced welding. Nor do any observations appear to exist of clustering of metal grains, characteristic of magnetic accretion. In contrast, such clustering is indeed observed for ferromagnetic iron oxide crystals (magnetite) accreted in space and subsequently aggregated into carbonaceous chondrites (Figure 15.3; Jedwab, 1967; Kerridge, 1970; Brecher, 1972a). Arguments have been given by Banerjee (1967) against magnetostatic accretion of multidomain grains of nickeliron. Finally, runaway accretion in the Earth's jet stream would take place at about $\frac{1}{10}$ of the present mass of the planet, corresponding to the mass of the core. Even if it had been possible to selectively accrete metal and leave silicate material behind in the jet stream during the formation of the inner core, all of the material orbiting in the source region of the Earth regardless of composition, would be swept up at the exponential runaway accretion coinciding with the formation of the outer core.

(d) *The differentiation took place in conjunction with the gas emplacement and condensation processes.* A suggestion of this kind, now mainly of historical interest, was made by Eucken (1944). It has recently been revived in modified form by Turekian and Clark (1969) but without application of the physical constraints of condensation (Arrhenius and De, 1973) or accretion dynamics (Sections 9.1–9.7). This type of hypothesis could in principle be made physically and chemically consistent if it is assumed *ad hoc* that the composition of condensable impurities in the region of the inner terrestrial planets changed with time, having higher iron content during the first $\sim 3 \times 10^7$ yr of infall (the order of magnitude of time required for accretion of the Earth's core; see Section 9).

If it were conclusively demonstrated that the high densities of the Earth and Venus are due to high content of iron, this fact would lend observational support to an assumption of a change with time of the composition of the source materials of these planets. At the present time such an assumption must however be regarded as speculative.

14.5.2. *Mercury*

Mercury with a radius of 0.38 $R_\oplus$ has a pressure at the center which is as low as that in the Earth's upper mantle (Lyttleton, 1969). In spite of this Mercury has a density as high as 5.46 g cm^{-3}. This can be understood in terms of the general mechanism for fractionation in the inner solar system discussed in Section 14.5.1.

14.5.3. *Venus*

The discussion of the composition of the Earth in Section 14.5.1 applies also to Venus which has 85.5% of the volume and 81.6% of the mass of the Earth. Its density, estimated at 5.25 g cm^{-3}, is only 5% less than that of the Earth. With the assumption of a core of densified silicate, Venus could have the same composition as the Earth, the Moon and Mars. If, on the other hand, as is likely, excess iron is needed to account for the high bulk density in both Earth and Venus, these two planets, together with Mercury, would be distinctly different from the Mars-Moon group (Figure 14.8a).

14.5.4. *Moon and Mars*

Since there are strong indications that the Moon is a captured planet (Alfvén, 1942–45, 1954; Urey, 1952; Gerstenkorn, 1969; Alfvén and Arrhenius, 1972) it is here included in the discussion of planetary compositions.

The observed chemical composition of the lunar surface cannot be characteristic of the interior. If the high thorium and uranium contents of the surface rocks persisted at depth, the lunar interior would be extensively melted, but seismic observations indicate possible partial melting only in the central region below 10^3 km (Toksőz *et al.*, 1972).

Furthermore rocks of the observed surface composition of the Moon would, in the interior and in a limited zone in the lower crust, seem to transform to high density assemblages (seismic data may indeed indicate such a transformation in the lower crust (Toksőz *et al.*, 1972)). If these high density phases prevailed throughout the interior of the Moon its average density would be considerably higher than the observed value, 3.35 g cm^{-3} (Wetherill, 1968). Therefore, the higher content of radioactive elements in the outer crust as well as its basaltic-anorthositic composition suggests either that the Moon accreted sequentially from materials of different chemical compositions, (Arrhenius, 1969; Arrhenius *et al.*, 1970; Gast, 1971) or that a differentiation process selectively removed the critical components from the interior to the surface.

The latter explanation would appear possible since it is difficult to escape the conclusion that an accretional front of hot spots has swept through the mantles of the terrestrial planets including that of the Moon (Sections 9.9–9.11 and Alfvén and Arrhenius, 1972). Such a progressive zone melting would be likely to cause removal to the planetary surface region of components with low melting temperature range, low density or large ionic radius (Vinogradov, 1962; Vinogradov and Yaroshevsky, 1967). The crusts of the Earth and the Moon consist of such materials except that much of the volatile components appear to have escaped thermally in the low gravitational field of the Moon (See Arrhenius *et al.*, 1974).

The former suggestion, namely that the source material for the lunar interior differed in composition from the material that formed the outer layer of the Moon, may seem more *ad hoc*. However, support for such an assumption can be drawn from the closeness and possible overlap of the A and B regions where the source materials of the terrestrial planets condensed. These relationships are discussed in Section 16.

Regardless of the cause of the lunar differentiation, the low mean density of the Moon (Table 14.1) makes it clear that it differs chemically from Mercury, most likely by having a lower iron content. It is also possible that the Moon differs substantially from the Earth and Venus in bulk chemical composition. This possibility becomes certainty if it can be verified that the latter two planets owe their high densities to a high content of iron (see Section 14.5.1).

The bulk density of Mars, 3.92 g cm^{-3}, suggests that the bulk proportion of heavy to light elements is similar to that of the Moon, and hence lower than those of Venus and the Earth (see Figure 14.8a).

14.5.5. *Asteroids*

These bodies are of sufficiently small size that pressure effects can be neglected. On the other hand, asteroids of a size larger than about 100 km have gravitation which is probably large enough to compact fluffy material. Hence the uncertainties in data interpretation discussed in Section 14.4 do not apply to such large asteroids. Their densities, in the few cases where they are known at all, furnish suggestive information on gross chemical composition.

Mass determinations from gravitational perturbation of the orbits of other asteroids exist only for Vesta and Ceres (Schubart, 1971). These values, combined with the most accurate measurements of radii (Morrison, 1973) give a density of 1.6 ± 0.5 for Ceres and 2.5 ± 0.7 for Vesta. In bodies like these, several hundred kilometers in size, porosity can probably only be maintained in a small surficial region. The low densities, if correct, therefore suggest the presence of hydrous minerals or ice in the interior, or (less likely) rocks virtually free of iron. Optical measurements by Chapman *et al.* (1971) indicate that the surface layer of Vesta consists of material with absorption properties closely similar to the meteorites known as calcium rich eucrites (density 3.4–3.7 g cm^{-3}) which are also similar to some common lunar rocks. Ceres, in contrast, has a lower albedo and more bluish color than Vesta and lacks diagnostic absorption bands; it does not bear close resemblance to any known type of meteorites (Chapman, 1972a).

The optical properties of the dusty surface material of the near Earth object 1685 Toro (Gehrels, 1972b) are similar to those of the most common type of chondritic meteorites (Chapman, 1972b). In general, however, the asteroids show widely differing optical surface properties, (Chapman *et al.*, 1971). We do not yet know to what extent, if any, there is a corresponding variation in their bulk composition.

14.5.6. *Jupiter and Saturn*

These planets are so massive that our lack of knowledge of matter at high pressures precludes any detailed speculation about their chemical composition. Not even a meaningful comparison between Jupiter and Saturn can be carried out in view of the large difference in size between them.

Attempts have been made to construct models for the different giant planets (DeMarcus, 1958; DeMarcus and Reynolds, 1963; Reynolds and Summers, 1965) but the assumptions used are necessarily highly uncertain. Existing calculations are generally based on the arbitrary assumption that the composition of the source material of all planets and satellites is the same and, specifically, that of the solar photosphere. Such assumptions are in conflict with the wide variation in bulk densities observed among small bodies in the solar system and discussed in previous and subsequent sections.

Furthermore, in order to draw conclusions about the chemical composition from the average density of a body it is necessary to know the internal temperature distribution. However attempts to estimate interior temperatures are highly sensitive to the assumed composition and to the unknown properties of the elements in question at high pressure. If the interior of Jupiter is assumed to be at relatively low temperature and to consist of solid metallic hydrogen and helium, heat could then effectively be removed by conduction. The discovery of excess energy emission from Jupiter (Hubbard, 1969; Bishop and DeMarcus, 1970) has, however, shown that this commonly accepted picture is unrealistic. This leaves a wide range of uncertainty regarding interior temperature and chemical composition. It should be noted that during planetesimal accretion the primordial heat distribution probably differed substantially for the different planets (Sections 9.9–9.11). This distribution is likely to have affected the present day internal temperature profile.

Finally, a strong magnetic field, as existing in Jupiter and possibly also in the other giant planets, could profoundly affect the heat transfer in a liquid or gaseous interior by inhibiting convection. Hence the interior temperature of the giant planets may well be much higher than existing models have indicated, and the average atomic mass could also be correspondingly higher.

Although space emissions to the giant planets will certainly provide additional information with direct or indirect bearing on the problem of the interior state, this problem is likely to remain in a speculative state for a long time. Suggestive information on the completely unknown composition of the nonvolatile material in the giant planets could perhaps be obtained from residues of the source material in their regions of formation. Small bodies in the Lagrangian points L_4 and L_5 (Trojan asteroids in the case of Jupiter) may consist of such material.

14.5.7. *Uranus and Neptune*

The uncertainties in chemical composition further complicated by the unknown internal thermal states of Jupiter and Saturn apply also to Uranus and Neptune. However, because of the close similarity in size, (Neptune possibly being slightly smaller but definitely more massive than Uranus; see Table 14.1) comparison of physical properties of this pair is, to some extent, meaningful. It is interesting to note that the density of Neptune (1.66 g cm^{-3}) at a solar distance of 30 AU, is larger than that of Uranus (1.3 g cm^{-3}) at 19 AU and both are much denser than Saturn at 9 AU (see Figure 14.8a).

14.5.8. *Triton*

The retrograde orbit of this body, now a satellite of Neptune, indicates that it was captured from a planetary orbit (McCord, 1966) and underwent an evolution partially similar to that suggested for the Moon (Gerstenkorn, 1969; Alfvén and Arrhenius, 1972). Mass and radius for Triton have been measured with estimated errors of $\pm 18\%$ and $\pm 30\%$ respectively. A combination of the extremes would give a lower density limit of 1.6 g cm^{-3} and an upper exceeding 8 g cm^{-3}; the 'best' value is around 5 g cm^{-3}.

14.5.9. *Pluto*

Considering even the largest estimated 'possible' errors in the values for the mass and diameter of Pluto, it is difficult to escape the conclusion that its density considerably exceeds 2 g cm^{-3}. A density of 4.8 g cm^{-3} (calculated assuming a radius of 3200 km, a value close to the definitive upper limit of 3400 km set by occultation measurements) is regarded as the best estimate (Newburn and Gulkis, 1973 p. 253 and Seidelmann *et al.*, 1971). Combining the occultation volume limit with a negative mass error of 25% of the best estimate gives a 'minimum' density of 2.9 g cm^{-3}. In order to bring the mass estimate into lower values it would be necessary to assume much larger errors in the mass estimates for Neptune and Saturn (Halliday, 1969) than is presently believed to be feasible (Newburn and Gulkis, 1973). The lower limit for the radius is relatively uncertain but it cannot be much different from the estimated best value since lowering of the radius rapidly results in unreasonably high densities (Table 14.1).

Pluto, like Triton, is sufficiently small to rule out the possibility of unknown high density phases in its interior. The relatively large bulk density consequently indicates also for Pluto a substantial fraction of rocky material, and, if the best present estimate is close to reality, also a significant proportion of iron.

14.5.10. *Bulk Density in Relation to Planetary Mass*

The densities of the terrestrial planets, discussed above and summarized in Table 14.1, have been plotted against planetary mass in Figure 14.4. A regular increase in density with increasing mass is found in the series Moon-Mars-Venus-Earth. This density increase could possibly be due to compression including pressure induced phase transformations; if this were the case, the chemical composition of all these bodies could be the same.

On the other hand it appears more likely that the content of heavy elements in Venus and Earth is higher than in the Moon and Mars. In the case of Mercury it is in any case necessary to assume a difference in chemical composition, presumably a higher iron content.

The densities of the outer planets have been plotted as a function of their masses in Figure 14.5. Also in this group it is obvious that other factors than the mass determine the densities of the planets.

14.5.11. *Composition of Satellites*

Except for the Moon which is here considered as a planet (Section 14.5.4), satellite mass and radius values are most reliable for the Galilean satellites of Jupiter. The

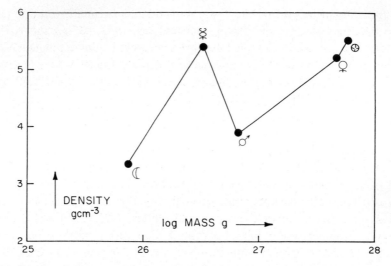

Fig. 14.4.   Density of the terrestrial planets as a function of their mass. A smooth curve could be drawn through Moon-Mars-Venus-Earth indicating that all may have a similar composition. This would require the assumption that Moon-Mars like material can be compressed to the high core densities indicated ($\sim 10\ \mathrm{g\,cm^{-3}}$) at the core pressures of Venus and the Earth ($\sim 1.5\ \mathrm{Mb}$). But it is also possible that Moon and Mars have a heavy element content entirely different from that of Earth-Venus. The composition of Mercury must in any case be different from all the other bodies.

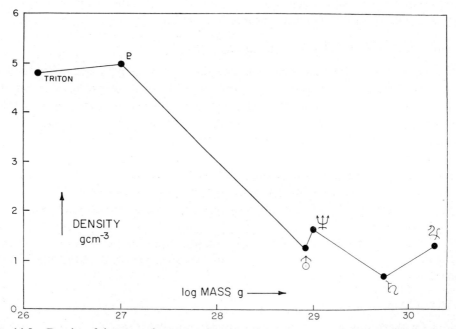

Fig. 14.5.   Density of the outer planets as a function of their mass. It is difficult to believe that the density variation can be due to only the difference in mass.

reported values of their densities display marked differences, the two smaller inner satellites (Io and Europa) consisting of more dense material (2.6–3.5 g cm^{-3}) than the outer ones (Ganymede and Callisto) (1.1–2.1 g cm^{-3}) (Table 14.1). This probably indicates differences in the proportion of light elements in icy or liquid compounds, to the heavier elements as found in earthy components (Lewis, 1971b), and demonstrate again the non-uniformity in composition of the source materials and bodies in the solar system.

The densities of the Saturnian satellites are poorly known except perhaps for Titan with a reported density of 2.3 g cm^{-3}. The estimated densities for the other satellites (Table 14.1), to the extent they can be relied upon, would suggest variations by a factor of four.

The densities of the Uranian satellites are completely unknown.

14.6. Composition of the Sun

14.6.1. *Spectrometric Analysis*

In principle, the composition of the polar photosphere, the chromosphere (including prominences) and the corona can be found by spectrometric analysis. This involves two steps, namely measurement of line intensity, profile, etc., which can be made with a high degree of accuracy, and secondly, calculations of abundances from the spectro-metric data based on models of the solar atmosphere. The models are usually homo-geneous in the sense that they assume that light received by the spectrograph emanates from a region with a density and temperature which are functions of only the height in the atmosphere. As pointed out in Section 10, homogeneous models are often mis-leading in astrophysics. In the case of the Sun, a homogeneous model is unrealistic, since we know that the solar atmosphere has a fine structure with elements of a size down to the limits of resolution and presumably still smaller. The differences in temperature and density between such elements are so large that the averaging introduced by the homogeneous model may cause gross errors. It is well-known that solar magnetograph measurements are seriously in error and in many cases it is even doubtful whether solar magnetograms can be interpreted at all. This is suggested by the fact that the 'magnetic field' derived from solar magnetograms does not obey Maxwell's equations (Wilcox, 1972). It is possible that the major uncertainties in chemical analysis by means of spectral analysis (Worrall and Wilson, 1972) are due to the same inhomogeneity effects. This must be clarified before we can rely on the spectrometric results for abundance estimates below the order of magnitude level.

14.6.2. *Analysis of Corpuscular Radiation from the Sun*

Space measurements of solar wind composition and of solar cosmic rays have given us qualitative information on the chemical composition of the upper corona and of the flare regions. However, the abundances obtained from these measurements have no simple relationship to the chemical composition of the regions from which they derive because of selective processes during emission (see Figure 14.6). For these

Fig. 14.6.   Coronal streamers, visible at solar eclipse. The photograph illustrates the inhomogeneous nature of emission of solar material. Homogeneous models of the Sun are often completely misleading.

reasons we know very little about the fractionation processes themselves, however, fluctuations in them are manifest by variations of two orders of magnitude in the helium content of the solar wind. In the range of heavier elements, variations in the solar flux of iron have been inferred from particle track studies (Price *et al.*, 1973). Selective emission of calcium has been postulated with support and evidence from apparent enrichment of this chemical in interstellar matter.

Long-term integration of the corpuscular flux may eliminate the effects of short term fluctuations in selective emission processes and give clues to their nature. However, they leave unknown any permanent differences between the composition of the Sun and the material that leaves it.

14.6.3. *Significance of Solar Abundance Data*

For the reasons outlined above, elemental abundances in the accessible layers of the Sun are known with much less accuracy than in samples of the Earth, Moon and meteorites analyzed under controlled conditions, and it is difficult to assign a probable error to any individual elemental abundance determinations (Urey, 1972).

It is often assumed that the bulk composition of the Sun is identical to some undifferentiated matter that was conjectured to be the source of other bodies in the solar system. This assumption derives from the Laplacian concept that all the matter of the solar system once formed a dense solar nebula. It was further assumed that throughout the presumed process of contraction and dynamic differentiation of such a nebula, the chemical composition somehow remained uniform.

As has been discussed in detail in other sections of this work, theories of this type are unrealistic since they ignore many of the important facts concerning the observed present state of the solar system and do not incorporate modern knowledge of the behavior of particles and fields. Hence there is no reason to believe *a priori* that the Sun has a composition which accurately corresponds to that of the bulk of any satellite, planet, or group of meteorites. Indeed, this is demonstrated already by the observed variability in composition of rocky components among various bodies in the solar system (Sections 14.5 and 14.7). Furthermore, we do not know whether the surface composition of the Sun is representative of its bulk composition. Theories of the solar interior are not very useful since they seem to be seriously out of line with observation (Fowler, 1972).

The range of actual variation in chemical composition is hard to specify because we have sampled only a few of the relevant bodies, and most of these are strongly differentiated. An indication of the variations in composition is given by the range of densities of the small bodies in the solar system (Section 14.5) and on a smaller scale, by the differences in composition between unmodified primordial condensate components in meteorites from different parent jet streams.

In order to place limits on the differences between accurately measurable materials such as meteorites and approximately measurable materials such as the solar photosphere, comparisons such as in Figure 14.7a are useful. Carbonaceous chondrites of Type I (Wiik, 1956) have been chosen for the comparison since there is general agreement that they consist of primary condensate material (one of the necessarily many different types) which does not seem to have been significantly modified with regard to elemental composition after condensation.

Elemental abundance data on this type of meteorites were obtained from a recent methodologically critical review compiled from work by a number of analytical experts (Mason (ed.), 1971). In order to avoid bias in selection of analyses, all reported measurements accepted in that review have been included without preferential selection. The solar abundances are taken from the recent evaluations by Müller (1968) and Grevesse *et al.* (1968). In the case of the solar abundances a potential bias may be caused by the presumption that the solar and meteoritic abundances ought to converge on a fictitious value, referred to as 'cosmic abundance'. The literature indicates that marked deviations from such agreement become subject to more extensive scrutiny, revision, rejection and exclusion than do the abundance ratio estimates which fall close to 1.0. The distribution shown in Figures 14.7a and b therefore probably represents a minimum dispersion.

As shown by Figure 14.7b, for about 50% of all abundance pairs determined, the solar and meteoritic values are within a factor five of each other. About 10% of all elements deviate by more than a factor of 60. The most extreme cases are the relative concentrations of the noble gases (not included in Figure 14.7a–b), mercury, thorium

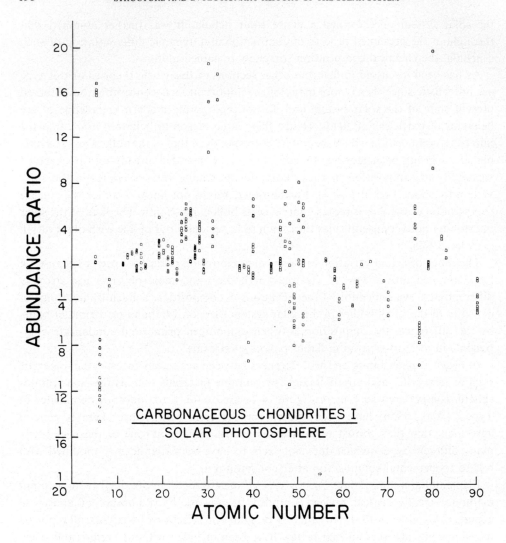

Fig. 14.7a. Comparison of solar photospheric abundance estimates with measurements on carbonaceous meteorites of Type I. Each analytical chondrite value, normalized to silicon, has been divided by each of the several current photospheric values normalized in the same way. Four of the ratio values for Mercury ($Z = 80$) exceed 20 and are not shown in the diagram. Data compiled by L. Shaw.

It has commonly been assumed that these two materials can be regarded as splits from a chemically homogeneous body 'the solar nebula' having 'cosmic abundances' of elements. Except for components with high vapor pressures or nuclear instabilities the compositions of these meteorites and of the solar photosphere then ought to approach identity, and the elemental abundance ratios should be close to 1. The often large deviations from a ratio 1:1 may be due to experimental uncertainties in the solar abundance values, actual differences in composition of the source materials and to a much smaller extent, errors in the chondrite analyses.

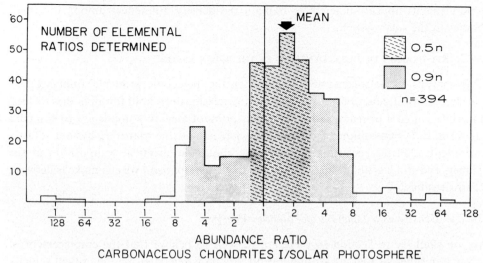

Fig. 14.7b. Frequency distribution of abundance ratios from Figure 14.7a. The diagram shows that on the average there is about 50% probability for solar photospheric observations to agree within a factor of five with their meteorite counterparts, and a 90% probability for agreement within a factor of sixty.

Ratios for elements with atomic number $\leqslant 10$ are not included in this diagram since they are affected by preferential nuclear instabilities or are highly volatile. Neither are the noble gases included because their abundances in solids are strongly permuted due to volatility and other factors; furthermore their photospheric abundances are not known.

The location of the mean and the mode of the distribution at values > 1 probably indicates an abundance difference or an observational error in the reference element, silicon.

Two abundance ratios exceed 128 and are not shown in the graph. Data compiled by L. Shaw.

and the rare earth elements. It is difficult to tell what fraction of these deviations reflect real differences and how much is due to solar observation and inversion errors. Particularly for thorium and the rare earth elements the oscillator strengths are very poorly known and the solar data for these elements may have large experimental errors. The noble gas anomalies on the other hand are based on implanted vs. occluded components in meteorites and implanted solar emissions in lunar materials. These anomalies would consequently seem to reflect real fractionation of the kind expected in the emplacement and condensation process of solids (Signer and Suess, 1963; Jokipii, 1964; Arrhenius and Alfvén, 1971; Arrhenius, 1972).

It is clear from the comparison that observational uncertainties leave room for considerable differences in composition between the solar photosphere on one hand, and various condensates such as the one represented by carbonaceous meteorites Type I on the other. As indicated above there is no particular *a priori* reason why there should be any close agreement in composition between these materials. The difference in bulk densities between the individual planets and satellites found in Section 14.5 are related to differences in abundances of the elements of which the bodies consist. Abundance differences of a factor of about four in the major condensable elements appear sufficient to explain the density differences among the small bodies in the solar system.

14.7. REGULARITY OF BULK DENSITIES IN THE SOLAR SYSTEM

Our analysis of the solar system is based on the 'hetegonic principle' implying that we should investigate to what extent the same relationships hold for all bodies formed in orbit around a primary body. From this point of view it is important to compare the chemical composition of the satellite systems and the planetary system. This is admittedly difficult because we know little about the chemical composition of the planets and still less about the satellites. The only comparison we can make is between their densities.

14.7.1. *Density as a Function of Potential Energy*

As we shall see in Section 16 there are reasons to believe that the emplacement of plasma in different regions around a central body is regulated by the critical velocity for ionization of the neutral gas falling toward the body. This implies that we should expect the abundances of elements in a system to vary with the gravitational potential energy. For this reason, it is useful to plot densities of the celestial bodies as a function of this gravitational potential energy, (the ratio of the mass, M_c, of the central body to the orbital radius, r_{orb}, of the body in question). In this way planets and satellites can be compared. Figure 14.8a shows gravitational potential energy as a function of density for the planets (including asteroids, Moon and Triton) Figure 14.8b, the satellite systems of Jupiter and Saturn, and Figure 14.8c, a composite of planets and satellites. The parameter, M_c/r_{orb}, allows a direct comparison of the planetary system and the different satellite systems.

   Looking at Figures 14.8a, b and c we can conclude that the bulk densities decrease from the high value for Mercury, Venus and Earth (at $M_c/r_{orb} = 3 \times 10^{20}$ g cm^{-1}) to a minimum at a gravitational potential energy of about 10^{19} g cm^{-1} (the region of Saturn in the planetary system) and then rise again to higher values with decreasing gravitational potential energy.

14.7.2. *Chemical Meaning of Bulk Densities*

The chemical meaning of the bulk densities of the large planets is rather uncertain. Because of the insignificance of pressure effects, the values for Mercury, Mars, Moon, Triton, Pluto, the asteroids, and the satellites are in principle more reliable, although possible measurement error is high in several cases.

   The interpretation of the densities of Uranus and Neptune also suffers from the uncertainties related to compression and temperature in the large planets but they can be better intercompared because of the closely similar size of these two planets.

   In the case of the least dense objects, namely Ganymede, Callisto, Tethys and the giant planets, it is likely that substantial amounts of volatile light elements in unknown proportions contribute significantly to the low density. This indicates that

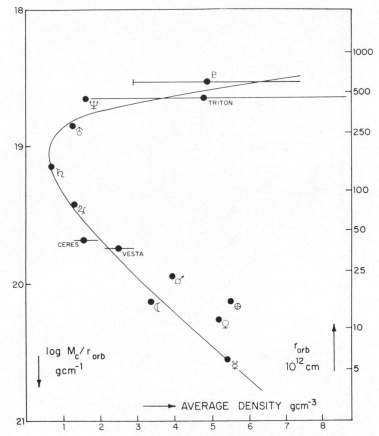

Fig. 14.8a. Average density of planets and former planets as a function of orbital distance $r_{\rm orb}$ from the Sun. The guideline through the population of density points is intended for intercomparison of this figure with Figs. 14.6b and c and does not indicate any functional relationship. The ordinate is also given in terms of gravitational potential energy (mass, M_c, divided by the orbital radius, $r_{\rm orb}$); this makes it possible to directly compare the distribution of satellites with that of the planets. The gravitational potential energy is also a parameter which enters in an important manner in the subsequent discussion of the critical velocity phenomenon (see Sections 16 and 17). Since the Moon and Triton are captured planets, the Sun is used as their central body. Hence the Moon and Triton have the approximate gravitational potential energy of the Earth and Neptune, respectively. The horizontal lines through the points for Ceres, Vesta, Triton and Pluto indicate the estimated range of uncertainty, with the vertical bar designating the lower limit for the density of Pluto as discussed in the text. Data are taken from Table 14.1.

heavy substances were accumulated both in the inner and the outermost regions of the systems, whereas light substances dominate in the intermediate region.

14.7.3. *Density as Influenced by Solar Radiation*

There is a common notion that the density of a body in the solar system is an inverse function of solar distance; this decrease in density is thought to be due to the decrease in radiation temperature at greater solar distances, which enhances capability for

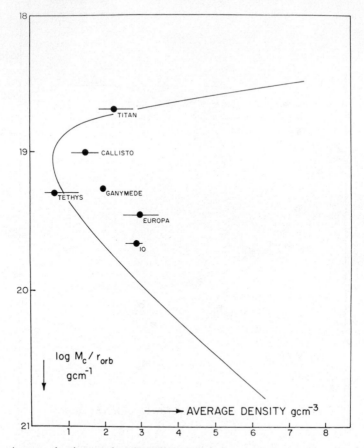

Fig. 14.8b. Average density as a function of gravitational potential energy, M_c/r_{orb}, for the regular satellite system of Jupiter and the two best known Saturnian satellites, Titan and Tethys. Solid circles denote density values based on the best estimates of radius and mass; horizontal lines indictae the estimated range of uncertainty. Data are taken from Table 14.1.

retaining lower density volatile elements and compounds. The fact that Neptune's density is higher than that of Uranus which in turn is higher than that of Saturn proves that this view is not correct and indicates together with the suggestive densities of Triton and Pluto that the chemical composition changes such as to give increasing density with increasing solar distance (increasing gravitational potential) in this part of the solar system.

Regardless of the fact that this parameter does not explain the major features of the density distribution among the planets and satellites it has long been recognized that the radiation of heat from the Sun might be one of the parameters influencing any condensation and evaporation process in the solar system. A quantitative treatment of this is given by Lehnert (1970). Since the state of the early Sun is unknown the magnitude of this heat contribution can not be directly established, but it must have been monotonically decreasing with r_{orb}^2. In this way it is possible to rationalize

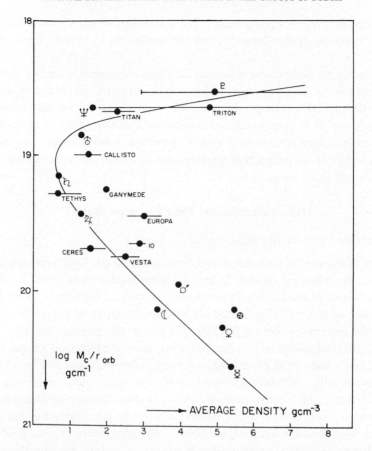

Fig. 14.8c.   Average density as a function of gravitational potential energy M_c/r_{orb} for the planets and better-known satellites. The distribution indicates that heavy substances accumulated both in the inner and outermost regions of the systems, whereas light substances dominate in the intermediate region. Symbols are those used in Figures 14.6a and b. Data are taken from Table 14.1.

the observation that among the small bodies in the outer regions of the solar system several have densities $2 \, \text{g cm}^{-3}$ or less (Ganymede, Callisto and Tethys) and that Titan retains a substantial atmosphere in spite of its relatively small mass.

14.7.4. *Theoretical Implications of Bulk Densities in the Solar System*

We have found (Section 14.5) that the bulk densities vary among the bodies of the solar system. This variation substantiates that the solar system did not form from a homogeneous medium. Hence it does not make sense to refer to any specific body in the solar system as representative of an average 'cosmic' composition of the source materials, and the Sun is no exception. Furthermore we know very little about the bulk composition of the Sun (see Section 14.6).

Another conclusion to be drawn from our survey of the bulk densities in the solar system is that density of a given body is not a function of mass (see Section 14.5.10) nor is it a function of the distance from the central body (see Sections 14.7.1 and 14.7.3).

Consequently an explanation is needed for these variations in density and presumably composition in different regions of gravitational potential. A theory making detailed predictions of composition, however, cannot yet be verified because such detailed data are not yet available. An explanation of the *variation* of densities and compositions throughout the solar system, however, follows from consideration of plausible courses of the primordial emplacement of matter around the central bodies, and is discussed in Sections 16 and 17.

15. Meteorites and Their Precursor States

15.1. INTERPRETATION OF METEORITE DATA

The studies of meteorites with increasingly sophisticated methods have supplied a rich collection of facts bearing on their history. This material is most promising for reconstruction of some of the conditions in the early history of the solar system. However, this potential cannot be fully utilized until we understand the processes which have produced the meteorites. We are far from this today. The physical models in terms of which the chemical and petrographical data are mostly interpreted are not internally consistent and some of the postulated processes violate fundamental laws of physics.

More specifically, the chemists who analyze the meteorites, commonly attribute their formation to the condensation of a gas in thermodynamic equilibrium with the growing grain. Such a process can be realized in the laboratory but not in space. The reason is that under wall free conditions the temperature of a solid normally is different from the kinetic temperature of the surrounding gas. Furthermore, the Laplacian concept of a homogeneous gaseous disc seems to be the general, although foggy, background for many speculations. The advent of magnetohydrodynamics and cosmic physics about 25 yr ago made this concept obsolete, but this seems not yet to be fully understood.

By analogy with familiar geological phenomena, the properties of meteorites are often thought to be due to processes in hypothetical planets which later have 'exploded' and thrown out the meteorites as debris. Such bodies are assumed to have had gravitational fields large enough to produce differentiation and to retain an atmosphere, in other words, to be larger than the Moon. It is not generally understood that there are no known processes by which such a large body can be blown up, especially as this must be done in such a delicate way as not to destroy the fragile structure of some meteorite materials.

In short, the most common current approach to interpreting meteorite data has led to contradictions and absurdities, which should be removed by a unified theory. Such a theory must account not only for the phenomena observed in meteorites but must at the same time be compatible with other observed properties of the solar system and it must take into account modern theory of particles and fields in space.

15.2. JET STREAMS AS METEORITE FACTORIES

In this section we shall embark on a new approach to meteoritic problems, which is in harmony with the general principles of this work. Realizing that grains in space under certain conditions become focussed into increasingly similar orbits, thus collectively forming jet streams, we shall study the possible processes in these and see to what extent one can interpret the empirical meteoritic properties as results of such processes. Because our knowledge of the structure of jet streams is in its infancy, our interpretations are necessarily tentative.

As we have found earlier (Sections 3 and 9) a jet stream consists of an assemblage of grains moving in similar Kepler orbits and interacting either by collisions or by a gas as an intermediary. The reason why a jet stream keeps together can be described in a number of ways: by the focussing action of a gravitational field, by diffusion with a negative diffusion coefficient, or by the action of an 'apparent attraction' or dynamic attraction between the grains.

Meteor streams are likely to be jet streams of this kind (Mendis, 1973) and a number of asteroidal jet streams have been discussed (Alfvén, 1969, 1970; Arnold, 1969; Lindblad and Southworth, 1971; Danielsson, 1971).

The 'profile' of an asteroidal jet stream is shown in Figure 15.1. In the focal regions

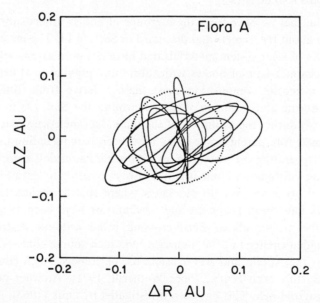

Fig. 15.1.   Profile of the Flora A jet stream. Diagram shows the intersections of the individual orbits of these asteroids with a heliocentric meridional plane as this plane is rotated one cycle around the ecliptic polar axis. The position of the orbits is shown relative to the mean orbit of the jet stream. (From Danielsson, 1971.)

The dotted circle shows the cross-section of the jet stream as calculated from Sections 9.2, Equation (2), and 9.8. Some observational support for the jet stream theory is given by the fact that most asteroid orbits in the Flora jet stream fall within the dotted circle. Profiles of other asteroidal jet streams show less concentration, probably indicating an earlier state of development.

the relative velocities are found to be as low as 0.2–1.0 km s^{-1}. This refers to visual asteroids with the size of 10 km or more. These are likely to be accompanied by a large number of small bodies. As they interact more frequently their relative velocities should often be much smaller; less than 100 m s^{-1}.

Hence, a jet stream should be depicted as a region where grains in similar orbits collide with a range of relative velocities. The average velocity is initially high, resulting in fragmentation, melting and vaporization. As a result of the gradual dissipation of energy by collision or gas friction the average internal velocity decreases with time.

When a certain fraction of the population has attained relative velocities of the order of 100 m s^{-1}, interparticle adhesion becomes effective and accretion into larger bodies can begin. During this evolution the grains and grain aggregates, forming, breaking up and regrouping while orbiting in free space, are exposed to irradiation by cosmic rays. Gas molecules may be retained in a jet stream by means of the apparent attraction.

As in meteor streams, density waves may produce local concentrations of particles, sometimes resulting in the formation of comets. Much of the history of meteorites should be studied with the formation and evolution of comets as background.

15.3. CHARACTERISTICS OF THE COMPOSITIONAL INFORMATION FROM METEORITES, METEOROIDS AND COMETS

The information that we have from these groups of bodies differs substantially from what is known about the planets and discussed in Section 14. The meteorites provide tangible samples of solar system materials that have not been extensively modified by the processes characteristic of bodies with substantial gravitational fields. Hence the variability in meteorite composition must mainly derive from differentiation (1) before or (2) during emplacement of matter around the Sun, (3) in the course of condensation, (4) during the evolution of the individual meteorite parent jet streams of which the meteorite parent bodies may have been part including the most recent breakup event generating the meteorite, or (5) during its residence on Earth.

From existing information alone it is not always possible to unravel the effects of these individual processes. We do not know where specific meteorites come from. Several sources have been proposed and mechanisms have been suggested for the transport of the meteoroids to Earth-crossing orbits without destruction of the material by sudden change in orbital energy. One such source consists of near-Earth objects such as the Apollo and Amor asteroids, or comets which come close to or intersect the Earth's orbit (Öpik, 1966; Wetherill, 1971). Another possible source region is the asteroid belt. This has been investigated by Öpik (1963), Anders (1964), Arnold (1965), Zimmerman and Wetherill (1973), but conclusive arguments have yet to be produced demonstrating that meteorites can come from the asteroid belt. In addition to the difficulty of identifying individual meteorites or groups with specific source regions we lack information on the location of the meteoritic fragments in their many generations of predecessor bodies.

Finally it should be remembered that the meteorites found on the Earth are not at all representative samples of small bodies in space, since most of the meteoroids approaching the Earth are destroyed in the atmosphere. It is estimated that even of the big meteoroids entering the Earth's atmosphere, only one in several hundred has sufficient tensile strength to reach the surface. From the retardation of meteors which burn up in the atmosphere, it has been concluded that many of them have a low bulk density indicating that they are fluffy aggregates (Verniani, 1969; McCrosky, 1970). Estimates of the chemical composition of nine Giacobinid and two Perseid fluffy meteoroids indicate similarity in their element composition to that of the much more compact chondrite meteorites (Millman, 1972). The Giacobinid showers are associated with Comet Giacobini-Zinner and the Perseids with Comet Swift-Tuttle.

Measurement of the light emission from other comets, however, indicates considerable differences in chemical composition from any known type of meteorites. This is suggested particularly by the widely different intensities of sodium D emission in different comets at comparable solar distances. In several cases notable emission of sodium occurs at solar distances as large as 0.5–1.2 AU (e.g., Bobrovnikoff, 1942; Swings and Page, 1948; Greenstein and Arpigny, 1962).

These observations together with the widely varying surface reflection spectra from asteroids (Section 14.5.5) confirm that even within these groups of bodies, wide variations occur in chemical composition both of volatile and refractory materials.

15.4. UPPER SIZE LIMITS TO METEORITE PREDECESSOR BODIES

Most meteorites investigated are clearly fragments of larger bodies, generally referred to as meteorite parent bodies. Particularly by study of nuclear transformations, induced by cosmic rays and of radiation damage in the material, it has been possible to reveal facts bearing on geometry of shielding and duration of exposure of the material. Such measurements, which are discussed in more detail in Section 15.7, indicate that the size of many of the bodies in the chain of precursor stages of meteorites must have exceeded the order of a few meters.

The largest possible size of any of the members in the chain of meteorite predecessor bodies can be estimated in different ways. One requirement is set by the size at which fragments of such a body, if it could be fragmented, would remain held together by gravitation. It is doubtful if a body larger than about a thousand kilometers in size can ever be blown apart by collision using any other body in the solar system as a projectile. It is also clear from the spin distribution of asteroids that they can not originate by explosion of much bigger bodies. This is discussed in Sections 5.7 and 5.8.

A limit to precursor body mass is also imposed by the structural changes in the meteorite material accompanying instantaneous acceleration to escape velocity at collision. Meteorites with delicate, well preserved structures and low cohesive strength, such as many carbonaceous and ordinary chondrites (see Figure 14.2) can hardly have been explosively accelerated to more than possibly a few hundred meters per second and probably less.

A body with $R = 0.01 R_\oplus = 67$ km and $M = 10^{-6} M_\oplus$ has an escape velocity $v = 0.01 v_\oplus = 110$ m s^{-1}. To break up such a body requires an explosive event which, if on the Earth's surface, would throw debris up to a height of more than 650 m. It is questionable whether fragile structures observed could tolerate such accelerations. If not, we may conclude that any one of the series of precursor bodies of such a meteorite must have been less than some ten kilometers in size.

Sizes of a hundred or a few hundred kilometers have been inferred from current interpretations of diffusion controlled crystal growth features in iron meteorites. These were believed to indicate cooling slow enough to require insulation thicknesses ranging up to the size indicated. However, the thermal history recorded in iron meteorites does not necessarily reflect a monotonic cooling from a high temperature state. It is more likely to have been a long series of heating events ($n \sim 10^6$) due to gas friction in the A and B clouds in the inner solar system (see Section 16). Each such heating event would be of short duration (the inner solar system fraction of elliptic orbits with aphelia in transplanetary space), and the amplitude of the maximum heating would monotonically decrease in time. Under these circumstances the objects could have been less than a hundred meters in size. In conclusion, meteorite parent bodies are unlikely to have been more than some kilometers or possibly some ten kilometers in size. But there is no compelling argument why they should have been larger than the order of a hundred meters.

15.5. PREDECESSOR STATES OF METEORITE PARENT BODIES

From the fact that parent bodies of meteorites must have existed, mostly considerably larger than the meteorites themselves, the question arises how these parent bodies were generated. This is in principle the same problem as the early growth of planetary embryos. We have seen in Section 7 that the only physically acceptable mechanism so far specified for this is planetesimal accretion. This means that all composite bodies in the solar system must have formed by aggregation of aggregates.

15.6. PPOPERTIES OF JET STREAMS INFERRED FROM METEORITE DATA

We shall here discuss to what extent the properties of meteorites can be understood as results of the conditions in a jet stream. Some types of meteorites, particularly the carbonaceous chondrites have a high proportion of single crystals and crystal aggregates with high content of volatiles remaining occluded in the structure from the time of condensation and with delicate growth and irradiation features perfectly preserved (see e.g. Figure 14.2). It is obvious that these particles have not undergone hypervelocity collisions in the course of their aggregation into larger bodies. Hence it is likely that these bodies have accreted in parent jet streams with sufficiently high gas content to achieve equalization of energy mainly by gas friction, and only a limited extent by high energy grain collision.

The formation of chondrules, the most abundant meteorite component, is in all likelihood a result of hypervelocity collision between single particles or aggregates. The chondrules are rounded particles of silicate and other materials, with structure indicating rapid quenching from melt or vapor.

The proportion of chondrules and chondrule fragments in a meteorite in relation to components unmodified by collision after their primordial condensation (see Figure 14.2) is likely to be a measure of collision relative to gas friction as an energy equalizing process and hence of the gas content in any specific meteorite parent jet stream.

Other manifestations of the high velocity collisions in the parent jet streams are deformation, particularly noticeable in nickel-iron metal grains (Urey and Mayeda, 1959); fragmentation, shock phenomena (see e.g., French and Short, 1968; Neuvonen et al., 1972 and subsequent articles) and complete melting of sufficiently large volumes of material to form igneous rocks (Duke and Silver, 1967) reduced to rubble in subsequent collisions (see Figure 15.5). A wide range of examples of such collision effects are also found in the lunar surface material, where, however, the relative extent of the various phenomena differs from the meteorites. This is likely to be due to the substantial gravitational field of the Moon.

In all of these collision phenomena a high degree of inelasticity characterizes the encounter, i e. the structure of the material shows that a substantial fraction of the kinetic energy of the colliding bodies has been converted into heat by deformation, melting and vaporization. This is of interest, since the degree of inelasticity is a controlling parameter in the focusing of a jet stream (Baxter and Thompson, 1973).

The material in chondrites is found to be in various states of welding due to heating of the aggregate at the time of its formation (Reid and Fredriksson, 1967) or at some later time (Van Schmus and Wood, 1967). A natural reason for such heating is gas friction in the region of the terrestrial planets as discussed above, hence there is no need for *ad hoc* assumptions of enhanced emission of corpuscular radiation from the Sun or other heat sources without basis in observation.

15.7. COHESION FORCES IN ACCRETING EMBRYOS AND THE RECORD OF MOBILE PARTICLES AND AGGREGATES IN METEORITES

When the internal energy of a jet stream has decreased sufficiently, collisions on the average cease to be disruptive and statistical net growth of aggregate bodies (embryos) is in principle possible. For this to take place, however, a cohesive force must necessarily act between the particles; in view of the small masses involved, gravity is ineffective as such a force.

15.7.1. *Cohesion by Electrostatic Bonding and Vapor Deposition*

The exploration of the Moon, particularly the investigation of the bonding forces in particle aggregates on the lunar surface has pointed at two processes as being of importance for initiating cohesion in the space environment; electrostatic bonding (Figure 15.2) and vapor deposition. Aggregates established by these processes can subsequently be compacted by shock. The fluffy state achieved by the persistent internal polarization in lunar dust particles exposed to the space environment, and by vapor deposition has been described in Section 14.4.2. The cohesive force between the grains ranges between 10 and 170 dyn (Arrhenius and Asunmaa, 1973) with dipole moments averaging a few hundred coulomb·cm. For such forces to cause adhesion

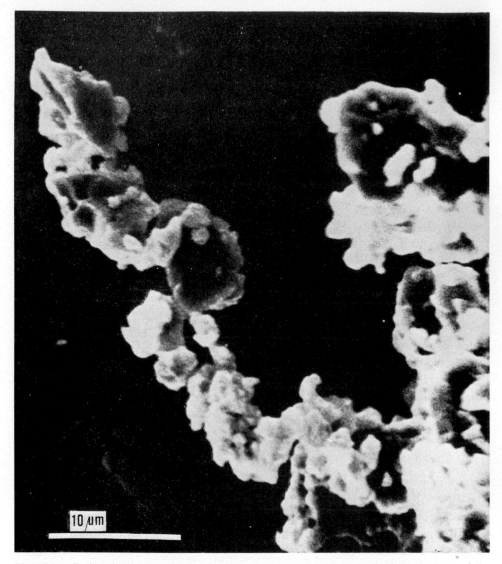

Fig. 15.2.   Grains clustering to form a flexible chain extending about 40 μm from the base of the aggregate. The chain structure illustrates the electric dipole nature of the individual microparticles. (From Arrhenius and Asunmaa, 1973.)

at grazing incidence of orbiting grains, their relative velocities need to be lowered into the range 10–100 m s^{-1} from the order of magnitude of a few thousand m s^{-1} characteristic of initial grains hitting a jet stream.

15.7.2. *Cohesion due to Magnetic Forces*

Magnetic forces of similar order of magnitude have led to clustering of magnetite (Figure 15.3; Kerridge, 1970; Jedwab, 1967; Brecher, 1972a). This is of subordinate

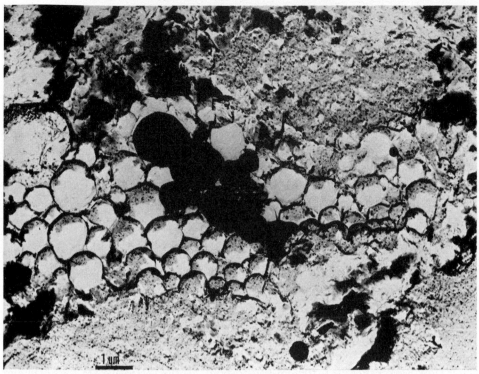

Fig. 15.3.   Section through a cluster of spheroidal magnetite crystals in the carbonaceous chondrite Orgueil. (From Kerridge, 1970; replica electron micrograph.)
    Assuming that the cluster is as high as it is wide, it consists of at least a thousand spheroids. Each of these spheroids appears to be a single magnetite crystal with a faceted surface. The crystals are easily detachable from each other and are presumably held together magnetically. Loosely bonded clusters like these are likely to have accumulated at orbital relative velocity of the individual spheroids in the range of 10–100 m s^{-1}.

importance in the main aggregation process since it affects only ferromagnetic solids, but it is a spectacular manifestation of the magnetic fields prevailing in the formative era (Brecher, 1971, 1972a, b; Banerjee and Hargraves, 1971, 1972; Brecher and Arrhenius, 1974; De, 1973). The observed magnetization cannot derive from planetary fields (Brecher, 1971, 1972a, b). To understand the origin and distribution of primordial fields and their effect on the distribution of matter it is necessary to consider the hydromagnetic processes active in the solar system today and in the past (see Part III and Sections 16 and 17).

    The magnetite grains in carbonaceous meteorites such as those shown in Figure 15.3 crystallized at grain temperatures below about 800 K (Brecher, 1972a, b) from a gas temperature an order of magnitude higher (Lehnert, 1970; Arrhenius, 1972; Arrhenius and De, 1973). The grains lack the microscopic inclusions of metallic nickel which are characteristic of oxidized nickel iron particles. Such a low nickel content is typical of magnetite in carbonaceous chondrites (Boström and Fredriksson, 1966). Hence this magnetite is not derived by oxidation of nickel iron metal as is sometimes assumed.

The hard component of the remanent magnetization and the magnetic alignment of the aggregates indicate that growth and aggregation took place in magnetic fields of the order of 0.1–1.0 G (Brecher, 1972a and b; Banerjee and Hargraves, 1971, 1972).

15.7.3. *Data from Meteorites*

For the reasons discussed in Sections 15.3 and 14.4.2 extra-terrestrial matter in the common loosely aggregated state is never sampled on Earth except in the rare cases where it is found preserved in cavities in meteorites. Examples of such materials are freely growing fibers of calcium silicate (wollastonite) formed late in the condensation sequence (Fuchs, 1971), and the magnetically adhering clusters of magnetic spheroids discussed above. However, even in the compacted material in meteorites, features are preserved from this precursor stage of low cohesion, where individual particles or aggregates were free to move relative to each other.

15.7.3.1. Particle track records in meteorites. This information obtained from compacted meteorite material is provided at a microscopic level by the cosmic ray particle track (Figure 15.4) and gas implantation record in the surface layer of exposed grains (Fleischer *et al.*, 1967a, b; Lal, 1972; Macdougall *et al.*, 1973; Eberhardt *et al.*, 1965; Wänke, 1965). The present day counterpart of this phenomenon has been extensively studied in lunar surface materials, where the source of the corpuscular radiation largely is solar wind (1 keV range; penetration about 10–1000 Å in silicates) and solar flares (low MeV range; 0.1–100 μm penetration). In the formative era the solar and planetary plasma clouds are likely to have provided major sources of accelerated particles in these energy ranges (Section 11.8) irradiating particles and aggregates also in the meteorite parent jet streams. Hence these phenomena are not necessarily related or only related to emission from the primeval Sun.

15.7.3.2. Conclusions from particle track data. Studies of dosage and gradient of this type of irradiation in meteorite grains and in lunar soil particles suggest the following conclusions:

(1) The proportion of track rich grains is variable, and often high; $\sim 30\%$ in Fayetteville, $\sim 6\%$ in Kapoeta (Wilkening *et al.*, 1971; Macdougall *et al.*, 1973). Erasure of the track record by shock makes these values lower limits. A major factor affecting the fraction of surface irradiated grains is the average size of the fluffy objects that must have evolved by a series of collision events resulting in alternating disruption and coalescence. The commonly high proportion of track rich grains suggests a high surface to volume ratio in most of these unconsolidated bodies, which consequently, must have been small. The insignificant escape velocity of small bodies indicates that throughout their history the material in any one of them has been redispersed in the jet stream and reaggregated into new configurations, probably many times.

(2) Chondrules as well as single crystals and rock fragments have generally been irradiated from all sides, with smooth variations of the dose as a function of direction; variations of the dose by a factor of 3 were observed by Macdougall *et al.* (1973).

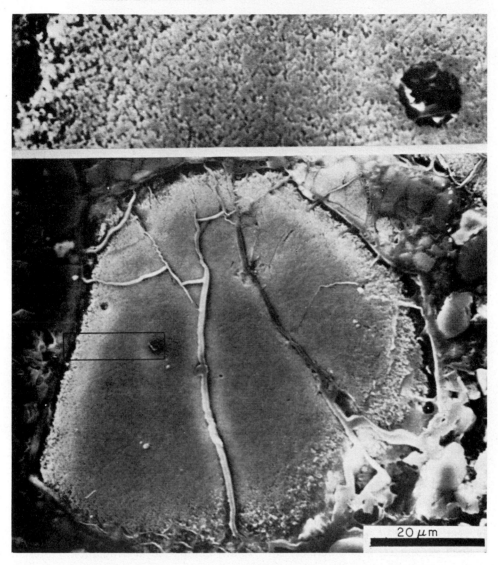

20 μm

Fig. 15.4.   Etched cross section through magnesium silicate (olivine) chondrule from the meteorite Fayetteville. The etching reveals a high concentration of cosmic ray particle tracks at the surface, rapidly falling off toward the interior, reflecting the hardness spectrum of the radiation.

The track distribution shows that the chondrule was turned around and irradiated from all sides before it was permanently embedded in the grain aggregate that now, in compacted form, constitutes the meteorite material. Before the preserved irradiation dose was received, the left part of the chondrule was broken off presumably in a collision event.

The area on the fracture edge, framed in the lower photograph, is shown in 5.1 times higher magnification in the upper photograph, illustrating the tracks and the radial track gradient in greater detail.

The track density at the edge is 10^{10} cm^{-2}. (From Macdougall et al., 1973.)

This demonstrates (a) that irradiation of fluffy aggregates persisted after the formation of the chondrules in question and (b) that a substantial fraction of the irradiation tracks, not annealed by shock, was received while the particles were loosely attached to at least one other particle. This is understandable on the basis that the initial single crystals formed by condensation necessarily had high relative velocities. Collisions between the particles in the early states of the jet streams consequently must have resulted in vaporization, melting, and shocking of grains and in the latter case to shock annealing of the tracks. When a substantial fraction of the relative velocities had been lowered into the range 10–100 m s^{-1}, clustering of particles into aggregates should have begun at a time not much different from that when shock annealing ceased to be dominant.

Lunar particles, in contrast, are much more frequently irradiated on one side only (Macdougall *et al.*, 1973; Macdougall, 1972; Arrhenius *et al.*, 1971). This presumably is due to the gravitational inhibition of free movement of material on the Moon, where sedimentary sequences characteristically are built up by blankets of material thrown out in collision events (Arrhenius *et al.*, 1971) or electrodynamically transported (Gold and Williams, 1972).

(3) The irradiation observed in meteorites requires screening less than 10^{-3} g cm^{-2}. This places limits on average density of gas and solids in the jet stream at any given time. However, the spread in screening during an orbital evolution must be considerable due to the eccentricities and inclinations of grains and aggregates. Furthermore neither the geometry nor the intensity of the radiation source is well known (see Section 11.8). Hence it is difficult to derive irradiation times from observed doses by comparison with the source that remains most active today, i.e., the Sun. The doses observed are characteristically of an order corresponding to 10^4 yr of present day solar flux at 1 AU.

(4) In addition to irradiated single crystals or crystal fragments, similarly surface irradiated aggregates of various sizes have recently been found in meteorites (Pellas, 1972; Macdougall *et al.*, 1972; Lal, 1972). It is likely that such aggregates are more common than the number so far discovered would indicate since their identification in the lithified meteorite material becomes more difficult with increasing size.

Such aggregates, which were certainly solidly compacted at irradiation, represent a stage in the evolution of the planetesimals where some material had been lithified, presumably by shock, fragmented again by collision, exposed to irradiation and reimmersed in fine grained, non-cohesive material, which later also was compacted.

15.8. Evolutionary Sequence of Predecessor States of Meteorites

The record in meteorites, discussed above, substantiates the self-evident but nonetheless commonly neglected fact that the immediate precursor bodies, from which the meteorites were derived, must have been aggregated from smaller bodies in a chain of collision events extending over a considerable period of time. Above some critical energy, depending on the material properties of the colliding bodies (see Sections 15.6 and 15.7), the collisions must be disruptive, below this level they would result in accretion. From the fact that a population of large bodies now exists it is clear that accretional collisions for some time have prevailed over disruptive ones. In the early part of the history of the jet streams the reverse must have been the case in order for orbital energies to equalize and to account for the record of aggregate disruption, particle fragmentation and extensive melt and vapor spray formation in meteorite material. A schematic representation is given in Figure 15.5 of the processes involved and the products observed.

15.9. AGE RELATIONSHIPS IN THE EVOLUTION OF METEORITE PARENT JET STREAMS (ASTEROIDAL AND COMETARY STREAMS)

The discussion in Section 9 shows that a satisfactory physical explanation of the accretion of the secondary bodies in the solar system requires continuous or intermittent emplacement of source gas in the circumsolar region during a time period of the order of 10^8 yr. Recent progress in radiochemical and mass spectrometric analysis of meteoritic materials has made it possible to resolve at an unprecedented level of precision the events that controlled the evolution of these materials in their formative era (Wasserburg *et al.*, 1969; Gopalan and Wetherill, 1969). Other measurements permit conclusions regarding subsequent events in individual meteorite parent jet streams, evolving into bodies, some of which yielded the meteorites by fragmentation. The observations of particular importance in this context are discussed in the following paragraphs.

15.9.1. *Closure of Chemical Reservoirs with Regard to Gain or Loss of Elements, Unperturbed by Subsequent Heating Events*

In this category belongs the establishment of proportions of elements, forming refractory oxides such as aluminum, calcium, strontium, titanium, and most rare Earth elements. Such separate reservoirs are represented by different meteorite groups, for example the various groups of chondrites. The fact that these groups are chemically distinct with regard to content of refractory elements shows that the finely disseminated material, from which the succession of meteorite predecessor bodies must have been formed, existed as separate streams in space. These streams were maintained as largely separate reservoirs during the orbital evolution of the meteorite material up to the most recent stage of formation of the meteorites in each stream.

Occasional exceptions to this rule are of equal interest, where an isolated chunk of material of one chemical type has been aggregated together with a major mass of material of another composition (Fodor and Keil, 1973). This indicates that separate jet stream reservoirs existed close to each other in velocity space, so that material could occasionally although infrequently be scattered from one stream to another.

The establishment of the distinct chemical characteristics of the material in any jet stream could have taken place (a) in the generation of individual source clouds (Section 16) (b) in the process of release of infalling matter from ionized source clouds by deionization (Section 16.11), (c) as a result of the critical velocity phenomenon at emplacement (Section 16.11), (d) in the process of condensation or (e) in the case of

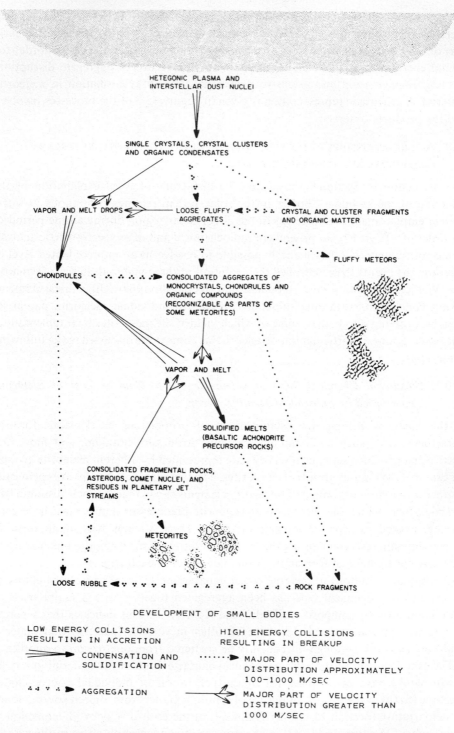

Fig. 15.5.   Flow diagram of collision processes (indicated by directional arrows) in assemblages of particles and particle aggregates moving in Kepler orbits (jet streams).

elements more volatile than those discussed here, by loss from the jet streams of a fraction of collision generated vapor.

Hence we do not know very well when and where the separate closed systems were established, which are suggested by groups of meteorites with different elemental and isotopic composition. Nonetheless, the fact is of great importance that such groups exist and have been maintained as largely separate entities.

A particularly interesting case of refractory reservoir closure is that of strontium, since the additional generation of the isotope Sr^{87} by radioactive decay of Rb^{87} provides chronological information. From highly precise measurements of the present contents of Sr^{87} and Rb^{87} (normalized to Sr^{86}) Wasserburg and his collaborators have demonstrated a range of initial Sr^{87}/Sr^{86} ratios in different meteorites and meteorite groups and in the lunar surface material. They have shown that this range represents a time interval of at least 10^7 yr and chose to interpret it in terms of condensation events (Papanastassiou, et al., 1973). However, in principle the concept can be expanded to refer to an interval of any of the events under (a) through (d) above.

15.9.2. Crystallization Ages

These are based on the accumulation of radioactive decay products inferred or measured in individual crystals. The system proven most practical and useful in space materials is that of Rb^{87}/Sr^{87}. The age defined is the latest event of crystallization. Petrological investigations sometimes make it possible to establish whether the crystallization took place from a low density gas and hence marks a primordial condensation event, from a dense gas or supercooled liquid, suggesting crystallization from collisional melt and vapor spray, or from a comparatively slowly cooling melt, presumably also generated by collision between the planetesimals in the jet stream where the parent planetesimal of the meteorite subsequently developed. Measured crystallization ages range over about 150×10^6 yr from a maximum 4700×10^6 yr (Wasserburg et al., 1969; Papanastassiou et al., 1973); 150×10^6 yr is consequently the known range within which initial crystallization and recrystallization events took place and within which the meteorite parent jet streams underwent their early development, some of them dissipating their initially high internal energies by gas friction, others by collisions with sufficiently high relative velocities to cause melting and recrystallization.

15.9.3. Gas Retention Ages

In decay systems where the daughter nuclide is a gas, the amount of this relative to the parent nuclide (inferred or measured) marks the time when the host solid was generated by condensation of cool grains in a hot gas or when hot solids became capable of retaining the gas in their structure. The decay system I^{129}/Xe^{129} with a half life of about 17×10^6 yr is of particular interest since it is capable of measuring events on the time scale of the formative era, the order of 10^8 years (Hohenberg and Reynolds, 1969; Podosek, 1970). Since the parent nuclide in this case is also relatively volatile, high energy grain collisions, such as in the early phase of development of the jet streams, are likely to largely remove both parent and daughter rather than just resetting the clock by selective removal of the daughter nuclide. In contrast, condensing crystals which by necessity must have been at a lower temperature than the surrounding gas (Arrhenius and De, 1973; Arrhenius, 1972) are bound to retain iodine and xenon already from the outset. This is illustrated by the record high I^{129} – generated Xe^{129} contents in the alkali halogenide silicate condensates characteristic of some carbonaceous meteorites (Marti, 1973). Hence the iodine-xenon ages should approximate the condensation ages.

Since neither the original abundance of I^{129} is known, nor the time scale of evolution of the source clouds or of the emplacement process, the iodine-xenon measurements yield only relative age values. The range found in the comparatively few meteorites analyzed as yet amounts to a few tens of million years (Podosek, 1970). Additional measurements are likely to expand this range, since plasma emplacement and condensation could be expected to continue over a substantial fraction of 10^8 yr.

15.9.4. *Degassing Ages*

The proportion of gaseous decay products such as He^4 relative to the refractory parent elements uranium and thorium or Ar^{40} in relation to primordial K^{40} would in ideal cases give the age of crystallization. Gas losses are, however, almost always indicated. Collisional heating appears to be the main cause for such loss, and the U-Th/He and K/Ar ages can consequently give some approximate information on the timing of such events, particularly when structural features suggest that shock is the main loss mechanism (Heymann, 1967).

15.9.5. *Cosmic Ray Exposure Ages*

Cosmic rays are largely absorbed in the surface one meter layer of meteorite material. Mainly due to spallation a wide variety of radioactive and stable nuclides are formed in the absorber. Measurements of these has provided insight into the total dose that any meteorite material has received at the surface of any of its predecessor bodies or at shallow depth below it. Detailed studies of the spatial distribution of different spallation products permit, in favorable cases, conclusions on gradients, distance to the surfaces existing during the periods of irradiation, and the shape of the body, if it remained unchanged during the period of irradiation (see, e.g., Fireman, 1958). The results also permit conclusions regarding the approximate constancy of cosmic radiation on a time scale of the order of 10^8 yr, hence dosages can be interpreted in terms of duration of irradiation (comprehensive reviews are given by Honda and Arnold, 1967; Kirsten and Schaeffer, 1971 and by Lal, 1972).

The cosmic ray induced radioactive nuclides used fall in two groups; the majority with half lives less than the order of 10^6 yr and a few with half lives exceeding 10^9 yr. Measurable activities of nuclides in the former group thus places the related irradiation in the recent history of the solar system. The activity of long lived species does, however, not provide any information on the period or periods when the irradiation was received.

Similarly most exposure dosages are, for practical reasons, not based on measurements of radioactive nuclides, but of stable spallation products, such as He^3. Also in these cases, which form the basis for statistical conclusions, there is mostly no evidence indicating when the irradiation was received, what part of the present meteoritic conglomerate was irradiated, and over how many separate intervals the exposure took place.

Nonetheless it is tacitly or explicitly assumed in most discussions of these matters, and indeed implied by the term 'exposure age', that the material was brought into exposure by one single breakup event and that the total observed dose was accumulated in the time period immediately preceding the fall of the meteorite. This view again seems to derive from the deistic tradition of regarding 'meteorite parent bodies' (which are likely to be asteroids or cometary nuclei) as having somehow been created large and then inexorably undergoing a one-way degradation process. As pointed out above, this belief ignores the need to build up the presently observed asteroidal bodies by a physically acceptable process, specifically by planetesimal aggregation. Hence it is necessary to assume that meteorite material was exposed to irradiation already in the early history of the solar system, and that a sequence of destructive and constructive collisions led to repeated burial and exposure events of which the latest fragmentation generating the meteorite, is only the last one. This is clearly reflected by the lower (MeV range) energy cosmic ray particle track record discussed in Section 15.7 above and by the distribution of keV range implantation products, deposited in a surface layer of the order 10^3–10^4 Å. Wänke (1966) has shown that material irradiated in this manner occurs much more commonly in some types of meteorites than in others (for example, in about 15% of all H-type chondrites investigated, compared to a few percent only of L and LL-type chondrites). This probably means that the planetesimals in the H-type parent jet stream spent a longer time in a relatively disseminated state, and that the L and LL-type material was focused more rapidly, possibly due to an initially narrower spread in velocity space. Similar conclusions can be drawn from the variability in frequency of grains with particle track irradiation (Macdougall *et al.*, 1973). Other observed phenomena which are probably due to this hierarchical exposure evolution are different exposure 'ages' found in different parts of the same meteorite (Zähringer, 1966) and the systematic discrepancies in K^{40} and Cl^{36} exposure ages (Voshage and Hintenberger, 1963).

15.10. CONCLUDING REMARKS

The time relationships of events recorded in meteorites need to be considered in the light of the fact that the potential source bodies must be products of both accretional and disruptive collisions. It is also necessary to take into account the fact that emplacement and condensation of matter in the circumsolar region was not instantaneous, but continued over an extended period of time.

Fractionation processes must be considered, taking place not only in the course of condensation events and during subsequent collisional heating but also in the preceding processes of emplacement of already fractionated portions of matter in different regions of the solar system and earlier still at release from interstellar source clouds and during other poorly known events further back in time. When these considerations ranging from observationally supported facts to generalized speculative reasoning, are taken into account, the record in the meteorites yields a picture consistent with that developed in other sections of this work. Remote analysis and return of samples from asteroids and comets, now in the stage of planning (Arrhenius *et al.*, 1973) are likely to increase the yield from the powerful analytical methods developed for probing the chronology of solar system history.

16. Mass Distribution and the Critical Velocity

16.1. MASS DISTRIBUTION IN THE SOLAR SYSTEM

16.1.1. *Inadequacy of the Homogeneous Disc Theory*

In theories of the Laplacian type it is assumed that the matter that formed the planets originally was distributed as a more or less uniform disc. The inadequacies of this type of approach have been discussed in Sections 2.12 and 10.6. For completeness a Laplacian type theory applicable to the planetary system must also prove applicable to the satellite systems. Hence let us turn our attention to the empirical aspects of Laplacian theories as applied to the satellite systems.

The distributed density (see Section 2.11) for the group of inner Saturnian satellites (Figure 2.8) is reasonably uniform from the ring system out to Rhea, and within this group a uniform disc theory might be acceptable But outside Rhea there is a wide region devoid of matter, followed by the giant satellite Titan, the very small Hyperion, and the medium-sized Iapetus. An even greater discrepancy between the homogeneous disc picture and the observed mass distribution is found in the Jovian satellite system (Figure 2.7). Although the Galilean satellite region is of reasonably uniform density there are void regions both inside and outside it This same general density pattern holds for the Uranian satellite system also (Figure 2.9).

Thus the distributed densities of the satellite systems of Jupiter, Saturn, and Uranus do not substantiate the homogeneous disc theory. Obviously the planetary system does not show a uniform distribution of density. In fact the distributed density varies by a factor 10^7 (Figure 2.6).

In spite of this there are many astrophysicists who believe in a homogeneous disc as the precursor medium for the planetary system. The low density in the asteroid region is then thought of as a 'secondary' effect, presumably arising from some kind of 'instability' caused by Jupiter. However, under present conditions several big planets (e.g. of 10 to 100 times the mass of Mars) moving between Mars and Jupiter would be just as perfectly stable in all respects as are the orbits of the present asteroids. And no credible mechanism has been proposed explaining how Jupiter could have prevented the formation of planets in this region.

In addition to these obvious discrepancies between the implied uniform and the actually observed distributions of mass in the solar system, the whole disc idea is tied to the theoretical concept of a contracting mass of gas which could collapse to form both the central body and the surrounding secondaries via the intermediate formation of the disc. As has been pointed out in Section 7.1, small bodies cannot be formed in this way and it is questionable whether even Jupiter is large enough to have been formed by such a collapse process. Another compelling argument against a gravitational collapse of a gas cloud is found in the isochronism of spins (Sections 5.7 and 8.3). We have also found in Section 14 that the chemical composition of the celestial bodies speaks against a Laplacian homogeneous disc. Other arguments against it are found in the detailed structure of the Saturnian rings and the asteroidal belt (see Sections 13.6 and 13.8). It is very unlikely that these features can be explained by a Laplacian model or by gravitational collapse.

16.1.2. *Origin and Emplacement of Mass*

Ejection of mass. Since the concept of homogeneous disc consequently is unrealistic when applied to any of the actual systems of central bodies with orbiting secondaries, we must look for other explanations of how the mass which now forms the planets and satellites could have been emplaced in the environment of the central bodies.

In principle, the mass which now constitutes the planets and satellites could either have been ejected from the central body or could have fallen in toward the central body from outside the region of formation. It is difficult to see how a satellite could have been ejected from its planet and placed in its present orbit. Such processes have been suggested many times, but have always encountered devastating objections. Most recently it has been proposed as an explanation of the origin of the Moon, but has been shown to be unacceptable (see Kaula, 1971).

This process would be still less attractive as an explanation of e.g. the origin of the Uranian satellites In fact, to place the Uranian satellites in their present (almost coplanar circular) orbits would require all the trajectory control sophistication of modern space technology. It is unlikely that any natural phenomenon, involving bodies emitted from Uranus could have achieved this result.

An ejection of a dispersed medium which is subsequently brought into partial corotation is somewhat less unnatural, but it also requires a very powerful source of energy, which is hardly available on Uranus, to use the same example. Moreover, even in this case, the launch must be cleverly adjusted so that the matter is not ejected to infinity but is placed in orbit at the required distances. Seen with the Uranian surface as launch pad, the outermost satellites have gravitational energies which are more than 99% of the energy required for escape to infinity.

16.1.3. *Infall of Matter*

Hence it is more attractive to turn to the alternative that the secondary bodies derive from matter falling in from 'infinity' (a distance large compared to the satellite orbit). This matter (after being stopped) accumulates at specific distances from the central body. Such a process may take place when atoms or molecules in free fall reach a kinetic energy equal to their ionization energy. At this stage the gas can become ionized by the process discussed in Section 16.4 and the ionized gas can then be stopped by the magnetic field of the central body.

16.2. THE BANDS OF SECONDARY BODIES AS A FUNCTION OF GRAVITATIONAL POTENTIAL ENERGY

If the latter hypothesis is correct then the matter which has fallen into the solar system would have accumulated at predictable distances from the central body. This distance is a function of the kinetic energy acquired by the matter during free fall under the gravitational attraction of the central body. Let us consider the positions of a group of secondary bodies as a function of their specific gravitational potential $\kappa\Gamma$, where

$$\Gamma = M_c/r_{orb} \tag{16.1}$$

and κ is the gravitational constant, M_c is the mass of a central body, and r_{orb} is the orbital radius of a secondary body. The gravitational potential Γ determines the velocity of free fall and thus the kinetic energy of infalling matter at a distance r_{orb} from the central body. In Figure 16.1 we have plotted this energy as a function of M_c for the Sun-planet system as well as for all the planet-satellite systems.

We see from Figure 16.1 that *the secondary bodies of the solar system fall into three main bands* and that *whenever a band is located far enough above the surface of a central body there is a formation of secondary bodies in the region*. These two important observational facts will be discussed in this and the next section.

Although there are some apparent exceptions to the general validity of these conclusions, cogent explanations can be offered for each discrepancy. Venus has no satellites, probably because of its extremely slow rotation and lack of a magnetic field. Both these properties, rotation and magnetic field of the central body, are the pre-requisites for formation of secondary bodies, as was discussed in Section 11.1. Further we find no satellite systems of the normal type around Neptune and the Earth. The reason for this seems to be straightforward; both of these bodies might very well have once produced normal satellite systems, but they have been destroyed by the capture of Triton (McCord, 1966) and of the Moon (Alfvén and Arrhenius, 1972). Mercury has a very slow rotation, probably no magnetic field, and is perhaps not massive enough for satellite formation. Whether Pluto has any satellite is not known.

We have not yet discussed the Martian satellites which fall far outside the three bands. From a formal point of view they may be thought to indicate a fourth band. However the Martian satellite system is rudimentary compared to the well-developed satellite systems of Jupiter, Saturn and Uranus, and the Martian satellites are the smallest satellites we know. In view of the rudimentary character of the Martian satellite system, we do not include this in our discussion of systems of secondary bodies. This question is further discussed in an earlier paper (Alfvén and Arrhenius, 1972).

In Figure 16.1 the satellite systems are arranged along the horizontal axis according to the mass of the central body. Groups of secondary bodies belonging to a particular band are generally located somewhat lower if the central body is less massive, thus giving the bands a slight downward slope. As a first approximation, however, we can consider the bands to be horizontal. (The reason for the slope is discussed in Section 17.9.2). We conclude from the gravitational energy diagram that *groups of bodies are*

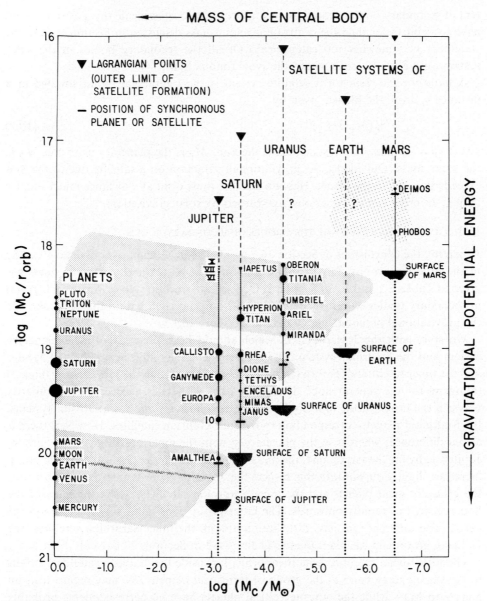

Fig. 16.1. Structure of the solar system in terms of the mass of the central bodies and the gravitational potential energy of the bodies orbiting around them. For a detailed analysis refer to Sections 16.2, 16.3 and 17.9.2.

formed in regions where the specific gravitational energy has values in certain discrete ranges.

In Figure 16.1 we have also plotted the positions of synchronous secondary orbits as well as those of the Lagrangian points for the satellite systems. The position of a synchronous secondary orbit around a primary body is a *natural inner limit* to a system of secondary bodies, since any secondary body located inside this position would have to orbit faster than the central body spins. (As discussed in Section 17.9.1, this statement is somewhat too categorical.) Of all the secondary bodies in the solar system only Phobos orbits within the synchronous limit.

A *natural outer limit* for a satellite system is the Lagrangian point situated at a distance r_L from the planet, given by

$$r_L = r_P (M_P/3M_\odot)^{1/3}, \tag{16.2}$$

where r_P is the planetary distance from the Sun, M_P is the planetary mass and $M_\odot$ is the solar mass. Outside r_L, the gravitational attraction on a satellite due to the Sun exceeds that due to the planet. Hence a satellite must orbit at a distance much smaller than r_L in order not to be seriously perturbed by solar gravitation.

16.3. COMPARATIVE STUDY OF THE GROUPS OF SECONDARY BODIES

Accepting the conclusions of Section 16.2 we shall now attempt a more detailed study of the different groups. Physical data for both the planet and satellite systems are given in Tables 2.1 and 2.2 of Section 2. Our general method is to compare each group of secondary bodies with its neighbors to the left and right within the same band of the gravitational potential energy diagram, Figure 16.1.

We start with the Jovian system which should be compared with the planetary system and the Saturnian system. The giant planets, the Galilean satellites of Jupiter and the inner satellites of Saturn (Janus through Rhea) fall in the same energy interval (allowing for the general slope discussed earlier). There is a conspicious similarity between the group of the four big bodies in the planetary and in the Jovian systems, the four giant planets corresponding to the four Galilean satellites. However, there is also a difference; whereas in the planetary system the innermost body in this group, Jupiter, is by far the largest one, the mass of the bodies in the Galilean satellite group increases slightly outward. In this respect the Jovian system is intermediate between the group of giant planets and the inner Saturnian satellites, where the mass of the bodies increases rapidly outwards. The latter group consists of six satellites and the rings. (The difference in mass distribution among the inner Saturnian satellites, the Galilean group and the giant planets is discussed in Sections 17.6–17.8).

The fifth Jovian satellite, Amalthea, orbits far inside the Galilean satellites. It falls in the same energy band as the group of terrestrial planets. We may regard it as an analog to Mars while the other terrestrial planets have no correspondents probably because of the closeness to the surface of Jupiter. The mass of Amalthea is unknown. Its diameter is estimated to be about 160 km. As the diameter of Io is about 3730 km its volume is about 10^4 times that of Amalthea. The mass ratio of these two satellites is unknown. The volume ratio of Io to Amalthea is of the same order as the volume ratio of Jupiter to Mars which is 9000 but the close agreement is likely to be accidental.

The outermost group of Jovian satellites, Jupiter VI, VII and X, is rudimentary. One may attribute the rudimentary character of this group to its closeness to the outer stability limit r_L for satellite formation, which is closer to this group than to any other group in the diagram. Although this group of Jovian satellites falls within the band including the outer Saturnian satellites and the Uranian satellites, it has no other similarity with these two groups.

In the planetary system the same band may have given rise to Pluto and Triton (the latter later captured by Neptune in a similar way as the Moon was captured by the Earth).

The Uranian satellites form the most regular of all the groups of secondary bodies in the sense that all orbital inclinations and eccentricities are almost zero, and the spacings between the bodies are almost proportional to their orbital radii ($q = = r_{n+1}/r_n \approx$ constant). The group is situated far outside the synchronous orbit and far inside the Lagrangian point. It should be noted that this is also the case for the Galilean satellites which also form a very regular group. In fact, these two groups should be studied as typical examples of satellite formation in the absence of disturbing factors.

Titan, Hyperion, and Iapetus are considered as a separate group which we refer to as the 'outer Saturnian satellites'. The assignment of these three bodies to one group is not altogether convincing and the group is the most irregular of all with regard to the sequence of satellite orbital radii and masses. However, it occupies a range of Γ values which closely coincides with that of the Uranian satellites. Furthermore, if we compare the group with both of its horizontal neighbors, we find that the irregular group in the Saturnian system constitutes a transition between the rudimentary group in the Jovian system and the regular group in the Uranian system. In this respect there is an analogy with the Galilean group in which the almost equal size of the bodies is an intermediate case between the rapid decrease in size away from the central body in the giant planets and the rapid increase in size away from the central body in the inner Saturnian system. The probable reason for these systematic trends is discussed in Sections 17.6–17.8.

16.4. THEORETICAL BACKGROUND FOR THE BAND FORMATION

Attempts to clarify the mechanism which produces the gravitational potential energy bands (Section 16.2) should start with an analysis of the infall of the gas cloud toward a central body. In order to avoid the difficulties inherent in all theories about the primitive Sun, we should, as stated in Sections 1.3 and 11.9, base our discussion primarily on the formation of satellites around a planet.

The gas cloud we envisage in the process of satellite formation is a local cloud at a large distance from a magnetized, gravitating central body. This cloud, called the source cloud (see Section 16.11.1) is located within the jet stream in which the central body has formed or is forming and is thusly part of the gas content of the jet stream itself (see Figure 16.2). This cloud also contains grains from which the central body is accreting. For the sake of simplicity let us assume that initially the cloud is at rest at such a low temperature that the thermal velocity of the particles can be neglected compared to their free-fall velocity. Then every atom of the cloud will fall radially toward the center of the gravitating body. If the gas cloud is partially ionized, the ions and electrons, which necessarily have a Larmor radius which is small compared to the distance to the central body, will be affected by the magnetic field already at great distances from the central body with the general result that their free fall will be prevented. Hence in our idealized case only the neutral component, the grains and gas, will fall in. The infall of grains is the basic process for the formation and growth of

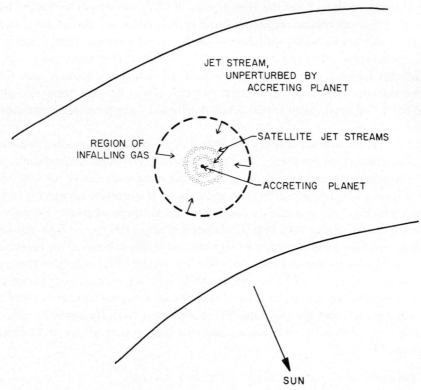

Fig. 16.2.   Qualitative picture of the infall of gas from the jet stream toward a planet. The gas becomes ionized, is brought into partial corotation and eventually forms satellite jet streams.

the central body (which acquires spin as a result of the asymmetry of this infall; for a detailed discussion see Section 8).

Let us now consider the infall of gas in an idealized case in which the gas is not disturbed by the infall of grains. Probably such a situation occurs when the accretion of the central body is near completion. Hence we assume that for a certain period of time there is a constant infall of gas toward the central body.

Suppose that at some distance r from the central body there is a very thin cloud of plasma which also has negligible thermal velocity and which, due to the action of the magnetic field, is at rest. (The effect of rotation of the central body is neglected here; it will be introduced in Section 16.13. The plasma density is assumed to be so low that the mean free path of the atoms exceeds the dimensions of the cloud. (For densities $\lesssim 10^3$–10^4, the mean free path is larger than the dimension of the satellite formation regions ($\leqslant$ planet-outermost satellite distance). However, the dimension of the cloud in question may be an order of magnitude or so smaller, allowing somewhat higher densities but the mean free path would still be much greater than the dimension of the cloud.)

When the infalling atoms reach the plasma cloud, some will pass through it without colliding, some will make nonionizing collisions and be deflected, but neither of these processes will affect the conditions in the plasma cloud very much.

However, under the condition that the atoms arrive at the cloud with a sufficiently high velocity, atoms may become ionized at some of the collisions. Due to the magnetic field, the ions and electrons thus produced will rapidly be stopped and become incorporated in the plasma cloud. Hence the density of the plasma cloud will grow, with the result that it will capture infalling atoms at an increased rate. In an extreme case the density may become so high that the mean free path of atoms is smaller than the size of the cloud, resulting in a complete stopping of the infalling gas. (We assume that the magnetic field is strong enough to support the resulting dense plasma cloud, see Sections 11.3–11.5).

Basic theoretical analysis of electric breakdown in a gas treats the conditions under which the electric field will give sufficient energy to an electron to produce new electrons so that an avalanche may start. The 'original' existence of free electrons can be taken for granted. Our case is essentially similar. The existence of thin plasma clouds anywhere in space can be taken for granted. The question we should ask ourselves is this: What are the conditions under which the infalling atoms get ionized so frequently that the density of the original plasma cloud will grow like an avalanche? It is likely that the infall velocity is the crucial parameter. In our simple model the infalling gas cloud will be stopped at the distance r_{ion} where its velocity of fall reaches the value v_{ion}, such that

$$\frac{\kappa M_c m_a}{r_{ion}} = \frac{m_a v_{ion}^2}{2}, \tag{16.3}$$

where m_a is the mass of an atom. At this distance the specific gravitational potential energy $\kappa \Gamma$ will have the value $\kappa \Gamma_{ion}$ with

$$\Gamma_{ion} = \frac{M_c}{r_{ion}} = \frac{v_{ion}^2}{2\kappa}. \tag{16.4}$$

Hence, Γ_{ion} is a function only of v_{ion}. Because v_{ion} is the parameter which sets the lower limit for ionization of the infalling gas, v_{ion} may be considered as an analogy to the breakdown electric field in the theory of electrical discharges.

The analogy between the stopping of an infalling cloud and the electric breakdown is in reality still closer. In fact, seen from the coordinate system of the infalling cloud, there is an electric field

$$\mathbf{E} = -\mathbf{v} \times \mathbf{B}, \tag{16.5}$$

which increases during the fall because both the velocity of fall v and the dipole magnetic field B increase. If the electric field exceeds a certain critical value E_{ion}, a discharge will start via some (yet unspecified) mechanism for energy transfer to the electrons. This will lead to at least partial ionization of the falling gas cloud. In situations where the collision rate for the electrons is low the mechanism for transfer of energy from the electric field (i.e. from the falling gas) to the electrons is very complicated and not yet quite clarified (the electric field $-v \times B$ in the coordinate system of the gas cloud cannot directly accelerate the electrons; in a magnetic field the electron cannot gain more than the potential difference over a Larmor radius for every collision and this is very small). It has, however, been demonstrated and proven to be very efficient in a variety of plasma experiments (see Section 16.8 and references). Under certain (rather general) conditions, this will lead to a braking of the velocity of the cloud, and possibly to stopping it. The discharge will occur when v exceeds the value v_{ion} which is connected with E_{ion} by

$$\mathbf{E}_{ion} = - \mathbf{v}_{ion} \times \mathbf{B}. \tag{16.6}$$

Hence the ionization of the infalling cloud may also be thought of as being due to the electric field exceeding E_{ion}.

16.5. ATTEMPTS TO INTERPRET THE BAND STRUCTURE

If we equate the ionization energy eV_{Ion} of an atom of mass m_a to its gravitational energy in the presence of a central body of mass M_c, we have

$$eV_{Ion} = \frac{\kappa M_c m_a}{r_{orb}} \tag{16.7}$$

or

$$\Gamma = \frac{M_c}{r_{orb}} = \frac{eV_{Ion}}{\kappa m_a}. \tag{16.8}$$

As we will see later there is a mechanism which converts the kinetic energy of an atom falling toward a central body to ionization energy. Hence, Equation (16.8) allows one to determine for an atom of known mass and ionization potential the orbital radius from the central body at which ionization can take place.

In Table 16.1 we list a number of elements of cosmochemical importance along with their relative abundance, average atomic mass, ionization potential, eV_{Ion}, gravitational energy as given by Equation (16.8), and critical velocity which will be discussed in a later section. Just as the Γ values for the bodies in the solar system, as given by Equation (16.1), were plotted in Figure 16.1, so the Γ values for the elements as given by Equation (16.8) are plotted in Figure 16.3. Looking at this plot of gravitational potential energy versus ionization potential we find that *all the elements fall in one of three bands*. Hydrogen and helium give a value for Γ which falls in the region of the lowest band (which will be referred to as Band I, since this is comprised of elements of the first row of the periodic table). All the elements in

TABLE 16.1

Element[a]	Ionization potential V_{Ion} (volt)	Average atomic mass (a.m.u.)	Log gravitational potential energy $\log \Gamma$ (g cm^{-1})	Atomic abundance[b] relative to Si $= 10^6$	Critical velocity v_{crit} (10^5cm s^{-1})	Band
H	13.5	1.0	20.29	2×10^{10}	50.9	I
He	24.5	4.0	19.94	2×10^9	34.3	I
Ne	21.5	20.2	19.18	2×10^6	14.3	II
N	14.5	14.0	19.18	2×10^6	14.1	II
C	11.2	12.0	19.11	1×10^7	13.4	II
O	13.5	16.0	19.08	2×10^7	12.7	II
(F)	17.42	19.0	19.11	4×10^3	13.3	II
(B)	8.3	10.8	19.08	1×10^2	12.1	II
[Be]	9.32	9.0	19.18	8×10^{-1}	14.1	II
[Li]	5.39	6.9	19.04	5×10^1	12.2	II
Ar	15.8	40.0	18.78	1×10^5	8.7	III
P	10.5	31.0	18.70	1×10^4	8.1	III
S	10.3	32.1	18.70	5×10^5	7.8	III
Mg	7.6	24.3	18.60	1×10^6	7.7	III
Si	8.1	28.1	18.60	1×10^6	7.4	III
Na	5.12	23.0	18.30	6×10^4	6.5	III
Al	5.97	27.0	18.48	8×10^4	6.5	III
Ca	6.09	40.1	18.30	7×10^4	5.4	III
Fe	7.8	55.8	18.30	9×10^5	5.2	III
Mn	7.4	54.9	18.30	1×10^4	5.1	III
Cr	6.8	52.1	18.30	1×10^4	5.0	III
Ni	7.6	58.7	18.30	5×10^4	5.0	III
(Cl)	13.0	35.5	18.70	2×10^3	8.4	III
(K)	4.3	39.1	18.30	2×10^3	4.6	III

[a] Minor elements (10^2–10^4) are indicated by parentheses; trace elements ($< 10^2$) are indicated by brackets.

[b] The very fact that separation processes are active in interstellar and circumstellar space makes it difficult to specify relative abundances of elements except by order of magnitude and for specific environments (such as the solar photosphere, the solar wind at a given point in time, the lunar crust, etc.). This is further discussed in Section 14.

The abundances are the averages estimated by Urey (1972). Most values are based on carbonaceous chondrites of type II which form a particularly well analyzed set, apparently unaffected by the type of differentiation which is characteristic of planetary interiors. Supplementary data for volatile elements are based on estimates for the solar photosphere and trapped solar wind. All data are normalized to silicon, arbitrarily set at 10^6.

[c] All values are calculated from Equation (16.9), using the data presented in this table.

the second row of the periodic table (Li-F), including C, N and O, have values around $\Gamma \approx 10^{19}$ falling in the intermediate band (Band II) whereas all the common heavier elements found in the third and fourth row of the periodic table fall in the upper band (Band III). This means that if a gas dominated by any one of the most abundant elements falls in toward the central body, its kinetic energy will just suffice to ionize it when its gravitational potential energy reaches the values indicated by its appropriate band. For our discussion it is decisive that the values of the cosmically most abundant

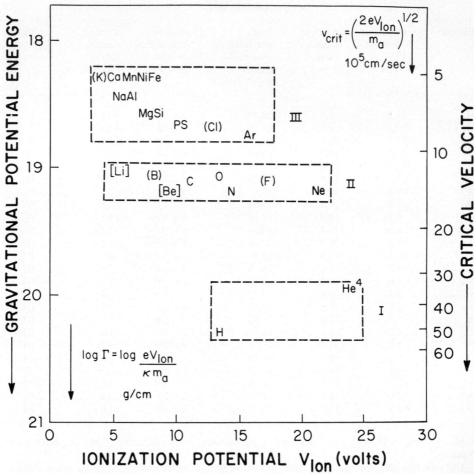

Fig. 16.3. The critical velocity v_{crit} and ionization potential of the most abundant elements. Roman numerals refer to row in the periodic table, with 'III' including the fourth row. All elements in a band have approximately the same critical velocity or v_{ion}, as discussed in Sections 16.4 and 16.5. Minor and trace elements are indicated, respectively, by parentheses and brackets.

elements fall in a number of discrete bands rather than forming a random distribution.

This ionization is a collective phenomenon dependent upon the gas mixture in the source cloud. The gas as a whole will tend to be stopped in one band. In the light of the above discussion, we note that because of the discrete regions where the Γ values of the most abundant elements fall, the discrete bands of gravitational energy discussed in Section 16.2 may be explained by *the hypothesis that they are related to these Γ values.* This relationship is discussed in detail in Sections 16.7–16.13.

16.6. THREE OBJECTIONS

When the preceding analysis was first made (Alfvén, 1942) there were three objections to the ensuing hypothesis:

(a) There was no obvious mechanism for the transfer of the kinetic energy into ionization. The requirement that Γ_{ion} of Equation (16.4) and Γ of Equation (16.8) should be equal, i.e.,

$$v_{ion} = \left(\frac{2eV_{Ion}}{m_a}\right)^{1/2}, \tag{16.9}$$

was crucial to the hypothesis, but no reason was known for this equality to be true.

(b) There was no empirical support for the hypothesis that masses of gas falling in toward central bodies would have differing chemical compositions.

(c) The chemical compositions of the bodies found in each gravitational potential energy band are not characterized by the elements giving rise to those bands. For example, the terrestrial planets fall in a band which corresponds to the Γ values for hydrogen and helium, but they contain very little of these elements. The band of the giant planets corresponds to C, N and O, but these planets were believed to consist mainly of hydrogen and helium.

However, the above situation has changed drastically over three decades of theoretical studies and empirical findings. Although we are still far from a final theory, it is fair to state that objection (a) has been eliminated by the discovery of the critical velocity phenomenon as discussed in Section 16.7–16.10. With reference to (b) we now know that separation of elements by plasma processes is a common phenomenon in space. We shall discuss such separation and variance of chemical compositions in Section 16.11. Objection (c) will be considered in Section 16.12.

In the meantime, no alternative theory has been proposed which in terms of known physical processes explains the positions of the groups of bodies and which at the same time is consistent with the total body of facts describing the present state of the solar system.

16.7. Search for a 'Critical' Velocity

Early attempts to theoretically analyze the stopping of an infalling cloud were not very encouraging. Equating the gravitational and ionization energies has no meaning unless there is a process by which the gravitational energy can be transferred into ionization. Further, in an electric discharge the energy needed to actually ionize an atom is often more than one order of magnitude greater than the ionization energy of that atom, because in a discharge most of the energy is radiated and often less than 10% is used for ionization.

In view of the fact that, as stated in Section 10, all theoretical treatments of plasma processes are very precarious unless supported by experiments, it was realized that further advance depended on studying the process experimentally. As soon as the advance of thermonuclear technology made it possible, experiments were designed to investigate the interaction between a magnetized plasma and a non-ionized gas in motion relative to that plasma. Experimental investigations have now proceeded for more than a decade. A survey has recently been given by Danielsson (1973).

16.8. EXPERIMENTS ON CRITICAL VELOCITY

Independently many experimental measurements of the burning voltage in magnetic fields were made. They demonstrated the existence of a limiting voltage V_L which if introduced into Equation (16.6) with $E_{ion} = V_L/d$ (d being the electrode distance) gives almost the same values v_{ion} as are calculated from Equation (16.9). This upper limit of the burning voltage is directly proportional to the magnetic field strength but independent of gas pressures and current in very broad regions. The presence of neutral gas, however, is a necessity for this effect to occur; once a state of complete ionization is achieved these limiting phenomena no longer appear.

Of the first observations most were accidental. Indeed the effect sometimes appeared as an unwanted limitation on the energy storage in various plasma devices, for example thermonuclear machines like the Ixion, the early homopolars and the F-machines (Lehnert, 1966).

16.8.1. *Homopolar Experiments*

One of the earliest experiments which was especially designed to clarify the phenomena occurring when a natural gas moves in relation to an ionized gas was performed by Angerth *et al.* (1962). The experimental apparatus, a homopolar device, is shown in Figure 16.4. In a vessel containing a gas at a pressure of the order of 5×10^{-3}–0.2 torr, or 10^4–10^{16} atoms cm^{-3}, a radial electric field is established by connecting a condenser bank between two concentric cylindrical electrodes. There is an almost homogeneous magnetic field of up to 10 000 G perpendicular to the plane of the lower figure. In order to have any reference to our problem, the gas density in the experiment should be scaled down in the same relation as the linear dimension is scaled up. As the densities during the formation of the planetary system should have been of the order of 10^1–10^5 atoms cm^{-3}, and the scaling factor is 10^{10}–10^{13}, the experiment is relevant to the astrophysical problem. The temperature is determined by the plasma process both in the experiment and in the astrophysical problem and should therefore be equal.

A portion of the gas is ionized by an electric discharge. This ionized component is acted upon by a tangential force, resulting from the magnetic field and the radial electric field, and begins to rotate about the central axis. The non-ionized component remains essentially at rest because of the friction with the walls. Hence there is a relative motion between the ionized part of the gas and the non-ionized gas. If the relative motion is regarded from a frame moving with the plasma, there is a magnetized ionized gas at rest which is hit by non-ionized gas. We can expect phenomena of the same general kind as when a non-ionized gas falls toward a central body through a magnetized gas (a plasma).

The experiment shows that the ionized component is easily accelerated until a certain velocity, the 'critical' velocity v_{crit}, is attained. However, this critical velocity cannot be surpassed as long as there is still nonionized gas. Any attempt to increase the burning voltage V_b above the limiting value V_L in order to accelerate the plasma results in an increased rate of ionization of the gas, but not an increase in the relative

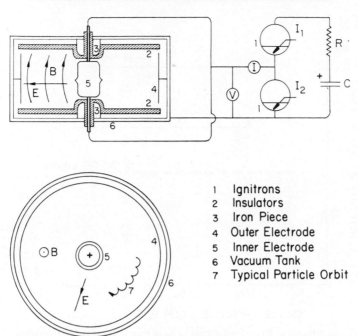

1	Ignitrons
2	Insulators
3	Iron Piece
4	Outer Electrode
5	Inner Electrode
6	Vacuum Tank
7	Typical Particle Orbit

Fig. 16.4.   Homopolar apparatus. A voltage V is applied across an inner electrode (5) and an uter electrode (4) to give a radial electric field. The electric field in the presence of an axial magnetic field B acts on the ionized portion of the gas to set into rotation (7). The interaction between the rotating magnetized plasma and the non-ionized, non-rotating gas (in contact with the wall) produces a voltage limitation indicating that the relative velocity of the two components attains a critical velocity v_{crit}. (From Angerth *et al.*, 1962.)

velocity between the ionized and nonionized components. From the theoretical point of view the phenomenon is rather complicated. The essential mechanism seems to be that kinetic energy is transferred to electrons in the plasma and these electrons produce the ionization (see Section 16.9).

   The limiting value of the burning voltage was found to be independent of the gas pressure in the whole range measured (Figure 16.5) but dependent on the magnetic field (Figure 16.6) as one would infer from Equation (16.6). Further the burning voltage was independent of the applied current, i.e. was equal to V_L, until this exceeded a certain value (which is related to the degree of ionization, see Figure 16.7). Given the relationship of the burning voltage to the radial electric field and the value of the axial magnetic field, one can, from Equation (16.6), determine the critical velocity from the measured value of the limiting voltage V_L. The dependence of the critical velocity on the chemical composition of the gas was also investigated and found to agree with Equation (16.9). Within the accuracy of the experiment, this equation has been checked experimentally for H, D, He, O and Ne (and also for Ar, but with less accuracy). The experimental results are shown in Figure 16.8, where one can observe that the plasma velocity remains rather constant while the applied current (and thus

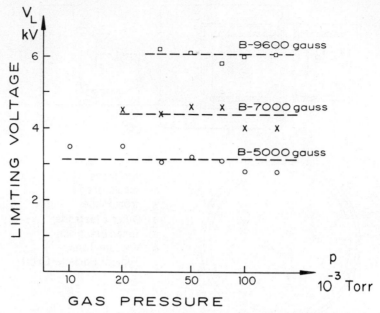

Fig. 16.5.   The limiting value V_L of the burning voltage as a function of gas pressure p for hydrogen in the homopolar experiment. V_L is independent of pressure, but proportional to the magnetic field B. (From Angerth *et al.*, 1962.)

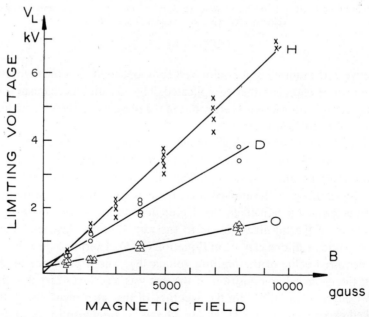

Fig. 16.6.   Limiting voltage V_L versus the magnetic field B in the homopolar experiment. V_L is proportional to B, and depends also on the chemical composition (O, D, H) of the gas being studied. (From Angerth *et al.*, 1962.)

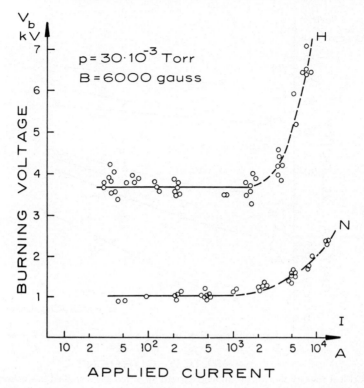

Fig. 16.7.   Burning voltage V_b versus applied current I, for hydrogen and nitrogen in the homopolar experiment. V_b is independent of current (degree of ionization) up to a maximum value related to the complete ionization. The plateau defines the limiting voltage V_L related to the critical velocity. (From Angerth *et al.*, 1962.)

the energy input and degree of ionization) is changed over almost two orders of magnitude.

16.8.2. *Plasma Beam Hitting Neutral Gas Cloud*

The experiment most directly related to the cosmic situation was carried out by Danielsson (1969, 1970). The experimental arrangement is shown in Figure 16.9. The hydrogen plasma is generated and accelerated in an electrodeless plasma gun (a conical theta pinch) and flows into a drift tube along a magnetic field. The direction of the magnetic field changes gradually from axial to transverse along the path of the plasma. As the plasma flows along the drift tube much of it is lost by recombination at the walls. A polarization electric field is developed and a plasma with a density of about $10^{11}-10^{12}$ cm^{-3} proceeds drifting across the magnetic field with a velocity up to 3×10^7 cm s^{-1}. In the region of the transverse magnetic field the plasma penetrates into a small cloud of gas, released from an electromagnetic valve. This gas cloud has an axial depth of 5 cm and a density of 10^{14} cm^{-3} at the time of the arrival of the plasma. The remainder of the system is under high vacuum. Under these conditions

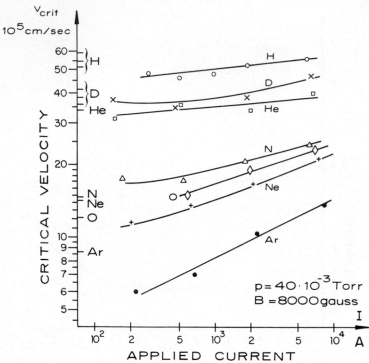

Fig. 16.8.   Critical velocity v_{crit} versus applied current for seven gases studied in the homopolar experiment. (The slope of the Ar curve is related to the magnetic field's being too weak to make the ion gyro-radii small enough). The theoretical v_{crit} for each gas, as calculated from Equation (16.9), is indicated on the ordinate. (From Alfvén and Wilcox, 1962.)

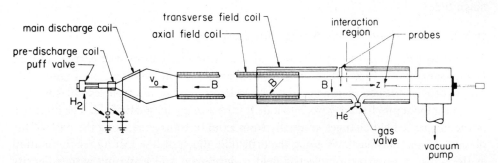

Fig. 16.9.   Experimental arrangement for critical velocity measurement used by Danielsson. The left part is a plasma gun, emitting a magnetized plasma with a velocity v_0. In a long drift tube, the longitudinal magnetization is changed to transverse magnetization. A thin cloud of gas is injected through the gas valve. If v_0 is below the critical velocity, the plasma beam passes through the gas cloud with very little interaction because the mean free path is long. If v_0 is above the critical velocity, there is a strong interaction, bringing the velocity to near the critical value. At the same time, the gas cloud becomes partially ionized. (From Danielsson, 1969.)

the mean free path for direct, binary collisions is much longer than 5 cm so that the interaction according to common terminology is collisionless.

In the experiment it was observed that the velocity of the plasma was substantially reduced over a typical distance of only 1 cm in the gas cloud (see Figure 16.10). It was also found that this reduction in plasma velocity depends on the impinging

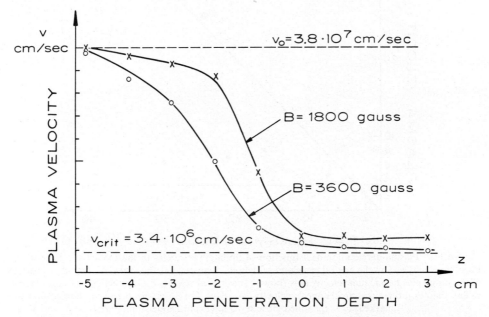

Fig. 16.10.   Velocity retardation to near the critical value in the Danielsson experiment. Plasma deceleration with depth of penetration z in a neutral gas cloud of helium is shown. The front of the cloud is located at $z = -5$ cm and the center, at $z = 0$ cm. The plasma undergoes deceleration from the impinging velocity v_0 to near the critical velocity v_{crit} of helium. Data for two values of the magnetic field B are shown. (From Danielsson, 1969.)

velocity as shown in Figure 16.11. If the neutral gas was helium there was no change in velocity for the smallest impinging velocities (below $\sim 4 \times 10^6$ cm s^{-1}) as the plasma penetrated the gas. For higher impinging velocities there was a relatively increasing deceleration of the plasma.

By investigation of the emission of radiation from the plasma and neutral gas it was found that the electron energy distribution changed drastically at the penetration of the plasma into the gas and that the ionization of the gas atoms was many orders of magnitude faster than anticipated from the parameters of the free plasma stream. The characteristic electron energy was found to jump from about 5 eV to about 85 eV at least locally in the gas cloud. This was inferred to be the cause of the ionization and deceleration of the plasma.

So far Danielsson's experiment has demonstrated that even in a situation where the primary collisions are negligibly few there may be a very strong interaction between a moving plasma and a stationary gas. In helium this interaction is active above an

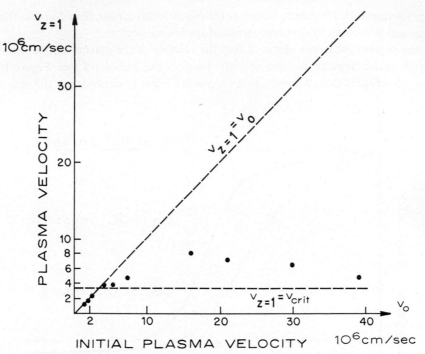

Fig. 16.11.   Plasma deceleration as a function of impinging velocity in the Danielsson experiment. Plasma velocity $v_{z=1}$ in the neutral gas cloud of helium, 1 cm beyond the cloud center, as a function of the initial plasma vacuum velocity v_0 is shown. The critical velocity v_{crit} for helium is indicated. For v_0 less than v_{crit} there was no change in velocity; $v_{z=1}=v_0$. For v_0 greater than v_{crit} deceleration was marked; $v_{z=1}$ remained close to v_{crit}. (From Danielsson, 1969.)

impinging velocity of 3.5×10^6 cm s^{-1} and it leads to

  (a) local heating of the electrons;
  (b) ionization of the neutral gas;
  (c) deceleration of the plasma stream.

16.8.3. *Other Experiments*

Analysis of a number of other experiments confirm these conclusions. In some of the experiments the critical velocity is much more sharply defined and hence better suited for a detailed study of the phenomenon. The experiment described above has the pedagogic advantage of referring most directly to the cosmic situation.

16.8.4. *Conclusions*

Experiments investigating the critical velocity or voltage limitation phenomenon have been conducted under a wide variety of experimental conditions (see Danielsson, 1973). These experiments have demonstrated that as the relative velocity increases a *critical velocity* v_{crit} is reached. When $v < v_{crit}$ there is a small and often negligible interaction between gas and plasma. With $v > v_{crit}$ very strong interaction sets in, leading to ionization of the gas. The onset of ionization is abrupt and discontinuous. The value of v_{crit} for a number of gases has been measured. Although under certain conditions there are deviations up to perhaps 50% the general result is that v_{crit} is the same as v_{ion} as given by Equation (16.9).

16.8.5. *Possible Space Experiments*

Experiments on the critical velocity phenomenon carried out in space are of particular interest since they give more certain scaling to large dimensions. The upper atmosphere provides a region where plasma-gas interaction of this kind could suitably be studied in the Earth's magnetosphere. The first observation of the critical velocity effect under cosmic conditions was recently reported by Manka *et al.* (1972) from the Moon. When an abandoned lunar excursion module was made to impact on the dark side of the Moon not very far from the terminator, a gas cloud was produced which when it had expanded so that it was hit by the solar wind gave rise to superthermal electrons.

16.9. THEORY OF THE CRITICAL VELOCITY

A considerable number of experiments representing a wide variety of experimental conditions have each demonstrated an enhanced interaction between a plasma and a neutral gas in a magnetic field. However, the theoretical understanding of the process is not yet complete although much progress has been made. A review of present theories has recently been given by Sherman (1973). An initial theoretical consideration might reasonably suggest that an ionizing interaction between a gas and a plasma should become appreciable when the relative velocity reaches a value of $(2\,eV_{Ion}/m_a)^{1/2}$, as noted in Section 16.6, Equation (16.9), because the colliding particles then have enough energy for ionization. However, two serious difficulties soon become apparent: (1), the kinetic energy of an electron in the plasma with the above velocity is only $m_e/m_a \cdot eV_{Ion}$ (where m_e is the electron mass) or just a few millivolts, and (2), ionizing collisions between the ions and the neutrals will not occur unless the ion kinetic energy in the frame of reference of the neutral gas exceeds $2\,eV_{Ion}$. This second difficulty follows from the fact that, assuming equal ion and neutral masses and negligible random motion of the neutrals, the maximum inelastic energy transfer equals the kinetic energy in the center of mass system of the colliding particles. It is then evident that any theoretical justification of the critical velocity hypothesis must explain how the ion and/or electron random velocities are increased.

Different theories have been suggested by Sockol (1968), Petschek (1960), Hassan (1966), Lin (1961), Drobyshevskii (1964), Lehnert (1966, 1967) and Sherman (1970a, b). They all refer to different mechanisms of transfer of energy from the atoms/ions to the electrons. We shall not discuss these theories here, but only cite the rather remarkable conclusion which Sherman (1973) draws from his review. He states that for the most part the theories discussed are internally self-consistent. The different theories give a good description of those situations which satisfy the assumptions on which the theories are based. It is remarkable that several widely different theoretical models should all predict the values of E/B near to $(2\,eV_{Ion}/m_a)^{1/2}$. Correspondingly the experiments show that values of E/B near the critical value are observed over a wide range of experiments. The critical velocity hypothesis is then an experimentally confirmed relationship which is valid over a wide range of conditions, but it seems likely that more than one theoretical model is necessary to explain it.

If the atomic mass in Equation (16.9) is replaced by the electron mass m_e we have a result which is a formal analog to the well-known law discovered by Franck and Hertz: $1/2 m_e v_{ion}^2 = e V_{Ion}$. The experimental and theoretical investigations demonstrate that a number of mechanisms exist which make the results of the Franck and Hertz experiment for pure electron interaction valid also for a plasma. The only difference is that here one additional step in the interaction is needed which transfers the energy from the atoms (or the ions, depending on the choice of coordinate system) to the electrons.

Hence the critical velocity experiment may be considered as the 'plasma version' of the classical Franck-Hertz experiment.

16.10. Conclusions about the Critical Velocity

The conclusion from our survey of the experimental and theoretical investigations is that the critical velocity v_{crit} at which a neutral gas interacts strongly with a magnetized plasma is

$$ v_{crit} = v_{ion} = \left(\frac{2 e V_{Ion}}{m_a} \right)^{1/2}. \tag{16.10} $$

Hence, if a gas of a certain chemical composition is falling toward a magnetized central body from a cloud at rest at infinity, it will become ionized when Γ has reached the value

$$ \Gamma_{ion} = \frac{M_c}{r_{ion}} = \frac{v_{crit}^2}{2\kappa} = \frac{e V_{Ion}}{\kappa m_a}. \tag{16.11} $$

Consequently objection (a) of Section 16.6 is not valid and Equation (16.9) is validated.

16.11. Chemical Composition of Infalling Gas

Objection (b) of Section 16.6 states that there is no empirical support for the hypothesis that masses of gas falling toward a central body would have differing chemical compositions. In this section we shall discuss the theoretical conditions under which such chemical differentiation and fractionation could occur.

16.11.1. *The Basic Model*

Let us return to the simple model of Section 16.4 which refers to a jet stream, partially ionized either by radiation or more important by hydromagnetic effects. We assume that the source cloud which contains all elements (e.g., in an abundance relationship more similar to some average 'galactic' composition than now found in the satellites and planets) is partially ionized to such an extent that all elements with ionization potential higher than a certain value $V_{\text{Ion}}(t)$ are neutral, but all with an ionization potential lower than $V_{\text{Ion}}(t)$ are ionized. The Larmor radii of electrons and ions are all assumed to be negligible, but all mean free paths are assumed to be larger than the dimension of the source cloud. The region which we call 'source cloud' may be so defined. All the neutral atoms will begin to fall toward the central body.

Let $V_{\text{Ion}}(t)$ decrease slowly with time (for example through a general cooling of the plasma by radiation or a change in current such as that discussed by De (1973) in the case of solar prominences). When it has fallen below the ionization potential of helium, helium will begin to recombine to form a neutral gas which falls in toward the gravitating central body. Helium reaches its critical velocity of 34.4×10^5 cm s^{-1} at a Γ_{ion} value (which we now realize, recalling Equation (16.11)), to be equivalent, to the Γ value) of 0.9×10^{20} (the upper region of Band I of Figure 16.12). The gas will at this point become ionized, forming a plasma cloud which will be referred to as the A cloud. When $V_{\text{Ion}}(t)$ decreases further, and has passed the ionization potential of hydrogen (which is nearly equal to the ionization potentials of oxygen and nitrogen), hydrogen, oxygen and nitrogen will start falling out from the source cloud. Because hydrogen is by far the most abundant element we can expect the infalling gas to behave as hydrogen and to be stopped at $\Gamma = 1.9 \times 10^{20}$ (the lower region of Band I) forming what we shall call the B cloud. In a gas, consisting mainly of H, the elements O and N will not be stopped at their critical velocities because of the quenching effect of the hydrogen on the acceleration of electrons that would lead to ionization in pure oxygen or nitrogen gas.

Next will follow an infall dominated by carbon, which is stopped at a v_{crit} of 13.5×10^5 cm s^{-1} and a Γ value of 0.1×10^{19} (Band II), forming the C cloud, and finally the heavier elements, mainly silicon, magnesium and iron, will fall in to $\Gamma = 0.3 \times 10^{18}$ (Band III), producing the D cloud with weighted mean critical velocity of 6.5×10^5 cm s^{-1}.

16.11.2. The A, B, C and D Clouds in the Solar System

From the above discussion one can consider the solar system as forming from four plasma clouds. The planets would form by accretion of planetesimals and grains, the matter condensing from the plasma cloud in the specific region of gravitational potential of each planet. The location of each plasma cloud is determined by the critical velocity of the controlling elements in it as depicted in Figure 16.12. Hence each plasma cloud can be characterized by a dominant critical velocity. Figure 16.13 shows the gravitational potential energy bands labeled as plasma clouds A, B, C and D with their respective critical velocities indicated. We see from Figure 16.13 and the discussion in the previous section that Mercury, Venus and Earth formed from the B cloud, while Moon and Mars accreted within the A cloud. As indicated in Figure 16.13, there was probably an overlap and possibly an interchange of matter between the A and B clouds in the region of the Earth and the Moon. The giant planets formed within the C cloud while Pluto and perhaps Triton accreted within the D cloud. Referring to Figure 14.8a, we can see that, although there is a wide range of densities

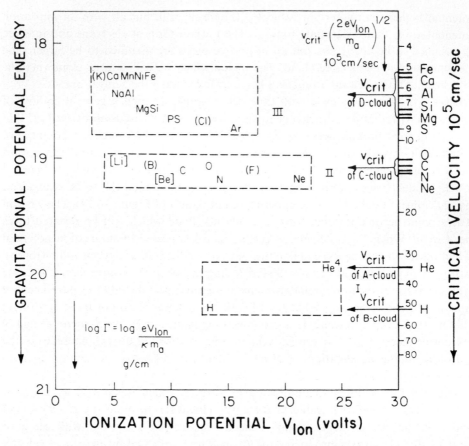

Fig. 16.12.   Critical velocity and ionization potential of the most abundant elements. The left hand ordinate showing gravitational potential energy and the right hand ordinate showing the critical velocities of elements and the *A, B, C* and *D* clouds allows a comparison of Γ values and v_{crit} values.

in the solar system, the bodies which formed in the same cloud have similar densities. This pattern can be understood on the basis of relatively constant composition within each cloud, but variance of composition among the A, B, C and D clouds.

Returning to Figure 16.13 we see that there were plasma clouds formed around each planet. Our hetegonic principle (Section 1.3) stresses that the same processes which formed the planetary system should also prove capable of forming the satellite systems. As depicted in Figure 16.2, the jet stream formed within a plasma cloud will provide material for a planet and will function as the source cloud for a series of plasma clouds which will form around that planet by the processes discussed in Section 16.11.1. Thus, each planet will act as the gravitating (and magnetized) rotating body around which A, B, C and D clouds will form. Formation of the plasma clouds is determined by the attainment of critical velocity by the element which determines the orbital distance of the cloud to the central body. For planets of less mass, the inner clouds cannot form due to inadequate acceleration of the infalling gas from the source

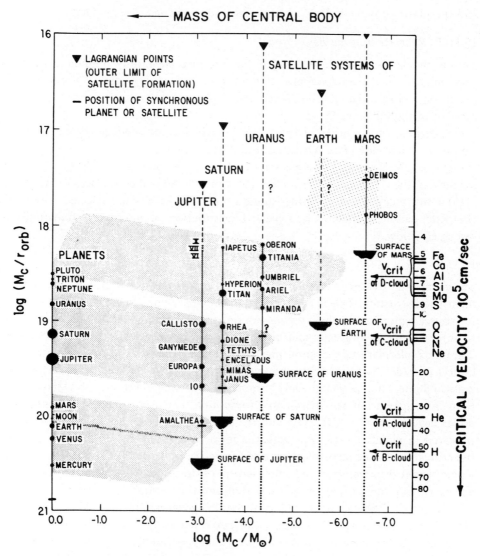

Fig. 16.13. Gravitational potential energy as a function of the mass of the central body for the planetary and satellite systems. The right hand ordinate showing critical velocity affords comparison of Γ values for the planets and satellites with v_{crit} values for abundant elements and the A, B, C and D clouds.

clouds. We see in Figure 16.13 that Jupiter is massive enough to have had an A cloud, but not a B cloud form. The Galilean satellites of Jupiter formed in the Jovian C cloud. The Saturnian inner satellites formed in the Saturnian C cloud, while the outer satellites formed in the D cloud around Saturn. The satellites of Uranus accreted in the Uranian D cloud.

Therefore all discussion of band formation, gravitational potential energy bands and the plasma clouds A, B, C and D refer both to planetary and satellite systems.

16.11.3. *Refinement of the Basic Model*

This is the most simple model which can produce chemically differentiated mass accumulation in the observed gravitational potential energy bands. Of course it is much too simple to be realistic. When discussing and developing it we have to take account of the following facts:

(1) There are a number of plasma processes which could produce chemical separation in a cosmic cloud (see Arrhenius and Alfvén, 1971).

(2) The critical velocity of a gas mixture has not yet been thoroughly studied. We expect that the value v_{crit} is determined by the most abundant constituent in the cloud.

(3) Other charged species besides single atomic ions have been neglected. The more complete picture including the expected distribution of molecular compounds is discussed below.

16.11.4. *Effect of Interstellar Molecules*

The elementary treatment given above suggests only the gross features of the emplacement band structure. This is modified to some extent by the fact that the elements in the source regions are likely to prevail not only as monatomic species but also, at least to some extent, as molecules and molecular ions. The experiments carried out with diatomic molecular gases (Section 16.8) indicate that ionization at the critical velocity limit is preceded by dissociation and that the limit hence is determined by the atomic mass and ionization potential. Only homonuclear molecules (H_2, D_2, N_2) have so far been investigated but it is reasonable to assume that in the case of heteronuclear molecules such as CH, CH_4, OH and the multitude of other polynuclear molecules observed in clouds the element with the lower ionization potential will determine the critical limit. The main effect expected from the presence of molecular precursors would therefore be transport and emplacement of stoichiometric amounts of hydrogen, accompanying carbon, oxygen and nitrogen into the C band.

In the case of the commonly observed simple hydrides (CH, NH, OH, OH_2, CH_2, NH_2, NH_3), the ligated hydrogen contributes relatively little to the mass of the molecule. Furthermore the molecular ionization potential is similar to or slightly lower than that of the core atom. Hence, even if there is a so far undetermined effect of the molecular state, we would expect the critical velocity to remain close to that of the core atom.

In the case of molecules containing elements from rows II and III (SiO, AlO, MgO), the ionization potential of the molecule is substantially increased over that of the metal atom. Critical velocities (which are entirely hypothetical) calculated from mass and ionization potential of such molecules place them in the same band as the metals (the increased ionization potential is balanced by the mass increase). The effect, if any, would consequently be to contribute oxygen to the D band.

It is important to notice that in no case does molecular formation from abundant species of atoms lead to such an increase in ionization potential that penetration inside the C cloud is possible by this mechanism. Deposition in the A and B clouds therefore would depend entirely on transport of impurities together with major amounts of helium and hydrogen, and on evaporation of solid grains falling toward the Sun, as discussed in Section 14.7.

One can conclude from the above discussion that, although direct empirical evidence of source cloud composition during the formation era of the solar system is indeed lacking, there are many cogent theoretical possibilities to account for differing composition of the gravitational potential energy bands resulting from infall into the circumsolar region. Therefore objection (b) is relevant only in its emphasis on the need for continued observation and experimentation.

16.12. The chemical composition of the solar system and inhomogeneous plasma emplacement

Objection (c) of Section 16.6 states that the chemical composition of the bodies found in each gravitational potential energy band are not characterized by those elements which theoretically give rise to each specific band. In this section we shall consider a more detailed theoretical model of band formation.

16.12.1. *A Model of Band Formation*

We are certainly far from a consistent model of the infall of plasma. The discussion here should be confined to some basic principles.

As stated in Section 10 homogeneous models are of little value in astrophysics. Hence if we assume that the source cloud is a homogeneous shell from which there is a symmetric and time constant fall-down of gas (the simple model of the previous section), we may go completely astray. Inhomogeneous models are necessarily rather arbitrary and the final choice between possible models can be made only after extensive experiments in the laboratory and in space.

In almost any type of inhomogeneous model one should envisage a number of source clouds from which a gas is falling down during finite periods (see Figure 16.14). At a certain instant one or several clouds may be active. The chemical composition of the gas falling out from a certain source cloud may vary. For our model the most important question is to ask which element dominates in such a way that it determines the value of the critical velocity v_{crit} and hence the arresting value of the gravitational potential energy Γ_{ion}.

Suppose that after there has been no infall for a long time, gas with a certain value of Γ_{ion} begins to fall in from one source cloud. This gas will then accumulate in the band characterized by Γ_{ion}. If another cloud with a different characteristic Γ_{ion} begins to fall in, its gas will accumulate in the correspondingly different region, under the condition that when the infall of the second cloud starts the first infall has already ceased, and there has been enough time for the accumulated plasma to condense. However, if this condition is not satisfied, the plasma from the first infall may interfere with the second infall. Suppose for example that the first infall produced a plasma cloud in the C cloud band, and that the second gas infall has Γ value of the B band.

INFALL PATTERN FOR SOURCE CLOUDS ACTIVE
DURING DIFFERENT TIME PERIODS

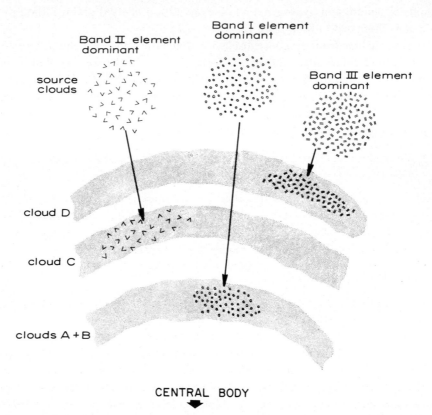

Fig. 16.14.   Infall pattern for source clouds active during different time periods. The infalling gas from one source cloud will be dominated by one element. The mass of infalling gas will be stopped when the dominant element is ionized, i.e. in the cloud corresponding to the critical velocity value of that element's band. For example, a Band II element reaches its critical velocity at a value r_{ion} which falls within the C cloud.

Then it can reach the B region only if there is no plasma in the C region, because, if there is, the infalling gas, normally penetrating to the B region will interact with the plasma in the C region (if it is dense enough; mean free path shorter than C-cloud thickness) and become ionized and hence stopped prematurely. Under certain conditions most of the new cloud will be trapped in the C region. See Figure 16.15.

Hence we see that an infall of hydrogen rich material may be trapped in any of the bands. It arrives to the B band only if it is not hindered by plasma in any of the upper bands, but if a recent infall of gas into e.g., the C cloud band has taken place, most of the gas which subsequently falls in may be trapped there. Then under certain conditions there may be, for example, more B band gas trapped in the C region than there is C band gas.

From this we can draw the important conclusion that although the stopping of

INFALL PATTERN FOR SOURCE CLOUDS
ACTIVE DURING ONE TIME PERIOD

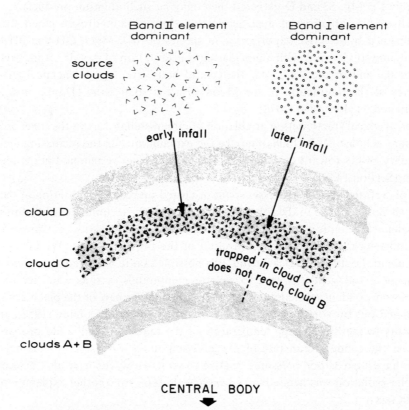

Fig. 16.15.   Infall pattern for source clouds active during one time period. A gas infall from a Band II element dominated source cloud will be ionized and stopped in Cloud C. If this plasma has not had time to condense, it will interact with any infall from a Band I element dominated source cloud. The Band I gas infall will be trapped in Cloud C by the plasma there and not reach its own r_{ion} in Cloud B.

injected gas in a certain band depends on the v_{crit} value of a controlling element, an inhomogeneous model need not necessarily predict that this element should dominate the ultimate chemical composition of the cloud. Although the trigger element should be enriched to some extent, the chemical composition of the bands need not deviate very much from that of the source clouds.

Furthermore the grains which have condensed in the source cloud will fall in through the regions where plasma gets accumulated. When penetrating these regions they may partially evaporate and inject part of their mass into the accumulated gas. This ablation effect would become important when the infalling grains have been accelerated to high velocities relative to the plasma clouds and in regions where the plasma density is comparatively high. Hence one would expect contamination by grain ablation to be most pronounced in the A-B region. Since the plasma in this region is dominated by helium and hydrogen, a major fraction of the condensable ions, probably Fe, Mg, Si and O gathered there may be such ablation products.

Furthermore, ablation of dust accelerated through a hydrogen cloud would be accompanied by selective vaporization of species of SiO, MgO, OH and SH leaving the infalling solid grains with an increasing concentration of metallic iron, vaporizing toward the end of the trail, near the central body. Analogous chemical effects are observed at the interaction of the Moon with the solar wind (Hapke *et al.*, 1970; Epstein and Taylor, 1970, 1972).

Such chemical fractionation at ablation of circumstellar dust in the inner solar and planetary nebulae (A-B clouds) may be the explanation for the increasing density of secondary bodies toward the central bodies in the Jovian system and in the planetary system (Section 14.7).

All of the fractionation processes so far discussed precede condensation of the solids from which the bodies in the solar system subsequently accumulated. In addition it is likely that fractionation processes associated with the condensation and later evolution have influenced the chemical composition of the present bodies. We do not know much about the state of the early Sun. It is possible that it had a radiation field at least as intense as today. If this were so, volatile compounds such as water ice were prevented from condensation and accumulation in the inner part of the planetary system, as pointed out by von Weizsäcker (1944), Berlage (1948) and Urey (1952). In close proximity to the Sun, a high temperature of the radiation field could probably also decrease the condensation rate of oxygen compounds with silicon and magnesium which have high vapor pressures relative to iron. However it is also possible that the solar radiation was negligible and we have to look for another explanation of the quoted facts.

Finally the embryos accreting to form the giant planets may have been massive enough to collect and retain substantial amounts of hydrogen and helium.

16.12.2. *Conclusions About the Chemical Composition of Celestial Bodies*

We are necessarily dealing with highly hypothetical phenomena which do not allow one to draw very specific conclusions. However, we here summarize the processes most likely to influence the bulk composition of the accumulated bodies:

(1) The critical velocities of the element groups corresponding to clouds A, B, C and D; this effect would also be responsible for the spacing of the groups of secondary bodies around their primaries.

(2) The vapor pressure of the solids that can form from the gases controlling the cloud formation; since hydrogen and helium are not condensable, the bodies formed in the A and B clouds consist largely of small amounts of 'impurities'.

(3) The fractional vaporization of infalling dust in the dense clouds, particularly bringing heavy elements to the central reactive hydrogen cloud (B-cloud).

(4) Trapping of infalling gases with high critical velocities in already established clouds.

(5) Fractionation at condensation, due to the gradient in the solar radiation field.

(6) Gravitational accumulation of hydrogen and helium by the giant planets.

It will require much work before we can decide between models giving similar composition to all the bands and models in which there are appreciable chemical differences among the regions. Such work should include both controlled interaction and fractionation experiments in hydrogen and in mixed plasmas as well as the sampling and analysis of materials as unaltered as possible from their primordial state.

What has been said in Chapter 14 and Section 16.12 shows the complexity of the problems relating to chemical composition of the celestial bodies. Hence objection (c) is not valid, but we are still far from a detailed theory of chemical composition, and we also lack empirical data. Indeed we are now far from the time when sweeping conclusions about the chemical compositions of the bodies in the solar system were considered acceptable.

16.13. MODIFICATION OF CRITICAL VELOCITY IONIZATION DISTANCE DUE TO INTERACTION WITH A PARTIALLY COROTATING PLASMA

The simple model of Section 16.4 could be developed in different directions. The falling gas need not necessarily interact with a plasma at rest. If, for example, the plasma is in the state of partial corotation (see Section 12) its tangential velocity is (from Table 12.1)

$$v_\phi = \left(\frac{2\kappa M_c}{3r}\right)^{1/2}.$$

(16.12)

Adding this vectorially to the velocity of fall

$$v = \left(\frac{2\kappa M_c}{r}\right)^{1/2},$$

(16.13)

we get a resulting relative velocity v_{rel}

$$v_{rel} = \left(\frac{8\kappa M_c}{3r_{rel}}\right)^{1/2}.$$

(16.14)

When v_{rel} reaches the critical velocity v_{crit} the infalling gas can become ionized. Let us determine the orbital radius r_{rel} at which ionization can take place.

From Equation (16.4) we have

$$v_{crit} = \left(\frac{2\kappa M_c}{r_{ion}}\right)^{1/2}.$$

(16.15)

Equating v_{crit} to v_{rel} we obtain

$$r_{rel} = \tfrac{4}{3} r_{ion}.$$

(16.16)

This relative velocity due to the corotation of the magnetized plasma attains the critical value v_{ion} at $\frac{4}{3}$ the orbital radius at which ionization would occur if the plasma were not in a state of partial corotation with the central body.

There is yet another effect seen when the interacting plasma is in a state of partial corotation. Condensation and accretion of matter reduces the orbital radius by a factor $\frac{2}{3}$ (as explained in Section 12). Combining the effects of the tangential velocity and the condensation characteristic of a corotating plasma we obtain the value for the effective ionization radius r'_{ion} for a gas falling through a corotating plasma as

$$r'_{ion} = \tfrac{2}{3} r_{rel} = \tfrac{2}{3} \times \tfrac{4}{3} r_{ion} = 0.89 \, r_{ion}. \tag{16.17}$$

Therefore, in Figure 16.13 the critical velocity scale should be displaced downward along the gravitational energy scale so that the value of r_{ion} is decreased to $0.89 \, r_{ion}$ and corresponds to r'_{ion} for the case of corotation of the plasma.

Yet another correction may be of some importance. If the central body is accreting mass during a period of plasma accumulation the angular momentum of the grains condensing in its environment will change during the accretion. Our present calculations are valid in the case that practically all the gas injection takes place when the central body is close to its final state of corotation. A refinement of the theory in this respect cannot be made before the variation of the gas content in the jet stream can be estimated. It should also be remembered that the formation of secondary bodies cannot start before the central body has acquired a magnetic field which makes possible a transfer of angular momentum.

17. The Structure of the Groups

17.1. IONIZATION DURING THE EMPLACEMENT OF PLASMA

In the preceding section we have discussed the hypothesis that the location of the different groups of secondary bodies is determined by the critical velocity phenomenon. However, the internal structures of the groups differ in the respect that in some of them (e.g., the giant planets) the mass of the bodies decreases rapidly with increasing distance from the central body, whereas in other groups (e.g., the inner Saturnian satellites) the reverse is true. In this section we shall show that this difference in structure among the groups probably is related to the total energy dissipated in the process of emplacement of the plasma. This leads to the conclusion that the structure of a group depends on the ratio τ/T between the typical orbital period τ of the secondary bodies of the group and the spin period T of the central body. There is observational support of this dependence (see Sections 17.5 and 17.6). In fact the mass distribution in the groups seems to be a function of τ/T.

As in some of the earlier sections we are obviously far from a detailed theory, and the aim of our discussion is essentially to call attention to what may be the basic phenomena determining the structure of the groups.

According to our model, a mass m, originally at rest at 'infinity', falls in to the ionization distance r_{ion} where it becomes partially ionized (Figure 17.1). By transfer of angular momentum from the central body this mass is brought into partial corotation.

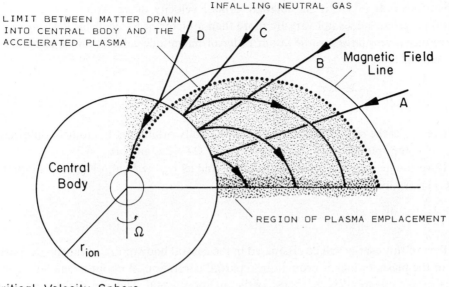

Fig. 17.1. Complete ionization of infalling gas. Gas falling in from infinity reaches the critical velocity at r_{ion} (the critical velocity sphere) and becomes partially ionized. It is rapidly included in 'superprominences' which, if $\zeta \tau_{ion}/T \gg 1$, are almost completely ionized. Matter falling in at low latitudes (A, B and C) will be emplaced in the equatorial plane and condense there. Matter arriving at the critical velocity sphere at high latitudes (D) will be drawn into the central body. Note that the processes A, B and C do not necessarily interfere because they may occur at different times or even simultaneously at different longitudes.

It condenses and through processes discussed in Sections 13.2 and 13.10 it is eventually placed in a circular orbit with the radius r. The total release of energy, $W = W_g + W_t$, during the process consists of two parts. The first one is

$$W_g = \frac{\kappa M_c m}{2 r_{ion}}, \tag{17.1}$$

deriving from the difference in energy between the initial position at infinity and the final state in orbital motion.

The second part W_t is the energy released at the transfer of angular momentum. When the spin of the central body is decreased by $\Delta\Omega$ from its original value Ω, the energy

$$W_t = \Xi\Omega\Delta\Omega \tag{17.2}$$

is released (if $\Delta\Omega \ll \Omega$). Ξ is the moment of inertia of the central body. Because

$$\Xi\Delta\Omega = mr^2\omega, \tag{17.3}$$

we find that

$$W = \frac{\kappa M_c m}{r}\left(\frac{\Omega}{\omega} + \frac{1}{2}\right), \tag{17.4}$$

where $\omega = (\kappa M_c/r^3)^{1/2}$ is the angular orbital velocity of m. As $r\omega \sim r^{-1/2}$ and as within a group r does not vary by more than a factor of 6 (see Table 2.4), we do not introduce a very large error if in our order-of-magnitude calculation we approximate Equation (17.4) by

$$W = \frac{\kappa M_c m}{r_{ion}} \frac{\tau_{ion}}{T},$$ (17.5)

where τ_{ion} is the orbital period of a fictitious body orbiting at the ionization distance r_{ion}, T is the spin period of the central body, $\Omega T = 2\pi$, and $\omega \tau_{ion} = 2\pi$.

If we equate m to the mass of an atom m_a and let $r_{ion} = \kappa M_c m_a/eV_{Ion}$ (from Equation (16.11)) we have

$$W = eV_{Ion} \frac{\tau_{ion}}{T}.$$ (17.6)

Part of this energy will be dissipated in the central body or in its ionosphere, part of it, in the plasma which is brought into partial corotation. Without a detailed analysis it is reasonable to guess that these parts are about equal. The energy delivered to the plasma goes primarily to increasing the electron temperature. When this has reached a certain value, most of the energy is radiated but a fraction ζ is used for ionization.

In laboratory studies of electric currents in gases it has been shown that ζ seldom exceeds 5%. For example, in a glow discharge the minimum voltage V_c between the electrodes (which actually equals the cathode fall) is usually 200–300 V, (essentially only pure noble gases have lower values). This holds for example for H_2, N_2, and air (v. Engel, 1955, p. 202) for which the voltage needed to produce ionization is in the range 10–15 V. Hence this ratio $\zeta = V_{ion}/V_c$, which gives the fraction of the energy which goes into ionization, is about 0.05. Even if the discharge in our case differs in certain respects, we should not expect ζ to be drastically different. Taking account of the fact that only a fraction of W is dissipated in the plasma we should expect ζ to be less than 0.05.

Hence, without making any detailed model of the process we may conclude that if ζW denotes the energy which goes into ionization of the plasma, ζ is not likely to exceed 0.05. This means that it is impossible to produce a complete ionization of the plasma if τ_{ion}/T is of the order 10 or less. A considerably higher value is probably needed for complete ionization to occur.

We then conclude:

(1) *Ceteris paribus* the degree of ionization at the emplacement is a function of τ_{ion}/T;

(2) We may have complete ionization if τ_{ion}/T is for example 100 or more, but probably not if it is of the order of 10 or less.

In Section 17.2 we shall treat the case

$$\frac{\tau_{ion}}{T} \gg \zeta^{-1},$$ (17.7)

which indicates complete ionization, reserving the case of incomplete ionization

$$\frac{\tau_{ion}}{T} < \zeta^{-1} \tag{17.8}$$

for Section 17.3.

17.2. COMPLETE IONIZATION

We shall now discuss the extreme case $\tau_{ion}/T \gg \zeta^{-1}$, implying that the plasma is completely ionized. The gas which falls in is stopped at the critical velocity sphere, which is defined by $r_{ion} = 2 \kappa M_c/v_{crit}^2$, where it immediately becomes partially ionized (see Figure 17.1). The transfer of angular momentum gives it an azimuthal velocity which increases until partial corotation is achieved. The energy release associated with this process ionizes the plasma completely.

As stressed earlier, it is important to note that homogeneous models are obsolete in cosmic plasma physics. In order to reduce the speculative element which hetegonic theories necessarily include, it is essential to connect the models as far as possible with laboratory experiments and such cosmic phenomena as we observe today. For the discussion references to magnetosphere and especially solar phenomena are essential. The transfer of angular momentum through a set of 'superprominences', as discussed in Section 11.7 and by De (1973), is the background for our present treatment (see Figure 11.2).

Hence we should consider the infall of gas as taking place in a series of intermittent infalls with a finite extension and a finite lifetime. Several infalls could very well be active simultaneously. The gas which arrives at the critical velocity sphere r_{ion} and becomes partially ionized, is rapidly incorporated in a superprominence which is almost completely ionized because $\zeta\tau_{ion}/T \gg 1$ guarantees that in the long run there is enough energy for ionization. The processes which the infalling gas is subject to at r_{ion} confine the gas to a magnetic field line. Its final destiny is either to fall along this field line to the central body, or to attain an increasing angular momentum so that it is brought to the neighborhood of the equatorial plane where it condenses. There are regions around the axis of the central body where the former process takes place, whereas the latter process occurs in a band near the equatorial plane.

Figure 17.1 depicts the projection on the meridional plane and should be interpreted with what is said above as a background.

As the average mass distribution is uniform over the surface of the sphere r_{ion}, the mass dM between the latitude circles at ϕ and $\phi + d\phi$ amounts to

$$dM = M_0 \cos \phi d\phi, \tag{17.9}$$

M_0 being a constant. The equation of the magnetic lines of force is

$$r_B = r \cos^2 \phi, \tag{17.10}$$

where r_B is the distance to the central body from a point on the line of force and r is the value of r_B at the equatorial plane. Putting $r = r_{\text{ion}}$ we obtain by differentiating Equation (17.10)

$$\mathrm{d}\phi = \tfrac{1}{2} \frac{\cot \phi}{r} \, \mathrm{d}r \tag{17.11}$$

and

$$\frac{\mathrm{d}M}{\mathrm{d}r} = \frac{M_0 r_{\text{ion}}}{2} \frac{1}{r^2 (1 - r_{\text{ion}}/r)^{1/2}}. \tag{17.12}$$

This function is plotted in Figure 17.2.

Let us now see whether it is possible that the outer planets have originated from a gas having the mass distribution given by Equation (17.12). We assume that r_{ion} coincides roughly with the present value of the orbital radius of Jupiter ($r_{\text{♃}}$) and that all gas situated between $r_{\text{♃}}$ and the orbital radius of Saturn ($r_{\text{♄}}$) is used to build up Jupiter. (The fact that according to Section 13, all distances are likely to decrease by a factor of $\tfrac{2}{3}$ is not crucial in this respect.) In the same way we assume that all matter between $r_{\text{♄}}$ and $r_{\text{♅}}$ (Uranus) is condensed to Saturn, etc. Thus we should expect the following masses of the planets:

Jupiter:

$$M_{\text{♃}} = A \int_{r_{\text{♃}}}^{r_{\text{♄}}} \frac{\mathrm{d}r}{r^2 (1 - r_{\text{ion}}/r)^{1/2}}, \tag{17.13a}$$

Neptune:

$$M_{\text{♆}} = A \int_{r_{\text{♆}}}^{r_{\text{♇}}} \frac{\mathrm{d}r}{r^2 (1 - r_{\text{ion}}/r)^{1/2}}. \tag{17.13b}$$

where $r_{\text{♇}}$ is the orbital radius of Pluto and A is defined by

$$M_{\text{total}} = A \int_{r_{\text{♃}}}^{r_{\text{♇}}} \frac{\mathrm{d}r}{r^2 (1 - r_{\text{ion}}/r)^{1/2}}, \tag{17.14}$$

The relative masses of the planets calculated from equations of the form (17.13) and the observed masses are given in Table 17.1. The calculated values agree with observations within a factor of 2. (The integral from Pluto to infinity is 32 units, but as this mass has become ionized near the axial region of the Sun, it is likely to have fallen directly into the Sun; note 'D' in Figure 17.1.)

The assumption that the gas is divided exactly at the present distances of the planets is, of course, arbitrary, and a more refined calculation has been given elsewhere (Alfvén, 1954, Chapter V). But if we go in the opposite direction, we can interpret the results as follows. Suppose that we 'smear out' the masses of the outer planets so

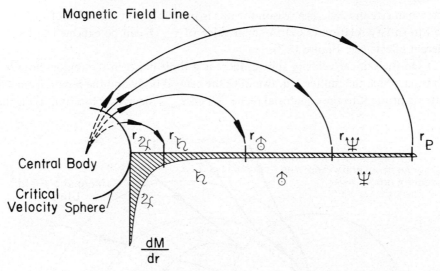

Fig. 17.2.   Matter stopped at the critical velocity sphere is displaced outward along the magnetic field lines and condenses in the region of the equatorial plane. For a rough estimate it is assumed, rather arbitrarily, that all matter between the present orbits of Jupiter and Saturn is now included in Jupiter, etc. As shown by Table 17.1 this gives roughly the observed mass distribution. The essence of the analysis is that the 'smeared out' density in the region of the giant planets is compatible with the model of Section 17.2.

TABLE 17.1

Planet	Mass (Earth = 1)	
	Calculated	Observed
Jupiter	320	317
Saturn	88	95
Uranus	26	15
Neptune	10	17

that we obtain a continuous mass distribution in the equatorial plane. A projection of this along the magnetic lines of force upon a sphere gives us an almost uniform mass distribution. Consequently, the mass distribution obtained in this way shows a reasonable agreement with the mass distribution among the giant planets.

We now turn our attention to the outer Saturnian satellites. This is a group which also has a very high value of τ_{ion}/T. The group is *irregular* (see Section 17.8) and it is difficult to conclude the original mass distribution from the three existing bodies. However, it is evident that in this group also most of the mass is concentrated in the innermost body, Titan, which is situated somewhat below ionization limit.

17.3. PARTIAL IONIZATION

It is only in these two groups – the giant planets and the outer Saturnian satellites – that the innermost body is the biggest one. In all other groups there is a slow or rapid decrease in size inward. The reason for this is probably that the value of τ_{ion}/T is too small to satisfy relation (17.7). A small value of τ_{ion}/T can be expected to have two different effects (see Figure 17.3):

(1) On the critical velocity sphere, there is a limit between the region close to the axis from which the matter is drawn in to the central body and the region from which matter is brought to the equatorial plane. When τ_{ion}/T decreases this limit is displaced

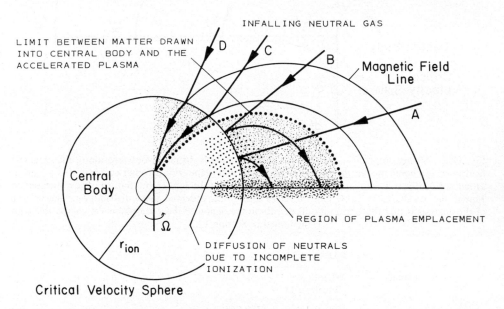

Critical Velocity Sphere

Fig. 17.3.   Partial ionization of infalling gas. Small values of $\tau_{ion}/T\,(<20)$ imply an increase of the region near the axis of the central body from which matter is drawn into the central body. Incomplete ionization at r_{ion} is also implied and diffusion of neutral gas toward the central body will take place. The result is a displacement inward of the region of plasma emplacement and a change in the mass distribution within a group of secondary bodies.

away from the axis. The result of this is that no matter is brought down to the equatorial plane at a large distance from the critical velocity sphere. Hence, in comparison with the case of very large τ_{ion}/T, the outer limit of the region where bodies are produced will be displaced inward.

(2) As all the gas is not ionized at the critical velocity sphere, part of it will fall closer to the central body where sooner or later a considerable part of it is collected in jet streams. Hence mass is collected even far inside the critical velocity sphere. These two effects are further discussed in Section 17.7.

17.4. CHANGE OF SPIN AT THE FORMATION OF SECONDARY BODIES

The result of our discussion is that we should expect the mass distribution within a group of bodies to depend on the value of τ_{ion}/T. However, the value of this quantity should not be the present value but the value at the time of formation. The angular momentum which Jupiter, Saturn and Uranus have transferred to orbital momenta of their satellites is small (of the order of 1%, see Tables 2.1 and 2.2) compared with the spin momenta of these planets, and no other mechanism by which they can lose a large fraction of their momenta is known (see Section 6.2). Hence, it is reasonable to suppose that they possessed about their present angular momenta at the time of formation of their satellite systems. Their moments of inertia may have changed somewhat during the planetary evolution, but this change is likely to be rather small. Hence, it seems reasonable to state that the axial rotations of these planets had approximately their present angular velocity at the time when their satellite systems were formed.

This conclusion does not hold for the Sun. Its present angular momentum is only 0.6% of the total angular momentum of the solar system. Hence, if the Sun has lost angular momentum only through transfer to planets, it has transferred 99.4% of its original angular momentum to the orbital momenta of the giant planets. This effect should make the value of τ_{ion}/T about 180 times larger at the beginning of the formation of the planetary system. However, the Sun may also have lost angular momentum to the solar wind. Whether this has been an appreciable amount or not is uncertain (see Section 11.2), but it is possible that this factor of 180 should be still larger.

On the other hand, the moment of inertia of the Sun may have changed. If, at a very early stage, the Sun was burning its deuterium, its radius would be about 16 times larger than now (Brownlee and Cox, 1961). Hence, for example, if the planets were formed around the deuterium-burning Sun, these two effects would approximately compensate each other, and the present values of τ_{ion}/T would be valid.

These considerations are not very important for the formation of the giant planets because this group would, for either extreme value of T, have values of τ_{ion}/T which satisfy Equation (17.7). On the other hand, it does not seem legitimate to use the present values of τ_{ion}/T for the terrestrial planets. Hence, we exclude them from our analysis.

17.5. OBSERVATIONAL VALUES OF τ_{ion}/T

Before calculating theoretically the values of τ_{ion}/T for the different groups, we shall plot the observational values of the ratio τ/T between the Kepler period τ of a secondary body and the period T of the axial rotation of its central body. This gives us Figure 17.4.

It appears that for the giant planets the value of τ/T is of the order of 1000, and for the outer Saturnian satellites about 100. The Galilean satellites and the Uranian satellites have similar values, ranging from about 5 up to about 50. The inner Saturnian satellites have values between 2 and 10. (The terrestrial planets, which should not be included in our analysis, lie between 3 and 30.)

In order to characterize each group by a certain value of τ/T we could take some sort of mean value of the values for its members. From a theoretical point of view the least arbitrary way of doing so is to use the value τ_{ion} of the Kepler motion of a mass moving at the ionization distance, as we have done in Section 17.1. Referring to Figure 16.13 we see that each group falls into one of the 'clouds' surrounding its

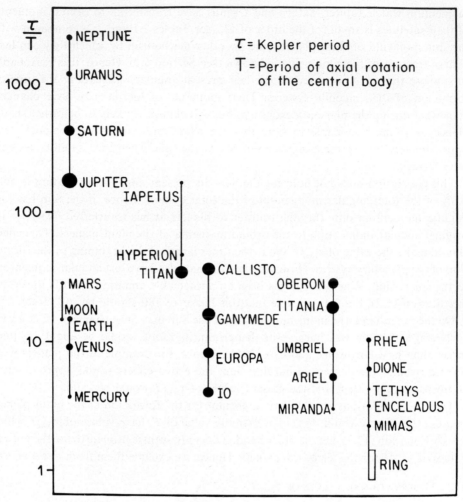

Fig. 17.4.   Ratio between the orbital period τ of secondary bodies and the spin period T of the prima-ry body. The latter quantity may have changed for the Sun, but not for the planets. The secondaries are grouped according to the 'cloud' in which they formed. From left to right are the terrestrial planets, the giant planets, outer Saturnian satellites, Galilean satellites of Jupiter, the Uranian satellites and the inner satellites of Saturn.

central body. To analyze a group in terms of τ_{ion}/T we must choose the ionization distance r_{ion} for the group as a whole. In this treatment we shall use the r_{ion} which corresponds to the critical velocity v_{crit} of each cloud as denoted in Figure 16.13.

Setting $r = r_{\text{ion}}$, we have

$$\tau_{\text{ion}} = \frac{2\pi}{\omega} = 2\pi \left(\frac{r_{\text{ion}}^3}{\kappa M_c}\right)^{1/2} ; \qquad (17.15)$$

and from Equation (16.11)

$$\frac{\kappa M_c}{r_{\text{ion}}} = \tfrac{1}{2}v_{\text{crit}}^2 ; \tag{17.16}$$

it follows that

$$\tau_{\text{ion}} = 2\pi \sqrt{2}\, \frac{r_{\text{ion}}}{v_{\text{crit}}} = 4\pi \sqrt{2}\, \frac{\kappa M_c}{v_{\text{crit}}^3}, \tag{17.17}$$

where v_{crit} is the velocity characterizing the cloud.

17.6. Mass distribution as a function of τ_{ion}/T

In Figure 17.5 the masses of the bodies are plotted as a function of the orbital distances. The distances are normalized with the ionization distance r_{ion} as unit: $x = r/r_{\text{ion}}$. This value for each body is called the 'normalized distance'. The normalized distances for the planets and their satellites are given in Table 17.2.

The values of the normalized distance are not rigorously obtained. As r_{ion} is a function of v_{crit} all the uncertainty introduced in assigning a characteristic v_{crit} to a specific cloud (see Sections 16.11–16.12) also pertains to the values of the normalized distance. Further (see Section 16.13) one ought to reduce the r_{ion} to $0.89\, r_{\text{ion}}$ to take account of the $\tfrac{2}{3}$ falldown process of condensation (see Section 12.5) and the corotation of the plasma. However, we attempt only a general understanding of the relationship of τ/T to the mass distribution. Thus the inaccuracy introduced in choosing r_{ion}, and hence τ_{ion}, for each group does not diminish the validity of the trends observed in each group.

For each group a straight line is drawn in Figure 17.5, and the slope of this line gives a picture of the variation of the average mass density of the gas from which the bodies condensed. Such a line can, in general, be drawn in such a way that the individual dots fall rather close to the line (mass difference less than a factor of 2). An exception is found in the outer Saturnian group, where Hyperion falls very much below the line connecting Titan and Iapetus.

The figure shows that the mass distribution within the groups depends in a systematic way on the value of τ_{ion}/T. Among the giant planets ($\tau_{\text{ion}}/T = 520$) the masses decrease outward, as discussed in detail in Section 17.2. The Jovian (Galilean) satellites with $\tau_{\text{ion}}/T = 29$ have almost equal masses. In the Uranian group ($\tau_{\text{ion}}/T = 12$) the masses increase outward, on the average, whereas the inner Saturnian satellites ($\tau_{\text{ion}}/T = 8$) show a rapid and monotonic increase outward. The outer Saturnian satellite group which has $\tau_{\text{ion}}/T = 80$ should be intermediate between the giant planets and the Jovian satellites. If a straight line is drawn between the dots representing Titan and Iapetus, the slope of this line is steeper than we should expect. However, Hyperion falls very far from this line, which hence does not represent the mass distribution within the group in a correct way. For reasons which we shall discuss later, this group is not so regular as the outer groups (see Section 17.8). Furthermore, the τ_{ion}/T value for the giant planets is uncertain because we do not know the spin period of the primeval Sun, which indeed must have changed when it transferred most of its angular momentum to the giant planets. An evolution of the solar size and spin as suggested by Alfvén

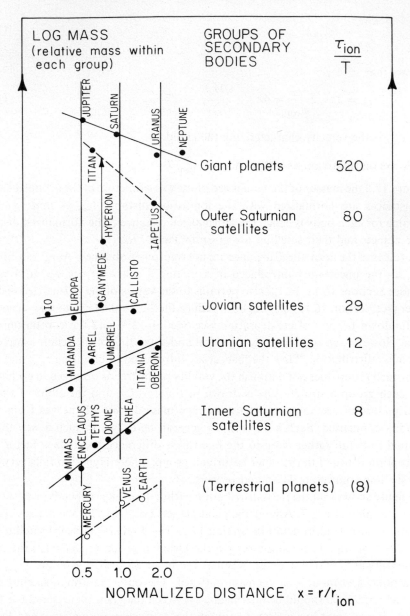

Fig. 17.5.   Mass distribution within the groups of secondary bodies as a function of their normalized distances $x = r/r_{ion}$. The figure shows that within a group characterized by a large value of τ_{ion}/T, the mass decreases outwards. For a value of τ_{ion}/T which is small, the mass decreases inwards.

(1963) should give an average value of τ_{ion}/T for the giant planets which may be smaller than the value for the outer Saturnian satellites. This would eliminate the only exception to the systematic trend in Figure 17.5.

It was found above that the Mercury-Venus-Earth group should not be included in the analysis because we could not be sure that the Sun has the same angular velocity now as when this group was formed, which means that its τ_{ion}/T value may not be the correct one. The present value is $\tau_{ion}/T=8.5$, close to the value of the inner Saturnian group. The mass distribution is also similar to the conditions in this Saturnian group (see Figure 17.5). Hence, if the present value of τ_{ion}/T for this group

TABLE 17.2

Normalized distance for secondary bodies in the solar system

Primary	Cloud	Secondary	Normalized distance r_{orb}/r_{ion}
Sun	B	Mercury	0.56
		Venus	1.05
		Earth	1.46
	A	Moon	0.67
		Mars	1.01
	C	Jupiter	0.49
		Saturn	0.89
		Uranus	1.79
		Neptune	2.81
	D	Triton	0.63
		Pluto	0.83
Jupiter	A	Amalthea	0.84
	C	Io	0.28
		Europa	0.44
		Ganymede	0.70
		Callisto	1.24
	D	rudimentary	
Saturn	C	Mimas	0.41
		Enceladus	0.52
		Tethys	0.65
		Dione	0.83
		Rhea	1.16
	D	Titan	0.60
		Hyperion	0.73
		Iapetus	1.75
Uranus	D	Miranda	0.42
		Ariel	0.61
		Umbriel	0.85
		Titania	1.40
		Oberon	1.87

is used, the terrestrial planets fit, though probably coincidentally, in the sequence of Figure 17.5. Likewise, the Moon and Mars are deleted from the discussion because of the uncertainty of the Sun's spin period in the formation era.

17.7. Discussion of the Structure of the Groups

In an earlier treatise (Alfvén, 1954) an attempt is made to develop a detailed theory of the variation of the mass distribution as a function of τ_{ion}/T. As this was done before experimental and theoretical investigations had clarified the properties of the critical velocity, the discussion must now be revised to some extent. We shall not try here to treat this problem quantitatively but confine ourselves to a qualitative discussion of the two effects which, according to Section 17.3 should be important. These are best studied for the C cloud (Section 16.11.1 and Figure 16.13 because this has produced three groups with very different values of τ_{ion}/T (giant planets with $\tau_{ion}/T=520$ Galilean satellites with $\tau_{ion}/T=29$, and inner Saturnian satellites with $\tau_{ion}/T=8$).

In the group of the giant planets the bodies have normalized distances $x=r/r_{ion}$, with a maximum of 2.81 (see Figure 17.5 and Table 17.2). In the two other groups the maximum value of x is 1.24 for the Galilean and almost the same (1.16) for the inner Saturnian satellites. The decrease in the outward extension may be caused by the first effect discussed in Section 17.3. Of the matter stopped at distance r_{ion}, that found in a larger region around the axis is drawn down to the central body (compare Figures 17.1 and 17.3). In this situation no matter is brought to the equatorial plane along these lines of force which intersect this plane at a large distance.

Further, the second effect discussed in Section 17.3 allows matter to become ionized closer to the central body, because not all the matter is ionized and stopped at the ionization distance r_{ion}. A result of this is that the innermost body of the Galilean group has a normalized distance of only $x=0.28$ compared to 0.49 for the giant planets. In the inner Saturnian group this effect is even more pronounced because of the smaller value of τ_{ion}/T. Certainly, the innermost body (Mimas) of this group has an x-value of 0.41, but the satellite group continues inside the Roche limit in the form of the ring system. In this we find matter collected almost down to the surface of Saturn, corresponding to an x-value as low as 0.1.

A similar effect, although less pronounced, is indicated in the D cloud, by the fact that the x-value of Miranda in the Uranian system is 0.42, and Titan, the innermost body of the outer Saturnian group, has $x=0.60$. However, there is no similar difference between the outer limits.

17.8. Complete List of τ_{ion}/T for All Bodies

Table 17.3 presents all the τ_{ion}/T values above unity for the A, B, C, and D clouds captured around the largest bodies in the solar system (see Figure 16.13). Also some values slightly below unity are given for comparison.

The six groups represented in Figures 17.4 and 17.5 all have τ_{ion}/T values $\geqslant 8$. As the process we have discussed has a general validity, we should expect similar groups to be produced in all cases where we find the same values of τ_{ion}/T, unless special phenomena occur which prevent the formation. In addition, to these six groups, we find high values of τ_{ion}/T also in three more cases. This means that we should expect groups of bodies also in these cases.

TABLE 17.3

Values of τ_{ion}/T where τ_{ion} is the Kepler period of a body at the ionization distance and T is the period of axial rotation of the central body

Central body	$T\ 10^{-5}$ s	τ_{ion}/T for secondary bodies in cloud:			
		B	A	C	D
Sun	21.3	8.5 ☿–⊕	28 ☾–♂	520 ♃–♆	5000 ♇
Jupiter	0.354	0.50	1.6 Amalthea	29 Galilean satellites	286
Saturn	0.375		0.45	8.4 inner satellites	outer satellites
Uranus	0.385			1.3	12 Uranian satellites
Neptune	0.568			1.0	9.5
Earth	0.14?				2.2?
(prior to capture of the Moon)					

According to the theory, bodies are produced only in the groups above the line in the table.

(1) *D cloud around the Sun.* We should expect a group of planets outside the giant planets. Pluto and probably also Triton may belong to this group. (Like the Moon, Triton was initially a planet which later was captured; see McCord (1966)). As the *D* cloud should contain heavy elements (see Section 16) the high density of Pluto, and possibly Triton (see Section 14.5), may be explained. It is possible that the extremely large distance to the Sun has made the hydromagnetic transfer of momentum inefficient or that an interstellar ('galactic') magnetic field has interfered with the solar field so that this group only has these two members. But there may also be yet undiscovered members of this group.

(2) *D cloud around Jupiter.* The absence of regular *D* cloud satellites around Jupiter may appear surprising. However, as has been shown elsewhere (Alfvén, 1954, p. 161), the solar magnetic field, if it is strong enough, should prevent, or interfere with, the production of satellites. The region which is most sensitive to this interference is the *D* cloud region around Jupiter; next is the *D* cloud region around Saturn. Hence, the solar magnetic field may have prevented the *D* cloud satellites around Jupiter and at the same time made the outer Saturnian satellites as irregular as they are with regard to the sequence of masses and orbital radii.

(3) *D cloud around Neptune.* We should also expect a *D* cloud group around Neptune. If such a group was once formed, it is likely to have been largely destroyed by the retrograde giant satellite Triton, when it was captured. The evolution of the Neptune-Triton system is likely to have been similar, in certain respects, to the Earth-Moon system (see Alfvén and Arrhenius, 1972).

This implies that Nereid is the only residual of an initial group of satellites, most of which may have impacted on Triton in the same way as the Earth's original satellites presumably impacted on the Moon, forming the maria relatively late in lunar history.

It should be added that the A cloud around the Sun probably has produced Mars and also the Moon as an independent planet, which later was captured as discussed above (Alfvén and Arrhenius, 1972).

So far we have discussed all the cases in which τ_{ion}/T has a value in the same range as the six groups of Figure 17.5. It is of interest to see what happens if τ_{ion}/T is smaller than this. From Table 17.3 we find that the next value ($\tau_{ion}/T = 1.6$) belongs to the A cloud around Jupiter. In the region where we expect this group, we find only one tiny satellite, the fifth satellite of Jupiter, which has a reduced distance $x = r/r_{ion} = 0.84$. This little body may be identified as the only member of a group which is rudimentary because of its small τ_{ion}/T value. If we proceed to the next value, which is $\tau_{ion}/T = 1.3$ for the C cloud around Uranus, we find no satellites at all.

Hence, the theoretical prediction that no satellite formation is possible when τ_{ion}/T approaches unity is confirmed by the observational material. The transition from the groups of Figure 17.5 to be absence of satellites is represented by Jupiter's lone satellite V.

17.9. COMPLETENESS

Summarizing the results of our analysis we may state that they justify our original assumption, namely, that it makes sense to plot the secondary bodies as a function of Γ. In fact, according to the diagram (Figure 16.1) a *necessary* condition for the existence of a group of secondary bodies is that its gravitational potential has specific values, and whenever this condition is fulfilled, bodies are present.

> *All the known regular bodies* (with a possible uncertainty in the identification of Pluto) *fall within three horizontal bands* – with a possible addition of a fourth band for the Martian satellites. *Groups of bodies are found wherever a band falls within the natural limits of formation of secondary bodies.*

There is no obvious exception to this rule but there are three doubtful cases:

(a) The band producing the Uranian, the outer Saturnian, and outermost Jovian satellites may also have produced bodies in the planetary system. It is possible that Pluto, whose density seems to be higher than those of the giant planets, is an example of such a group. However, it is also possible that the band falls outside the region of present detection capability for small planets. Moreover, the outer limit for formation of planets around the Sun cannot be calculated in the same way as for the satellite systems.

(b) From only looking at the observational diagram (Figure 16.1) we may expect a correspondence to Martian satellites in the outermost region of the Uranian system, and possibly also in the outskirts of the Saturnian system. However, we see from Figure 16.13 that there is no critical velocity which is so small, so we should not expect such bodies theoretically.

(c) It is likely that once a group of natural satellites were formed around the primeval Earth, but that this group was destroyed during the capture of the Moon.

Before the capture of the Moon the Earth had a much more rapid spin. A reasonable value for the spin period is 4 h. With a D cloud around the Earth this gives $\tau/T=2.2$. This value is intermediate between Amalthea and the inner Saturnian satellites. Hence we should expect that the Earth originally had a satellite system somewhat intermediate between Amalthea and the inner Saturnian satellites. The satellites were necessarily very small, and all were swallowed up or ejected by the much bigger Moon (see Alfvén and Arrhenius, 1972).

17.9.1. *Note on the Inner Limit of a Satellite System*

As derived in Section 12.3, the state of partial corotation is given by

$$v_\phi^2 = \frac{2\kappa M_c}{3r}$$

(17.18)

with

$$v_\phi = \omega r \cos \lambda.$$

(17.19)

As ω, the angular velocity of the orbiting body, cannot surpass the angular velocity Ω of the spinning central body we cannot have equilibrium unless $r > r_0$ with r_0 defined by

$$r_0^3 \cos^2 \lambda \geqslant \frac{2\kappa M_c}{3\Omega^2}.$$

(17.20)

Introducing the synchronous radius r_{syn} for a Kepler orbit when $\omega = \Omega$

$$r_{syn}\Omega^2 = \frac{\kappa M_c}{r_{syn}^2},$$

(17.21)

we find that

$$\frac{r_0}{r_{syn}} = \left(\tfrac{2}{3} \cos \lambda\right)^{1/3}.$$

(17.22)

The minimum distance r_{min} of condensed matter in circular orbit given by the $\tfrac{2}{3}$ law (Section 12.5) is

$$r_{min} = \tfrac{2}{3} r_0 \cos \lambda$$

(17.23)

and

$$\frac{r_{min}}{r_{syn}} = \left(\tfrac{2}{3}\right)^{4/3} (\cos \lambda)^{5/3} = 0.58 \, (\cos \lambda)^{5/3}.$$

Hence, within an order of magnitude, the synchronous orbit gives the inferior limit to the position of a satellite. Due to the nature of the condensation process (Section 12.5) $\cos \lambda$ approaches unity.

There are only two cases known where matter is orbiting inside the synchronous orbit:

(a) *Phobos.* The orbital radius of Phobos is 0.44 of the synchronous orbit. Matter could be brought into circular orbit at this distance only if $\cos\lambda=(0.44/0.58)^{3/5}$ or $\cos\lambda<0.85$ and $\lambda>31°$. There is no apparent reason why condensation should have taken place exclusively so far from the equatorial plane of Mars. Possible explanations for the small orbital radius of Phobos are:

(1) Mars might have slowed down its spin after the generation of Phobos. This is compatible with the fact that according to the law of isochronism Mars should have had an initial spin period of the order of five hours (as with the Earth before the capture of the Moon). This would leave Phobos far outside the synchronous orbit. However, it is difficult to see how the required slow-down could have occurred.

(2) Phobos might have been generated when Mars was much smaller than it is today. If the mass of a central body increases the angular momentum of an orbiting body remains constant. Hence the mass must have increased at least in the proportion $(0.58/0.44)^3=2.29$.

(3) It has sometimes been suggested that Phobos might be a captured satellite. Phobos' small eccentricity and inclination make this suggestion highly unlikely.

(b) *Saturnian rings.* The synchronous orbit is situated in the outer part of the B ring. The minimum value $0.58\ r_{syn}$ is very close to Saturn, being only 7% of Saturn's radius above the surface of the planet. The density in the C ring, which begins at 0.8 of the synchronous orbit, is very small, but this is due to the 'ring's own shadow' (see Section 13.6) and is not likely to be connected with the synchronous orbit. Hence in the Saturnian rings we see a confirmation that matter can also be accreted at some distance inside the synchronous orbit.

17.9.2. *Tilt of the Bands in the Gravitational Potential Energy Diagram*

In Section 16 we expected theoretically that the bands in which the secondary bodies are located should be horizontal, i.e., independent of the mass of the central body. In the diagram Figure 16.1 we observe a slight tilt of the bands. In fact, the gravitational energy at which the C groups are located is larger for Jupiter than for the Sun, and larger for Saturn than for Jupiter. From what has been discussed above, this tilt is likely to be due to the fact that τ_{ion}/T values for these three groups differ. The similar difference between the D cloud groups of Saturn and Uranus may be attributed to the same effect.

17.10. CONCLUSIONS ABOUT THE MODEL OF PLASMA EMPLACEMENT

The model of plasma emplacement which we have treated in Sections 16 and 17 must necessarily be more speculative than the theories in earlier chapters. The basic phenomenon, the critical velocity phenomenon, although well established, is not yet so well understood in detail that we know the behavior of gas mixtures in this respect. We do not know very well what abundance of a particular element is necessary in order to make the critical velocity of this element decisive for the stopping and ionization of the gas except that it presumably must be a major component. Nor is the distribution of elements between molecular ions sufficiently known. In connection with what has been found in Section 16.10 this means that we cannot predict the chemical composition of the bodies in a specific group.

Moreover, such predictions cannot yet be verified since the chemical composition of celestial bodies belonging to different clouds is not yet available. We are far from the days when it was claimed with certainty that Jupiter consisted almost entirely of pure solid hydrogen. It is now generally admitted that we do not know with certainty the bulk composition of the Earth and still less, of any other body (see Section 14). Hence, detailed, precise predictions will not be possible until the theory is refined under the influence of more adequate experimental and observational data.

The success of the model in giving a virtually complete and non-arbitrary classification of the bodies in the solar system weighs heavily in its favor. However, the processes considered in this section are obviously open to many refinements and modifications.

Acknowledgements

The present work was funded by the Planetology Program Office, Office of Space Science under grant NASA NGL 05-009-110, the Lunar and Planetary Program Division under grant NASA NGL 05-009-002 and the Apollo Lunar Exploration Office under grant NASA NGL 05-009-154. The generous support from these NASA offices is gratefully acknowledged.

We wish to thank Mr Bibhas R. De for crucial help and constructive discussion. We are also indebted to Drs L. Danielsson and D. Lal for fruitful discussions and suggestions.

The competent and devoted editorial help from Dawn S. Rawls and Marjorie Sinkankas has been of crucial importance; we also wish to acknowledge assistance from the Royal Institute of Technology in Stockholm.

References

Alfvén, H.: 1942–45, 'On the Cosmogony of the Solar System', *Stockholm Obs. Ann.* **14**, No. 2; **14**, No. 5; **14**, No. 9.

Alfvén, H.: 1954, *On the Origin of the Solar System*, Oxford Univ. Press, London.

Alfvén, H.: 1963, 'On the Early History of the Sun and the Formation of the Solar System', *Astrophys. J.* **137**, 981–990.

Alfvén, H.: 1969, 'Asteroidal Jet Streams', *Astrophys. Space Sci.* **4**, 84–102.

Alfvén, H.: 1970, 'Jet Streams in Space', *Astrophys. Space Sci.* **6**, 161–174.

Alfvén, H. and Arrhenius, G.: 1972, 'Origin and Evolution of the Earth-Moon System', *The Moon* **5**, 210–230.

Alfvén, H. and Wilcox, J. M.: 1962, 'On the Origin of the Satellites and the Planets', *Astrophys. J.* **136**, 1016–1022.

Allen, C. W.: 1964, *Astrophysical Quantities*, The Athlone Press, Univ. of London, London, 1964. 291 pp.

Aller, L. H.: 1967, in K. Runcorn (ed)., *Int. Dictionary of Geophysics*, vol. 1, Pergamon, New York, p. 285.

Anders, E.: 1964, *Space Sci. Rev.* **3**, 583.

Angerth, B., Block, L., Fahleson, U., and Soop, K.: 1962, 'Experiments with Partly Ionized Rotating Plasmas', *Nucl. Fusion Suppl. Pt. I*, p. 39.

Arnold, J. R.: 1965, *Astrophys. J.* **141**, 1536–1548.

Arnold, J.: 1969, 'Asteroid Families and Jet Streams', *Astron. J.* **74**, 1235–1242.

Arrhenius, G.: 1969, 'Kosmologisk revolution från månen', *Forskning och Framsteg* **7**, 2–5.

Arrhenius, G.: 1972, 'Chemical Effects in Plasma Condensation', in *From Plasma to Planet, Nobel Symp.* **21** (ed. by A. Elvius), Wiley, New York, pp. 117–132.

Arrhenius, G. and Alfvén, H.: 1971, 'Fractionation and Condensation in Space', *Earth Planetary Sci. Letters* **10**, 253–267.

Arrhenius, G. and Asunmaa, S. K.: 1973, 'Aggregation of Grains in Space', *The Moon* **8**, 368–391.

Arrhenius, G. and De, B.: 1973, 'Equilibrium Condensation in a Solar Nebula', *Meteoritics* **8**, 297–313.

Arrhenius, G., Asunmaa, S., Drever, J. I., Everson, J., Fitzgerald, R. W., Frazer, J. Z., Fujita, H., Hanor, J. S., Lal, D., Liang, S. S., Macdougall, D., Reid, A. M., Sinkankas, J., and Wilkening, L.: 1970, 'Phase Chemistry, Structure and Radiation Effects in Lunar Samples', *Science* **167**, 659–661.

Arrhenius, G., Liang, S., Macdougall, D., Wilkening, L., Bhandari, N., Bhat, S., Lal, D., Rajagopalan, G., Tamhane, A. S., and Venkatavaradan, V. S.: 1971, The Exposure History of the Apollo 12 Regolith, in *Proc. 2nd Lunar Sci. Conf.*, *3*, The M.I.T. Press, Cambridge, Mass., pp. 2583–2598.

Arrhenius, G., Alfvén, H., and Fitzgerald, R.: 1973, *Asteroid and Comet Exploration*, NASA CR-2291, NASA, Wash., D.C., 56 pp.

Arrhenius, G., De, B. R., and Alfvén, H.: 1975, 'Origin of the Ocean', in E. D. Goldberg (ed.), *The Sea* **5**, p. 837.

Banerjee, S. K.: 1967, 'Fractionation of Iron in the Solar System', *Nature* **216**, 781.

Banerjee, S. K. and Hargraves, R. B.: 1971, 'Natural Remanent Magnetization of Carbonaceous Chondrites', *Earth Planetary Sci. Letters* **10**, 392–396.

Banerjee, S. K. and Hargraves, R. B.: 1972, 'Natural Remanent Magnetizations of Carbonaceous Chondrites and the Magnetic Field in the Early Solar System', *Earth Planetary Sci. Letters* **17**, 110–119.

Baxter, D. and Thompson, W. B.: 1973, 'Elastic and Inelastic Scattering in Orbital Clustering', *Astrophys. J.* **183**, 323–336.

Berlage, H. P.: 1948, 'The Disc Theory of the Origin of the Solar System', *Proc. Koninkl. Ned. Acad. Wetenschap.*, **51**, 796.

Birch, F.: 1964, 'Density and Composition of Mantle and Core', *J. Geophys. Res.*, **69**, 4377–4388.

Birch, F.: 1965, 'Energetics of Core Formation', *J. Geophys. Res.* **70**, 6217–6221.

Bishop, E. V. and DeMarcus, W. C.: 1970, 'Thermal Histories of Jupiter Models', *Icarus* **12**, 317–330.

Boström, K. and Fredriksson, K.: 1966, 'Surface Conditions of the Orgueil Meteorite Parent Body as Indicated by Mineral Associations', *Smithsonian Miscellaneous Collections* **151**, 1–54.

Bobrovnikoff, N. T.: 1972, 'Physical Theory of Comets in Light of Spectroscopic Data', *Rev. Mod. Phys.* **14**, 168–178.

Brecher, A.: 1971, 'On the Primordial Condensation and Accretion Environment and the Remanent Magnetization of Meteorites', in C. L. Hemmenway, A. F. Cook, and P. M. Millman (eds.), *The Evolutionary and Physical Problems of Meteoroids*, NASA SP-319, NASA, Wash., D.C., p. 311.

Brecher, A.: 1972a, 'Memory of Early Magnetic Fields in Carbonaceous Chondrites', in H. Reeves (ed.), *On the Origin of the Solar System*, Centre Nationale de la Recherche Scientifique, Paris, p. 260.

Brecher, A.: 1972b, 'I. Vapor Condensation of Ni-Fe Phases and Related Problems. II. The Paleomagnetic Record in Carbonaceous Chondrites', Ph. D. Thesis, Dept. Applied Physics and Information Science, University of California, San Diego.

Brecher, A. and Arrhenius, G.: 1974, 'The Paleomagnetic Record in Carbonaceous Chondrites: Natural Remanence and Magnetic Properties', *J. Geophys. Res.* **79**, 2081.

Brownlee, R. R. and Cox, A. N.: 1961, 'Early Solar Evolution', *Sky Telesc.* **21**, 252–256.

Chapman, C. R.: 1972a, 'Surface Properties of Asteriods', Ph. D. Thesis, Massachusetts Institute of Technology.

Chapman, C. R.: 1972b, Paper presented at the Coll. on Toro, Tucson, Ariz., Dec., 1972.

Chapman, C., Johnson, T. V., and McCord, T. B.: 1971, 'A Review of Spectrophotometric Studies of Asteroids', in T. Gehrels (ed.), *Physical Studies of Minor Planets*, NASA SP-267, pp. 51–66.

Danielsson, L.: 1969, 'On the Interaction Between a Plasma and a Neutral Gas', Research Report 69-17, Division of Plasma Physics, Royal Institute of Technology, Stockholm.

Danielsson, L.: 1970, 'Experiments on the Interaction between a Plasma and a Neutral Gas', *Phys. Fluids* **13**, 2288–2294.

Danielsson, L.: 1971, 'The profile of a jetstream', in T. Gehrels (ed.), *Physical Studies of Minor Planets*, NASA SP-267, Government Printing Office, Washington, D.C., pp. 353–362.

Danielsson, L.: 1973, 'Review of the Critical Velocity of Gas-Plasma Interaction, Part I: Experimental Observations', *Astrophys. Space Sci.* **24**, 459–485.

De, Bibhas: 1973, 'On the Mechanism of Formation of Loop Prominences', *Solar Phys.* **31**, 437–447.

DeMarcus, W. C.: 1958, 'The Constitution of Jupiter and Saturn', *Astron. J.* **63**, 2.

DeMarcus, W. C. and Reynolds, R. T.: 1963, 'The Constitution of Uranus and Neptune', *Mém. Soc. Roy. Sci. Liège, cinquième série, VII*, 51–64.

Dobryshevskii, E. M.: 1964, 'The Volt-Ampere Characteristics of a Homopolar Cell', *Soviet Phys. – Techn. Phys.* **8**, 903–905.

Dollfus, A.: 1971, 'Physical Studies of Asteroids by Polarization of the Light', in T. Gehrels (ed.), *Physical Studies of Minor Planets*, NASA SP-267, NASA, Wash., D.C., pp. 95–116.

Drickamer, H. G.: 1965, 'The Effect of High Pressure on the Electonic Structure of Solids', in F. Seitz and D. Turnbull (eds.), *Solid State Physics*, **17**, 1–133.

Duke, M. B. and Silver, L. T.: 1967, 'Petrology of Eucrites, Howardites and Mesosiderites', *Geochim. Cosmochim. Acta* **31**, 1637–1665.

Eberhardt, P., Geiss, J., and Grögler, N.: 1965, 'Über die Verteilung der Uredelgase im Meteoriten Khor Temiki', *Tschermaks Min. Petr. Mitt.* **10**, 535–551.

Elsasser, W. M.: 1963, 'Early History of the Earth', in J. Geiss and E. D. Goldberg (eds.), *Earth Science and Meteoritics*, dedicated to F. G. Houtermans, North-Holland Publ. Co., Amsterdam, pp. 1–30.

v. Engel, A.: 1955, *Ionized Gases*, Oxford Univ. Press., London.

Ephemerides of Minor Planets for 1969. Institute of Theoretical Astronomy, Acad. Sci. USSR. Publication 'Nauka' Leningrad Department, Leningrad, 1968.

Epstein, S. and Taylor, H. P. Jr.: 1970, 'The Concentration and Isotopic Composition of Hydrogen, Carbon and Silicon in Apollo 11 Lunar Rocks and Minerals', in *Proc. Apollo 11 Lunar Sci. Conf.*, **2**, (ed. by A. A. Levinson), Pergamon Press, New York, pp. 1085–1096.

Epstein, S. and Taylor, H. P., Jr.: 1972, 'O^{18}/O^{16}, Si^{30}/Si^{28}, C^{13}/C^{12}, and D/H studies of Apollo 14 and 15 Samples', *Lunar Science* vol. III, (ed. by C. Watkins), Lunar Science Institute Contribution No. 88, pp. 236–238.

Euken, A.: 1944, 'Über den Zustand des Erdinnern', *Naturwissenschaften*, No. 14/26, 112–121.

Fireman, E. L.: 1958, 'Distribution of helium-3 in the Carbo Meteorite', *Nature* **181**, 1725.

Fleischer, R. L., Price, P. B., Walker, R. M., Maurette, M., and Morgan, G.: 1967a, 'Traces of Heavy Primary Cosmic Rays in Meteorites', *J. Geophys. Res.* **72**, 355–366.

Fleischer, R. L., Price, P. B., Walker, R. M., and Maurette, M.: 1967b, 'Origins of Fossil Charged-Particle Tracks in Meteorites', *J. Geophys. Res.* **72**, 331–353.

Fodor, R. V. and Keil, K.: 1973, 'Composition and Origin of Lithic Fragments in L- and H-Group Chondrites', *Meteoritics* **8**, 366–367.

Fowler, W. A.: 1972, 'What Cooks with Solar Neutrinos?', *Nature* **238**, 24–26.

French, B. and Short, N.: 1968, *Shock Metamorphism of Natural Meteorites*, Mono Book Corp., Baltimore, Md.

Fuchs, L. H.: 1971, 'Occurrence of Wollastonite, Rhönite, and Andradite in the Allende meteorite', *Amer. Mineralogist* **56**, 2053–2068.

Gast, P. W.: 1971, 'The Chemical Composition of the Earth, the Moon and Chondritic Meteorites', in E. C. Robertson (ed.), *The Nature of the Solid Earth*, McGraw-Hill, New York, pp. 19–40.

Gault, D. E., Quaide, W. L., and Oberbeck, V. R.: 1968, 'Impact Cratering Mechanics and Structures', in B. M. French and N. M. Short (eds.), *Shock Metamorphism of Natural Materials*, Mono Book Corp., Md., pp. 87–99.

Gehrels, T.: 1972a, 'Physical Parameters of Asteroids and Interrelations with Comets, in *From Plasma to Planet*, *Nobel Symp* **21** (ed. by A. Elvius), Wiley, New York, pp. 169–178.

Gehrels, T.: 1972b, Paper presented at the Coll. on Toro, Tucson, Ariz., Dec., 1972.

Gerstenkorn, H.: 1969, 'The Earliest Past of the Earth-Moon System', *Icarus* **11**, 189–207.

Gold, T. and Williams, G.: 1972, 'Secondary Emission Charging and Movement of Dust on the Lunar Surface', Sixth ESLAB Symp., *Photon and Particle Interactions with Surfaces in Space*. Noordwijk, Netherlands, Sept. 26–29, 1972 (abstract).

Gopalan, K. and Wetherill, G. W.: 1969, 'Rubidium-Strontium Age of Amphoterite (LL) Chondrites', *J. Geophys. Res.* **74**, 4349–4358.

Greenstein, J. L. and Arpigny, C.: 1962, 'The Visual Region of the Spectrum of Comet Markos (1957d) at High Resolution', *Astrophys. J.* **135**, 892–905.

Grevesse, N., Blanquet, G. and Boury, A.: 1968, 'Abondances solaires de quelques éléments représentatifs au point de vue de la nucléosynthese', in L. H. Ahrens (ed.), *Origin and Distribution of the Elements*, Pergamon, New York, pp. 177–182.

Halliday, I.: 1969, 'Comments on the Mean Density of Pluto', *Publ. Astron. Soc. Pacific* **81**, 285–287.

Hanks, T. C. and Anderson, D. L.: 1969, 'The Early Thermal History of the Earth', *Phys. Earth Planetary Interior* **2**, 19–29.

Hapke, B. W., Cohen, A. J., Cassidy, W. A., and Wells, E. N.: 1970, 'Solar Radiation Effects on the Optical Properties of Apollo 11 Samples', in *Proc. Apollo 11 Lunar Sci. Conf.*, **3** (ed. by A. Levinson), Pergamon Press, New York, pp. 2199–2212.

Harris, P. G. and Tozer, D. C.: 1967, 'Fractionation of Iron in the Solar System', *Nature* **215**, 1449–51.

Hassan, H. A.: 1966, 'Characteristics of a Rotating Plasma', *Phys. Fluids* **9**, 2077–2078.

Heymann, D.: 1967, 'On the Origin of Hypersthene Chondrites: Ages and Shock Effects of Black Chondrites', *Icarus* **6**, 189–221.

Hohenberg, C. M. and Reynolds, J. H.: 1969, 'Preservation of the Iodine-Xenon Record in Meteorites', *J. Geophys. Res.* **74**, 6679–6683.

Honda, M. and Arnold, J. R.: 1967, 'Effects of Cosmic Rays on Meteorites', in *Handbuch der Physik* **46**/2, 613–632, Springer-Verlag, Berlin-Heidelberg.

Hubbard, W. B.: 1969, 'Thermal Models of Jupiter and Saturn', *Astrophys. J.* **155**, 333–344.

Jedwab, J.: 1967, 'La magnetite en plaquettes des météorites carbonées d'Alais, Ivuna et Orgueil', *Earth Planetary Sci. Letters* **2**, 440–444.

Jokipii, J. R.: 1964, 'The Distribution of Gases in the Protoplanetary Nebula', *Icarus* **3**, 248.

Kaula, W. M.: 1971, 'Dynamical Aspects of Lunar Origin', *Rev. Geophys. Space Phys.* **9**, 217–238.

Kerridge, J. F.: 1970, 'Some Observations on the Nature of Magnetite in the Orgueil Meteorite', *Earth Planetary Sci. Letters* **9**, 299–306.

Kerridge, J. F. and Vedder, J. F.: 1972, 'Accretionary Processes in the Early Solar System: an Experimental Approach', *Science* **177**, 161–163.

Kirsten, T. A. and Schaeffer, O. A.: 1969, 'High Energy Interactions in Space', in L. C. L. Yuan (ed.), *Elementary Particles, Science, Technology and Society*, Academic Press, New York 76–157.

Kumar, S. S.: 1972, 'On the Formation of Jupiter', A*strophys. Space Sci.* **16**, 52–54.

Lal, D.: 1972, 'Hard Rock Cosmic Ray Archaeology', *Space Sci. Rev.* **14**, 3–102.

Lehnert, B.: 1966, 'Ionization Process of a Plasma', *Phys. Fluids* **9**, 774–779.

Lehnert, B.: 1967, 'Space-Charge Effects by Nonthermal Ions in a Magnetized Plasma', *Phys. Fluids* **10**, 2216.

Lehnert, B.: 1970, 'On the Conditions for Cosmic Grain Formation', *Cosmic Electrodyn.* **1**, 219–232.

Levin, B. J.: 1972, 'Origin of the Earth', in A. R. Ritsema (ed.), *Upper Mantle, Tectonophysics* **13**, 7–29.

Lewis, J. S.: 1971a, 'Consequences of the Presence of Sulfur in the Core of the Earth', *Earth Planetary Sci. Letters* **11**, 130–134.

Lewis, J. S.: 1971b, 'Satellites of the Outer Planets: Their Physical and Chemical Nature', *Icarus* **15**, 174–185.

Lin, S.-C.: 1961, 'Limiting Velocity for a Rotating Plasma', *Phys. Fluids* **4**, 1277–1288.

Lindblad, B. A. and Southworth, R. B.: 1971, 'A Study of Asteroid Families and Streams by Computer Techniques, in T. Gehrels (ed.), *Physical Studies of Minor Planets*, NASA SP-267, NASA, Wash., D.C., p. 337.

Lodochnikov, V. N.: 1939, 'Some General Problems Connected with Magma Producing Basaltic Rocks', *Zap. Mineral. O-va.* **64**, 207–223.

Lyttleton, R. A.: 1969, 'On the Internal Structures of Mercury and Venus', *Astrophys. Space Sci.* **5**, 18–35.

MacDougall, J. D.: 1972, 'Particle Track Records in Natural Solids from Oceans on Earth and Moon', Ph. D. Thesis, University of California, San Diego.

MacDougall, D., Martinek, B., and Arrhenius, G.: 1972, 'Regolith Dynamics', in C. Watkins (ed.), in vol. III *Lunar Science*, Lunar Science Inst. Contrib. No. 88, The Lunar Science Institute, Houston, Texas, pp. 498–500.

Macdougall, D., Rajan, R. S., and Price, P. B.: 1973, 'Gas-Rich Meteorites: Possible Evidence for Origin on a Regolith', *Science* **183**, 73–74.

Majeva, S. V.: 1971, 'Thermal History of the Earth with Iron Core', *Izv. Akad. Nauk SSSR, Fiz. Zemli*, 1971, No. 1, 3–12, English translation, *Physics of the Solid Earth*, 1971, 1–7.

Manka, R. H., Michel, F. C., Freeman, J. W., Dyal, P., Parkin, C. W., Colburn, D. S., and Sonett,

C. P.: 1972, Evidence for acceleration of lunar ions, in C. Watkins (ed.), *Lunar Science*, Vol. III, Lunar Science Contribution No. 88, The Lunar Science Institute, Houston, Texas, p. 504.

Marti, K.: 1973. 'Ages of the Allende Chondrules and Inclusions', *Meteoritics* 8, 51.

Mason, B. (ed.): 1971, *Handbook of Elemental Abundances in Meteorites*. Gordon and Breach Sci. Publ., New York.

Mendis, A.: 1973, 'Comet-Meteor Stream Complex', *Astrophys. Space Sci.* 20, 165–176.

McCord, T. B.: 1966, 'Dynamical Evolution of the Neptunian System'. *Astron. J.* 71, 585–590.

McCrosky, R. E.: 1970, 'Fireballs and the Physical Theory of Meteors', *Bull. Astron. Inst. Czech.* 21, 271.

McQueen, R. L. and Marsh, S. P.: 1960, 'Equations of State for Nineteen Metallic Elements from Shock-Wave Measurements to Two Megabars', *J. Appl. Phys.* 31, 1253–1269.

Millman, P. M.: 1972, 'Cometary Meteoroids', in *From Plasma to Planet*, *Nobel Symp.* 21, (ed. by A. Elvius), Wiley, New York, pp. 157–168.

Morrison, D.: 1973, 'New Techniques for Determining Sizes of Satellites and Asteroids', *Comments Astrophys. Space Phys.* 5, 51–56.

Müller, E. A.: 1968, 'The Solar Abundances', in L. H. Ahrens (ed.), *Origin and Distribution of the Elements*, Pergamon, New York, pp. 155–176.

Murphy, R. E., Cruikshank, D. P., and Morrison, D.: 1972, 'Radii, Albedos, and 20-Micron Brightness temperatures of Iapetus and Rhea', *Astrophys. J. Letters* 177, L93.

Neukum, G., Mehl, A., Fechtig, H., and Zahringer, J.: 1970, 'Impact Phenomena of Micrometeorites on Lunar Surface Material', *Earth Planetary Sci. Letters* 8, 31–35.

Neuvonen, K. J., Ohlson, B., Papunen, Heikki, Häkli, T. A., and Ramdohr, Paul: 1972, 'The Haverö Ureilite', *Meteoritics* 7, 515–531.

Newburn, R. L., Jr. and Gulkis, S.: 1973, 'A Survey of the Outer Planets Jupiter, Saturn, Uranus, Neptune, Pluto, and Their Satellites', *Space Sci. Rev.* 14, 179–271.

Öpik, E. J.: 1963, *Advan. Astron. Astrophys.* 2, 219.

Öpik, E. J.: 1966, *Advan. Astron. Astrophys.* 4, 302.

Orowan, E.: 1969, 'Density of the Moon and Nucleation of Planets', *Nature* 222, 867.

Papanastassiou, D. A., Gray, C. M., and Wasserburg, G. J.: 1973, 'The Identification of Early Solar Condensates in the Allende Meteorite', *Meteoritics* 8, 417–418.

Pellas, P.: 1972, 'Irradiation History of Grain Aggregates in Ordinary Chondrites. Possible Clues to the Advanced Stages of Accretion', *From Plasma to Planet*, *Nobel Symp.* 21 (ed. by A. Elvius), Wiley, New York, pp. 65–92.

Petschek, H. E.: 1960, 'Comment Following Alfvén, H., Collision between a Nonionized Gas and a Magnetized Plasma', *Rev. Mod. Phys.* 32, 710–712.

Podosek, F. A.: 1970, 'Dating of Meteorites by the High-Temperature Release of Iodine-Correlated Xe^{129} *Geochim. Cosmochim. Acta* 34, 341–365.

Price, P. D., Rajan, R. S., Hutcheon, I. D., Macdougall, D., and Shirk, E. K.: 1973, 'Solar Flares, Past and Present', in J. W. Chamberlin and C. Watkins (eds.), *Lunar Science*, vol. 4, Lunar Science Institute, Houston, Texas, pp. 600–602.

Rama Murthy, V. and Hall, H. T.: 1970, 'On the Possible Presence of Sulfur in the Earth's Core', *Phys. Earth Planetary Interior* 2, 276–282.

Ramsey, W. H.: 1948, 'On the Constitution of the Terrestrial Planets', *Monthly Notices Roy. Astron. Soc.* 108, 406–413.

Ramsey, W. H.: 1949, 'On the Nature of the Earth's Core', *Monthly Notices Roy. Astron. Soc., Geophys. Suppl.* 5, 409–426.

Reid, A. M. and Fredriksson, K.: 1967, 'Chondrules and Chondrites', in P. H. Abelson (ed.), *Researches in Geochemistry*, vol. 2, John Wiley, New York, pp. 170–203.

Reynolds, R. T. and Summers, A. L.: 1965, 'Models of Uranus and Neptune', *J. Geophys. Res.* 70, 199–208.

Ringwood, A. E.: 1959, 'On the Chemical Evolution and Densities of the Planets', *Geochim. Cosmochim. Acta* 15, 257–283.

Ringwood, A. E.: 1966, 'Chemical Evolution of the Terrestrial Planets', *Geochim. Cosmochim. Acta* 30, 41–104.

Samara, G. A.: 1967, 'Insulator-to-Metal Transition at High Pressure', *J. Geophys. Res.* 72, 671–678.

Schubart, J.: 1971, 'Asteroid Masses and Densities, in T. Gehrels (ed.), *Physical Studies of Minor Planets*, NASA SP-267, pp. 33–40.

Seidelmann, P. K., Klepczynski, W. J., Duncombe, R. L., and Jackson, E. S.: 1971, 'Determination of the Mass of Pluto', *Astron. J.* **76**, 488–492.

Sherman, J. C.: 1970a, Dept. of Electron and Plasma Physics Report 70-30, Royal Institute of Technology, Stockholm.

Sherman, J. C.: 1970b, Dept. of Electron and Plasma Physics Report 70-14, Royal Institute of Technology, Stockholm.

Sherman, J. C.: 1973, 'Review of the Critical Velocity Gas-Plasma Interaction, Part II: Theory', *Astrophys. Space Sci.* **24**, 487–510.

Signer, P. and Suess, H. E.: 1963, 'Rare Gases in the Sun, in the Atmosphere, and in Meteorites'. in J. Geiss and E. D. Goldberg (eds.), *Earth Science and Meteoritics*, North-Holland Publ. Co., Amsterdam, pp. 241–272.

Simakov, G. V., Podurets, M. A., and Trunin, R. F.: 1973, *Doklady Akad. Nauk SSSR* **211**, 1330.

Sokol, P. M.: 1968, 'Analysis of a Rotating Plasma Experiment', *Phys. Fluids* **11**, 637-645.

Suess, H. E. and Urey, H. C.: 1956, 'Abundances of the Elements'. *Rev. Mod. Phys.* **28**, 53–74.

Swings, P., and Page, T.: 1948, *Astrophys. J.* **108**, 526.

Toksöz, M. N., Solomon, S. C., Minear, J. W., and Johnston, D. H.,: 1972. *The Moon* **4**, 190.

Turekian, K. K. and Clark, S. P., Jr.: 1969, 'Inhomogeneous Accumulation of the Earth from the Primitive Solar Nebula', *Earth Planetary Sci. Letters* **6**, 346–348.

Urey, H. C.: 1952, *The Planets: Their Origin and Development*, Yale Univ. Press, New Haven, Conn.

Urey, Harold C.: 1972, 'Abundances of the Elements. Part IV: Abundances of Interstellar Molecules and Laboratory Spectroscopy', *Ann. New York Acad. Sci.* **194**, 35–44.

Urey, H. C. and Mayeda, T.: 1959, 'The Metallic Particles of Some Chondrites', *Geochim. Cosmochim. Acta* **17**, 113–124.

Van Schmus, W. R. and Wood, J. A.: 1967, 'A Chemical-Petrological Classification for the Chondritic Meteorites', *Geochim. Cosmochim. Acta* **31**, 747–765.

Verniani, F.: 1969, 'Structure and Fragmentation of Meteorites', *Space Sci. Rev.* **10**, 230.

Vinogradov, A. P.: 1962, 'Origin of the Earth's Shells', *Izv. Akad. Nauk SSSR, Ser. Geol.* **11**, 3–17.

Vinogradov, A. P. and Yaroshovsky, A. A.: 1967, 'Further Investigation of the Differentiation Mechanism of the Earth's Mantle: The Problem of Heat-Mass Transfer in Connection With Zone Melting in the Mantle', IUGG General Assembly, 14th, Abstracts of Papers, IIa, A-2.

Voshage, H. and Hintenberg, H.: 1963, 'The Cosmic-Ray Exposure Ages of Iron Meteorites as Derived from the Isotopic Composition of Potassium and Production Rates of Cosmogenic Nuclides in the Past', in *Radioactive Dating*, IAEA, Vienna, pp. 367–379.

Wänke, H.: 1965, 'Der Sonnenwind als Quelle der Uredelgase in Steinmeteoriten', *Z. Naturforsch.* **20A**, 946.

Wänke, H.: 1966, 'Meteoritenalter und verwandte Probleme der Kosmochemie', *Fortschr. Chem. Forsch.* **7**, 332–408.

Wasserburg, G. J., Papanastasssiou, D. A., and Sanz, H. G.: 1969, 'Initial Strontium for a Chondrite and the Determination of a Metamorphism or Formation Interval', *Earth Planetary Sci. Letters* **7**, 33–43.

von Weizsäcker, C. F.: 1944, 'Über die Entstehung des Planetsystems', *Z. Astrophys.* **22**, 319–355.

Wetherill, G. W.: 1968, 'Lunar Interior: Constraint on Basaltic Composition', *Science* **160**, 1256–1257.

Wetherill, G. W.: 1971, 'Cometary versus Asteoidal Origin of Chondritic Meteorites', in T. Gehrels (ed.), *Physical Studies of Minor Planets*, NASA SP-267, NASA, Wash., D.C., p. 447.

Wiik, H. B.: 1956, 'The Chemical Composition of some Stony Meteorites', *Geochim. Cosmochim. Acta* **9**, 279–289.

Wilcox, J. M.: 1972, 'Why Does the Sun Sometimes Look Like a Magnetic Monopole?' in *Comments on Modern Physics*, Part C', *Comments Astrophys. Space Phys.* **4**, 141–147.

Wilkening, L., Lal, D., and Reid, A. M.: 1971, 'The Evolution of the Kapoeta Howardite Based on Fossil Track Studies', *Earth Planetary Sci. Letters* **10**, 334–340.

Worrall, G. and Wilson, A. M.: 1972, 'Can Astrophysical Abundances Be Taken Seriously?', *Nature* **236**, 15–18.

Zähringer, J.: 1966, 'Die Chronologie der Chondriten aufgrund von Edelgasisotopen-Analysen', *Meteoritika* **27**, 25.

Zimmerman, P. D. and Wetherill, G. W.: 1973, 'Asteroidal Sources of Meteorites', *Science* **182**, 51–53.

SUMMARY AND CONCLUSIONS

As this fourth part completes our analysis of the origin and evolution of the solar system, it is appropriate to sum up the general results.

Our analysis is based on the following principles:

(1) We aim at a *general theory of the formation of secondary bodies* around a primary body. This 'hetegonic' theory should be equally applicable to the formation of planets around the Sun and the formation of satellites around a planet.

The results confirm that this approach is sensible. In fact it is shown that the properties of a system of secondary bodies is a unique function of the mass (Section 16) and the spin (Section 17) of the central body. No special assumption needs to be introduced concerning the Sun.

(2) In order to avoid the uncertainty concerning the state of the primeval Sun and its environment, the analysis should *start from the present state* of the solar system, *and systematically reconstruct increasingly older states*. Hence, the first part (Sections 1–6) is a critical review of those facts which are considered to be relevant for a reconstruction of the origin and evolution of the system.

(3) Before an analysis of the evolution of the solar system can be made, it is essential *to clarify what physical laws govern its evolution*. A lack of clarity in this respect has been disastrous to many other attempts at such analysis. More specifically the following mistakes have been made:

(a) Based on the pre-hydromagnetic Laplacian concepts, the importance of electromagnetic effects has been neglected. Studies have been made without any knowledge at all of plasma physics, or with erroneous concepts of its laws ('frozen-in' field lines, etc.). (Section 10),

(b) Condensation of grains has been thought to occur in a state of thermodynamic equilibrium, and it has not been realized that in space the solid grain temperature normally is an order of magnitude lower than the plasma or gas temperature under such conditions where condensation can take place (cooling of the medium). This has led to inferred plasma densities which are too high by orders of magnitude, and to chemical interpretations which are clearly unrealistic.

(c) The nature of collisions between grains has not been understood. It has been assumed that these result only in fragmentation, and the accretional processes which necessarily are more important have been neglected. Studies of electrostatic attraction and of collision involving fluffy aggregates are essential.

(d) The orbital evolution of a population of grains, although of obvious importance has not been properly considered. It is necessary to introduce a new concept 'jet streams' as an intermediate stage in the accretional process.

(4) The papers discussed, Parts I–IV, show that *the origin and evolution of the solar system can be reconstructed as a result of the following processes:*

(a) *Emplacement of plasma* in specific regions around the central bodies. The critical velocity phenomenon is essential for these processes. The resulting chemical differentiation produces substantial differences in the composition of the bodies (Sections 16 and 14).

(b) The transfer of angular momentum from the central body to the surrounding plasma: a *partial corotation* is established as demonstrated by the structure of the Saturnian rings and the asteroid belt (Section 13).

(c) The condensation from this state results in a *population of grains* which is *focussed into jet streams* in which the *accretion* of planets or satellites takes place.

(d) Whereas all these processes took place during a period of some hundred million years, there has been a *slow evolution* during 4–5 Gy until the present state is reached.

Following principle (2), (d) is analyzed in Part I, (c) and Part II, (b) in Part III and (a) in Part IV.

The *general conclusion* is that already with the empirical material now available *it is possible to reconstruct the basic events leading to the present structure of the solar system.* With the expected flow of data from space research the evolution of the solar system may eventually be described with about the same confidence and accuracy as the geological evolution of the Earth.

INDEX OF SUBJECTS

GEOPHYSICS AND ASTROPHYSICS MONOGRAPHS

AN INTERNATIONAL SERIES OF FUNDAMENTAL TEXTBOOKS